The World of Andrei Sakharov

The World of
ANDREI SAKHAROV

A Russian Physicist's Path to Freedom

Gennady Gorelik

with

Antonina W. Bouis

OXFORD
UNIVERSITY PRESS

2005

OXFORD
UNIVERSITY PRESS

Oxford University Press, Inc., publishes works that further
Oxford University's objective of excellence
in research, scholarship, and education.

Oxford   New York
Auckland   Cape Town   Dar es Salaam   Hong Kong   Karachi
Kuala Lumpur   Madrid   Melbourne   Mexico City   Nairobi
New Delhi   Shanghai   Taipei   Toronto

With offices in
Argentina   Austria   Brazil   Chile   Czech Republic   France   Greece
Guatemala   Hungary   Italy   Japan   Poland   Portugal   Singapore
South Korea   Switzerland   Thailand   Turkey   Ukraine   Vietnam

Published by Oxford University Press, Inc.,
198 Madison Avenue, New York, New York 10016

www.oup.com

Oxford is a registered trademark of Oxford University Press

Library of Congress Cataloging-in-Publication Data
Gorelik, G. E. (Gennadii Efimovich)
[Andrei Sakharov. English]
The world of Andrei Sakharov : a Russian physicist's path to freedom / Gennady Gorelik
with Antonina W. Bouis.
p.   cm
Includes bibliographical references.
ISBN-13 978-0-19-515620-1
ISBN 0-19-515620-X
1. Sakharov, Andreæ, 1921–   2. Physicists—Soviet Union—Biography.
3. Dissenters—Soviet Union—Biography.   4. Human rights workers—Russia
(Federation)—Biography.   5. Soviet Union—Politics and government—1953–1985.   I. Bouis,
Antonina W.   II. Title.
QC16.S255G67 2004
530'.092—dc22
[B]   2003016991

1 3 5 7 9 8 6 4 2
Printed in the United States of America
on acid-free paper

In memory of Lydia Chukovskaya

PREFACE

This book is about how a theoretical physicist and the acknowledged father of the Soviet hydrogen bomb became a human rights
activist and the first Russian to win the Nobel Peace Prize.

In order to understand this incredible transformation, we must examine how
several powerful forces—life-enhancing as well as death-dealing—intersected
in Andrei Sakharov's life. Within his family, he joined the enigmatic world of
the Russian *intelligentsia*. Just the fact that this word, so Western in appearance, is followed by the tag "*Russ*" in dictionaries around the world is an indi-
cation of its mystery. It was Sakharov's lot to live his life during the era of Soviet
civilization with its glaring contrasts: the first Sputnik in space while kerosene
burned in lamps in the villages, the heights of creativity in the arts against the
background of daily suppression of freedom.

The scientific school, or rather scientific family, where Sakharov began his
path in physics was a miracle, given the backdrop of the Stalinist era. In a
society in which conformity was a means of survival, the teachers of this
school contrived to follow the voices of their conscience. And finally,
Sakharov's life unfolded against the backdrop of the nuclear alchemy which
had, in just a few short years, jumped from the pages of physics journals,
understandable to very few people, to the front pages of newspapers around
the world.

Only by comprehending how these forces shaped Andrei Sakharov's life can
his role in history be grasped.

One of the main sources for this book was a collection of oral histories—about
fifty interviews with colleagues, friends, and family of Andrei Sakharov—that
the author began collecting in 1989. My direct contact with the people who
participated in and witnessed events helped me understand archival materials
and publications.

Sakharov's own eyewitness account is contained in his book, *Memoirs*.[1] Although he wrote most of it during his exile in Gorky, relying only on his memory and respecting the constraints of secrecy then in effect, it is truly an invaluable source. Quotations from his book in the body of this text are translated directly from Russian and cited without endnotes.[2]

Those who shared their reminiscences with the author include Leon Bell, Yury Zamyatin, Boris Erozolimsky, Mikhail Levin, Sofya Shapiro, and Akiva Yaglom, who remembered Andrei Sakharov from his student years; Izrail Barit, Vitaly Ginzburg, Moisei Markov, Pavel Nemirovksy, Iosif Shapiro, and Yevgeny Feinberg, who knew him as a graduate student; Mates Agrest, Viktor Adamsky, Lev Altshuler, Lev Feoktistov, Yefim Fradkin, German Goncharov, Mikhail Meshcheryakov, Vladimir Ritus, Yury Romanov, Yury Smirnov, and Isaak Khalatnikov, who worked with him on the Soviet atomic project; Boris Bolotovsky, David Kirzhnits, Lev Okun, and Vasily Sennikov, who knew the Sakharov who returned to theoretical physics; Lyubov Vernaya, Sakharov's daughter, and Maxim Frank-Kamenetsky, who told me about the lives of their families in the closed city of Sarov (aka Arzamas-16); Yakov Alpert, Boris Altshuler, Sarra Babenysheva, Natalya Dolotova, Aleksandr Esenin-Volpin, and Maria Petrenko,who knew Sakharov as a defender of human rights; and Elena Bonner, Sakharov's widow, who talked to me about the last twenty years of his life (I also relied on fascinating material that she collected about Sakharov's genealogy).

The photographs and autographs from personal collections appear in this book courtesy of Elena Bonner and Lyubov Vernaya, as well as Vladimir Kartsev and Maxim Frank-Kamenetsky.

I received enormous help in my archival research from Galina Savina. Irina Dorman helped conduct many of the interviews with participants in and eyewitnesses of the events described in this book. I acquired much understanding about Soviet history from my association with Pavel Rubinin. Priscilla McMillan helped me grasp the history of the American nuclear project. I am indebted to Helmut Rotter for a perspective of events on both sides of the Iron Curtain from the center of Europe. Friendship with these wonderful people was an important support in my work.

Bela Koval and Yekaterina Shikhanovich helped me a great deal in Sakharov's archive in Moscow. Boston University's Center of Philosophy and History, headed by Fred Tauber, extended me hospitality while I worked on the book,

1. Andrei D. Sakharov, *Vospominaniya*, 2 v., New York: Chekhov Publishing Corp., 1990. English editions: *Memoirs* (Richard Lourie, trans.), New York: Knopf, 1990; *Moscow and Beyond, 1986 to 1989* (Antonina W. Bouis, trans.), New York: Knopf, 1991.

2. All the original Russian versions of quotations presented here can be found in the Russian edition of this book—Gennady Gorelik, *Andrei Sakharov: nauka i svoboda* (Moscow: Vagrius, 2004), while the full references to the Russian-language sources as well as Sakharov's *Memoirs* are available at http://ADSakharov.narod.ru/.

and my contact with Bob Cohen, its director emeritus, was particularly inspiring. I am grateful to Anne Fitzpatrick and Tom Reed for acquainting me with the world of Los Alamos and Livermore.

I am very thankful to Dmitri Zimin for helping me to understand the problem of antiballistic defense that was so important for transforming Andrei Sakharov the scientist into the public figure and human rights advocate.

Boris Altschuler, Sarra Babenysheva, Leon Bell, Boris Bolotovsky, Elena Bonner, Elena Chukovskaya, Vitaly Ginzburg, German Goncharov, Boris Erozolimsky, Vladimir Kogan, Leonid Litinsky, Klara Lozovskaya, Lev Okun, Galina Shabelskaya, Sofya Shapiro, Lyubov and Aleksandr Vernyi, Akiva Yaglom, and Sergey Zelensky all read this book in manuscript form (complete or partial) and made stimulating comments. I am deeply grateful to all of them.

My work in the history of science would have been impossible without the support of people who believed in me. The first was my father, from whom I learned about life, with whom I discussed all the questions that interested me. For many years, my wife, Svetlana, was a major help in deciphering extensive interviews, while selflessly supporting the family hearth.

I am grateful to David Holloway for stimulating contact of many years' duration and for his support of my oral history program on Soviet physics. Loren Graham imbued me with the confidence that I needed to tackle Sakharov's biography. And it was their encouragement helped me to start this work at the Dibner Institute for the History of Science and Technology at MIT, thanks to a fellowship from the Bern Dibner Fund in 1993.

My work on this book was generously supported by a grant for research and writing from the John D. and Catherine T. MacArthur Foundation; a fellowship from the John Simon Guggenheim Memorial Foundation; grants-in-aid from the Friends of the Center for the History of Physics, the American Institute of Physics, and the International Research and Exchanges Board; and funds provided by the U.S. Department of State (Title VII program) and the National Endowment for the Humanities.

The generous support of the Alfred P. Sloan Foundation made it possible for Antonina W. Bouis to devote herself to the translation of this book.

While none of the providers are responsible for the views expressed here, we gratefully acknowledge this assistance.

Heather Hartman and her colleagues at Oxford University Press were incredibly thorough and patient with a complex text, and we would like to acknowledge their artistry and skill.

And we both are very thankful to Anya Kucharev for her felicitous translations of most of the verses cited in the text.

CONTENTS

Photo gallery appears after page 146.

INTRODUCTION

How I Came to Write This Book

A ndrei Sakharov was my contemporary, a fellow countryman and, you might say, colleague. In the 1970s I listened to him at the Physical Institute of the Academy of Sciences (known by its Russian acronym—FIAN) in Moscow. The topic of these seminars was theoretical physics, and Sakharov seemed so totally engrossed in science, so open and gentle, that it was hard to reconcile this image with the recklessly brave words and actions of Sakharov the academician, which the "voices of the enemy," as Muscovites used to dub Western radio in those days, discussed at night over the howling of Soviet jamming.

The last time I was in the same room with him was in late 1979 in a small auditorium of the Theoretical Department. Sakharov was lecturing on very nonmaterial matters: the early universe and how the symmetrical laws of nature could have led to the puzzling asymmetry of the observable cosmos. The talk had not been announced anywhere, so only those in the know were there. Academician Yakov Zeldovich, a stocky, quick man with a shiny bald head, wearing a heavy sweater, was also present. As usual, Sakharov's voice lacked confidence; he sounded like he was thinking aloud. When he ended the talk, Zeldovich rushed to the blackboard and began speaking in a very confident voice about the difficulties of the cosmological blueprint under discussion. He deftly wrote formulas and drew graphs on the blackboard with swift, athletic movements.

Sakharov was an entirely different type: much taller than Zeldovich and slightly stooped, and he spoke slowly, almost haltingly. There was nothing athletic about his movements. One thing, however, made him clearly superior to his opponent—he held chalk and wrote on the blackboard equally well with his right or left hand.

They were discussing the concentration of unseen bosons in a young, super-dense universe. At that point I was becoming aware of the unusually high concentration of invisible stars at the blackboard—six stars of the Hero of Socialist Labor award per square meter. I thought about how these two theorists had met and become friends, then academicians and three-time Heroes, while they were creating Soviet nuclear weapons. They both had emerged from the closed scientific-military world into the open quest for the physics of the universe. But although they shared a passionate interest in the secrets of the universe, they lived very differently in the everyday world.

A few weeks later, Soviet troops invaded Afghanistan. Sakharov spoke out in strong condemnation. He was exiled in January 1980, without a trial, to the city of Gorky, which was closed to foreigners—probably so that nothing would distract him from the problems of the early universe. Zeldovich, who kept silent about such matters, had to contemplate lofty subjects in the bustle of the capital.

As it happened, the physics that interested me most of all in 1980 dragged me into the history of Russian science. This area of physics was directly related to the nature of the early universe. I was studying the prophetic and underappreciated work done by a young Soviet theorist, Matvei Bronstein, in 1936. This physicist remained forever young. He was thirty when he was arrested in August 1937, at the height of the Great Stalinist Terror. A bullet to his head in the cellar of a Leningrad prison ended the life of a theorist who understood the physics of the earliest period of the universe better than any of his contemporaries.

The more I immersed myself in his yellowed pages, the more I felt attracted to the author. And on October 18, 1980, this attraction led me to a room in the center of Moscow, just five minutes from Red Square. Lydia Chukovskaya, the widow of that forever-young physicist, lived there.

I spent many evenings in that room, learning more and more about the hero I had unexpectedly discovered. The unfolding picture of the amazing, amusing, and terrible events of the 1930s was turning me into a historian and biographer. And I acquiesced in this transformation.

I also began to appreciate the kind of eyewitness that life had connected me with. The walls of this room were covered with photographs of people who were the pride of Russian literature—Anna Akhmatova, Boris Pasternak, Kornei Chukovsky, Aleksandr Solzhenitsyn. I began to understand how their lives were intertwined with the life of Lydia Chukovskaya and her late husband.

There was only one person in the photographs whom I did not recognize, until she told me it was Sakharov—the smile of this man with a child in his arms was just too carefree and cozily happy at home. It was unthinkable that anyone with such a smile would be forcibly taken from his own house and isolated with round-the-clock surveillance by the KGB. I learned that the seditious academician visited this room often. A common cause united the writer Chukovskaya and the physicist Sakharov: the defense of the hurt and the humiliated, the defense of human rights.

I am indebted to Lydia Chukovskaya for my first impressions of Andrei Sakharov outside the realm of physics, the jammed Western radio station voices, and the loud braying of Soviet newspapers. She spoke of Sakharov with tenderness and pain. At that time, the details of his life in exile in Gorky were unknown, and this tormented her.

Sakharov had endured thirteen years of official persecution and seven years of exile before Gorbachev's new leadership of the country finally allowed him to return to Moscow at the very end of 1986, the second year of *perestroika* and *glasnost* (these Russian words turned into political terms for the policies of social reform and open discussion). The new Soviet leaders allowed Sakharov to be himself. They allowed him to say what he thought.

It was incredible, but incredible things were beginning to happen in the nation. For the first time in Soviet history, people were given the right—albeit limited—to choose among several candidates in elections for some seats in Parliament. Sakharov, who was neither an orator nor a politician, became a People's Deputy. For the first time, his countrymen could see and hear him on Soviet television and find in him the personification of conscience.

For a Soviet historian of science, walking into the main KGB building on Lubyanka Square was just as incredible. I was able to enter simply in my capacity as a historian of science!

In the fall of 1990 Soviet perestroika was still under the control of the Politburo of the nation's only party. The sword and shield of the Politburo—as the KGB liked to see itself—was still called by its old name, but the people wielding the sword were trying to change the organization's image. I took advantage of this. I also took advantage of the fact that I had just been hired by the new director of the History of Science and Technology Institute, Nikolai Ustinov, whose late father, a former defense minister, had held an important post in the Politburo. Despite his background, the younger Ustinov was an unusually mild and polite man.

I asked him to sign a carefully composed letter, addressed to one of his father's former colleagues who had been the head of the KGB. In the letter were the names of some physicists who had been arrested in the 1930s and a request for access to their files "for the creation of a complete and objective history of Soviet science in a social context." A few months later I got the call, "This is the Committee for State Security," with an invitation to come over.

And here I was sitting in an office paneled in solid dark wood. The five case files on the seven defendants lay on the desk in front of me. The man who had "prepared" cases like these had once sat in this very chair and had perhaps seen these very same "enemies of the people"—in the language of the Great Terror—in front of him. Through the window I could see the notorious Lubyanka inner prison, which is not visible from the street because it is completely surrounded by the monumental building of the KGB's headquarters. Those being investigated were brought from this prison for interrogation in "my" office. I learned from the archival files that three of the seven were executed (shot). One physicist was released from prison after only a year, on the orders of the head of the KGB, despite being guilty as charged: he had dared to compare Stalin with Hitler, not just in his thoughts, but on paper. This released criminal—Lev Landau—would later win the Nobel Prize for Physics.

I was overwhelmed by thoughts and feelings. Two main questions had arisen in my mind as I walked into the terrifying Lubyanka building and while the guard carefully compared my face with my photograph receded: why did *they* grant me permission to come—what was the reason? And when did *they* prepare the papers I was being allowed to see?

I was able to answer the second question, an archival and historical one, once I studied the papers—apparently, they had been written in the years that were indicated on them. All their documentary fictions and incongruities, as well as precious traces of reality, also dated from that time.

As for the question regarding my own personal, rather than state, security, it diminished on the first day of my unusual archival work. It began with an interrogation, so to speak. Two KGB men had a conversation with me for about an hour and a half. One was somewhat grim and world-weary; the other was younger, kind, and curious. They wanted to know what I, frankly, hoped to find in such documents and what my findings could contribute to the history of science.

I immediately began sincerely testifying, explaining with concrete examples, how much the simple exact date of an event can sometimes yield. I had a lot of examples in reserve, and my KGB officers visibly softened. The conversation loosened up, with history converging with contemporary life at times. In closing, they asked me a question that both upset and amused me: "Was Sakharov really a good physicist?"

How could the members of such a competent or, in any case, well-informed institution imagine that the dissident academician had been a physicist in name only? You see, by that time it had been more than year since the famous People's Deputy had died. The genuine public mourning had been shown on television. And how many publications had there been about him?!

After that ingenuous question from the KGB officer, I almost completely stopped worrying about being used by them in some way for purposes unrelated to my work in the history of science. I realized that they were simply obeying orders from their superiors to assist historians. Had their orders been different, they would have docilely obeyed those orders.

Afterward, I revisited that odd question on many an occasion. I admitted to myself that I, too, didn't understand how such disparate things could be contained in a single life: the hydrogen bomb and the Nobel Prize for Peace, his mourning for Stalin in 1953 and his staunch opposition to the Soviet system created by Stalin, and last but not least, the physics of the early universe. I knew that all of this was true—the most powerful thermonuclear explosion in history and the brave defense of human rights before the powers-that-be, the gentle nature and the symmetries of the universe—but how could it all come together in a single person, a single life? And so, having begun with the young physics of the early universe, I ended up reflecting upon one of the most humanitarian of physicists on this planet.

Six months after my visit to the bowels of the KGB, the head of the organization, Comrade Kryuchkov, took part in a coup against the state. He wanted to save the Soviet regime. The outcome was just the opposite—the Soviet state collapsed, and the last Soviet head of the KGB ended up in prison (not Lubyanka). But before that, alas, his institution managed to destroy hundreds of volumes of materials relating to Sakharov, including his manuscripts, which had been stolen by KGB agents.

However, the highly visible demise of the Soviet regime allowed even people of older generations to discover freedom of speech. Along with my work in the archives, I began collecting oral histories from people in Sakharov's circle. I interviewed people who had shared his life's journey: his university classmates, those with whom he began his scientific path and work on nuclear weapons, and those who knew him when he returned to pure physics, when he came out into the world of human rights and entered world history.

The archival materials helped me ask the eyewitnesses the proper questions, while their stories helped me ask myself new questions and seek out new documentary evidence. The documents from the KGB archives, which I studied at the actual scene of the crime, also turned out to be useful.

I discovered much that was unexpected. For instance, I learned that Lev Landau, the theorist who had compared Stalin to Hitler in 1938, did calculations for the Sakharov hydrogen bomb in the 1950s, for that same Stalin. And I realized that when Sakharov returned to pure science after working on the bomb, he answered the very same question about the physics of the early universe that had made me a historian of science—the question asked by Matvei Bronstein, my first hero, who was executed in 1938.

As a result, I think I began to understand the connection between the incredible contrasts in Andrei Sakharov's world, the world of Russian physics, Soviet unfreedoms, and personal aspirations for knowledge and freedom. I realized that in order to tell the story of Andrei Sakharov I must explain the "physics" of this strange world. And I set out the results of my research in my hero's native tongue, Russian.

In order to bring this story to the American reader, I turned to Antonina W. Bouis. Her numerous translations include books by Sakharov and his wife, Elena Bonner. Nina was among the first Americans to see Sakharov when he returned to Moscow, and she was a frequent guest at his home, working on various human rights and literary projects. She observed his scientific curiosity, manifested in his responses to the most mundane, everyday experiences. Once, Nina's husband offered to bring the right kind of adhesive compound for the tiles that had started to fall off after a bathroom refurbishment at the Sakharov apartment. "How interesting," Sakharov mused several minutes later, although the general conversation over tea had turned to other topics, "that you put glue only in the corners and not over the entire tile. So, it's the surface tension . . ."

I am grateful to Nina Bouis not only for her translation and additions to the text but for keeping it focused on the human side of Andrei Sakharov.

The World of Andrei Sakharov

PROLOGUE

Pyotr Lebedev: The Pressure of Light and the Pressure of Circumstances

---

At one time Lebedev measured the pressure of light radiation in the most refined, for that time, experiments, but here [in the physics of the thermonuclear explosion] it was enormous and definitive. [. . .]

Has our intelligentsia really become that much more insignificant since the time of Korolenko and Lebedev? To be sure, Pyotr Lebedev loved science no less than our contemporaries, was no less connected to the University when he left after Minister of Education Kasso's decision to allow the police onto University grounds (most likely, only Andropov knows the number of KGB agents at Moscow University right now).

—Andrei Sakharov, *Memoirs*

It's worth beginning the story of Andrei Sakharov's life with the events that took place a decade before he was born. At the center of these events was Pyotr Lebedev (1866–1912), the first Russian physicist of world stature. Lebedev received a European education and recognition in the international community of physicists; he also belonged to the Russian intelligentsia. His life was proof of the second affiliation, as was his death. When Russian history challenged him with the choice between science and his moral duty, he sacrificed his beloved profession and, given his heart ailment, the sacrifice proved too much for him to bear.

Half a century later, Russian history challenged Sakharov with a similar choice.

There were other reasons why Sakharov felt linked to Lebedev. Sakharov's first physics teacher, his own father, had studied with Lebedev at Moscow University.

3

The scientific home in which Sakharov began his path in science, the Physical Institute of the Academy of Sciences, was named after Lebedev for good reason. The institute building itself was constructed for him when he was forced to leave the university in 1911. Lebedev, it turns out, has something to do with Sakharov's military physics. The main scientific outgrowth of Lebedev's scientific work became an element in the making of the thermonuclear bomb. The most powerful of nature's forces awakened by man is related to an elusively tiny force that Lebedev succeeded in capturing.

Light Exerts Pressure

Lebedev was first to discover and measure the pressure exerted by light. He did this through an exceedingly difficult experiment. You might doubt its difficulty, if you have seen the amusing scientific toy that imitates Lebedev's apparatus. A small vane, covered by a glass bell jar, begins rotating continuously when an ordinary desk lamp is turned on next to it. It's hardly a surprise to anyone when wind causes this type of vane to rotate, but here we have a glass bell jar that shuts out the slightest puff of air. Only light can penetrate it. Does it therefore exert pressure on the blades just as well as a stream of air does? It's an interesting toy, but does one really wind up in the annals of science with toys like this?

The history of science itself is more interesting. Lebedev was only seven years old when the English physicist William Crookes inadvertently constructed the first light vane while pursuing other goals. Physicists understood that the reason for the vane's rotation indeed was light, but not that it was due to its pressure. It's easy to feel the warmth of the sun's rays, but not their pressure. This perceivable warmth is precisely what rotates the vane, warming up the air near it under the bell jar. Theorists calculated that these weak "warm air" forces are thousands of times greater than the predicted forces of the pressure of light.

That light exerts a pressure was predicted a year before Lebedev was born by the great James Clerk Maxwell (1831–1879) in his electromagnetic theory of light—a very unusual theory for its time. Matter, electricity, and light are so obviously different that, for a long time, physicists investigated them separately. Michael Faraday conjectured that they were related, and Maxwell fleshed out the conjecture to form a precise theory. Some of the conclusions of this theory, however, turned out to be so strange that hardly anyone believed them. However, no one is obliged to believe a theory until an experiment verifies its prediction.

From his equations, Maxwell concluded that electromagnetic impulses could travel without wires, and that their speed is equal to the speed of light. Based on this, he proposed that light itself was a sort of electromagnetic wave. According to these same equations, a stream of light should not only warm up the surface it illuminates but also exert pressure on it. Maxwell calculated this pressure and realized that it must be very small—simply because the speed of light, c, is unusually large, much larger than all other measured speeds. The calculations of the great Maxwell can be accepted on faith. Or one can become convinced of their correctness by means of the only equation permitted in the company of nonphysicists:

$$E = mc^2$$

Even those who do not know what the letters in this famous equation stand for—that E is energy, m is mass, and c is the speed of light—are familiar with it. And anyone who has ever thrown a ball knows without any equations that the greater the speed and mass of the ball, the greater its impact (or pressure, p) will be on the person hit by it. In other words,

$$\text{pressure} \sim \text{mass} \times \text{speed}$$

(the reader acquainted with physics will easily make this more precise by adding the words "per unit of area per unit of time"). Keeping this in mind, let's rewrite the famous equation slightly:

$$E = mc^2 = mc \times c = pc$$

Since $E = pc$, it means that

$$p = E/c$$

Thus, in order to calculate the pressure of light, p, the energy of light must be divided by the speed of light, which is an enormous quantity, nearly 300,000 kilometers per second. Since it must be divided by such a large number, the pressure of light turns out to be very small. This was the root of all the difficulties facing experimenters, including Lebedev.

Theorists were experiencing difficulties because new ideas did not fit into the framework of the existing scientific concepts. The British idea of the electromagnetic field, or a force that was somehow dispersed in space, was foreign to continental physics—first and foremost to German physicists, who recognized only electrical particles and the forces between them. Confusion reigned over the course of several decades: there was no basis for refuting Faraday; as for Maxwell's ideas, there was yet not enough courage to believe in them.

There is only one reliable path to truth in physics—the experiment. The first support for Maxwell's theory came from the experiments of the German scientist Heinrich Hertz (1857–1894). Initially Hertz regarded the British theory with skepticism, but in 1888 he succeeded in grounding Maxwell's equations in reality. As a result, he became convinced and persuaded others that electromagnetic oscillations can travel without wires and actually do so at the speed of light.

As for the pressure of light, Maxwell's predictions remained in question. Even his compatriot Lord Kelvin (1824–1907) did not believe them, even though he received the rank of Lord mainly for his accomplishments in the science of electricity.

Crookes's light vane might have revealed the pressure of light if it were sophisticated enough. To make their measurements, physicists tried to remove the air from the bell jar. By the time Lebedev became familiar with the problem, his experienced colleagues had learned to pump almost all the air out of the jar, leaving only one hundred thousandth part in it. However, even this amount was too large. Its lightest zephyrs were still many times greater than the force of light radiation pressure.

And so a Russian who had received an excellent German education in thoroughly French Strasbourg took on the job. At that time, the turn of the 20th century, thirty-year-old Lebedev was at the height of his powers in his Moscow laboratory—and he needed them, plus the acquired experience, enthusiasm, and persistence of a young researcher not lacking in healthy ambition.

And Lebedev managed to do what had eluded the very experienced Crookes—he devised a way of decreasing the remaining amount of air in the bell jar by a factor of 100 and finally succeeded in making its interference smaller than the pressure of light. It took several years to measure such a tiny force, as tiny as the weight of a flea compared with that of a horse. This was, of course, amazing—but what did it really matter?

The preceding paragraphs mention British experts, a skilled Russian craftsman, and a flea—everything you need to recall Leskov's famous tale about the skilled left-handed craftsman. In the tale the Russian craftsman shod the British-made mechanical flea in order to show off and to humiliate the Brits. With such intentions, it's no wonder that the windup flea could no longer jump once it was shod.

Lebedev shod his flea so it could jump much better. And he would not have tried so hard if weren't so important for science, for world science. The outcome of his experiments, despite the smallness of the quantity he measured, answered a big question posed by science at that time. That is why Lebedev quickly made a name for himself with a lecture about his experiments at the First International Congress of Physicists in Paris in August 1900, and his publications in the major scientific journals.[1] Furthermore, Lebedev's experiments forced the renowned Kelvin to capitulate to Maxwell's theory.

If science is about prediction, history of science is mostly about mispredictions. Hertz did not believe that the electromagnetic waves he discovered could be used for long-distance communication. Lebedev thought that his experiments would explain the forces between molecules. However, in science, achievements take on a life of their own, independent of their authors' hopes and intentions. Radio communication was born seven years after Hertz's experiments. Five years after Lebedev's experiments, the theory of relativity completed Maxwell's revolution. The shortest exposition of the theory of relativity and its main outcome was that same equation, $E = mc^2$, which helped elucidate Lebedev's experiments. The actual sequence of events was the opposite: at the very beginning of the twentieth century, Lebedev's experiment, which had definitively convinced physicists of the correctness of Maxwell's electro-dynamics, firmly established the experimental basis for the physics of the new century, first of all for the theory of relativity. The edifice of physics was to be built and repeatedly rebuilt on this foundation.

But wasn't this really too little to be considered a great scientific achieve-ment—the proof of one theory and the foundation for others? Since 1901, in addition to history's judgment, an authoritative people's court has been oper-ating in science: its judgments are called Nobel Prizes. The Nobel Committee begins its annual selection by asking leading scientists to nominate candidates. Lebedev had already received such a request in 1902.[2]

Lebedev himself was nominated a candidate in 1912. His name was proposed by Wilhelm Wien, who had received the Nobel Prize the previous year, for the discovery of the laws of heat radiation (in the middle of the 1890s). Wien also nominated Einstein. The Russian physicist had more of a chance to win—not because his contribution to science was more significant, but because of the Nobel Committee's very cautious attitude toward theoretical works. They always await very solid experimental proof. So great was this caution that Einstein did not receive the Nobel Prize until 1921, a decade and a half after the work that immortalized his name, whereas Lebedev's work was experimen-tal and had been widely recognized by 1912.

But the 1912 Nobel Prize in Physics was awarded to the Swedish engineer Nils Gustaf Dalen for the invention of an automatically regulated acetylene burner for use in lighthouses. This invention did not enter the history of phys-ics. Why wasn't the Nobel Prize awarded to Lebedev? Forty-six-year-old Pyotr Lebedev died in March 1912, and Nobel Prizes are not awarded posthumously.

Pressure of Circumstances

The circumstances that preceded Lebedev's death, and became one of its causes, gave Lebedev a role in Russia's social history in addition to the history of sci-ence. Lebedev did not aspire to this at all—he didn't feel cramped in the phys-ics world. By the beginning of the twentieth century, physics had already developed into a worldwide enterprise. While he worked in Moscow, Lebedev took part in the life of this international community through his research, cor-respondence, and personal contacts, and in the spirit of his times, he hoped, along with this community, that science could improve people's lives. He had no other particular social interests. However, circumstances in Russia did not allow him to remain just a scientist.

This is how he himself described these circumstances in letters to his West-ern colleagues:

In January of this year [1911], student uprisings began, and the police took the initiative of maintaining order in the university buildings, without subordinating themselves to the Rector. Under the circumstances, the Rector was unable to assume responsibility for the normal course of academic life at the university, as granted to him by law, and the Rector and his two assistants submitted petitions resigning from their positions to the university Board. The Board was in agreement with both the reasons for these petitions and with the resignations themselves. The ministry accepted the resignations of these individuals as officials of the university, but it also fired them from their professorial and teaching positions at the university, without indicating the reasons why. Then many of the colleagues of the fired professors considered it their moral duty to resign too.

We were faced with the alternatives of either distancing ourselves in a cowardly way from the Rector and his assistants, whom we had elected and who acted on our behalf, or expressing our protest by resigning.

I was forced to abandon my professorship in Moscow, shut down my laboratory, where independent research was now being conducted at full speed, and am now with my family, left without a position and without the hope of completing the work which I had planned.[3]

What did all this mean to him, a physicist with a sense of moral duty? The famous Russian biologist Kliment Timiryazev (1843–1920) talked about this in an article:

Lebedev has died . . . Could I, old enough to be his father, imagine that my trembling elderly hand would ever write these words? There was a time when I spoke out as his only defender, a time when he was ready to leave Moscow University and flee to Europe. Time and time again I have repeated with pride that I had saved him for Russia, and now I repeat with horror: would it not have been better to save him for science?

At this enormous institute, on whose establishment he had expended a considerable amount of his powers, a wretched apartment, a work room on another, higher, floor, and a dark basement for his students to work in were found for him, and this—even as he was increasingly showing signs of heart disease. Young strength overcame everything; a powerful spirit was still stronger than the body. Work went into full swing, and with it came fame and glory—at first, of course, in foreign lands, then at home. The last Congress in Moscow was a triumph for Lebedev. It seemed as if a long series of years of feverish activity for the good and glory of the nation lay ahead; but those who are in charge of destinies decided otherwise.

A wave of Stolypin's "pacifications" rolled its way over to Moscow University and took Lebedev away to eternal rest.

This is not a mere phrase—it is a naked fact. Do I mean by this that he was one of those hotheads who get swept away by the facts of the political life around them and torn away from their ordinary favorite occupations? Not at all. I have not met anyone of his generation with such a chilling and skeptical distrust of the capacities of Russians, of the "Slavic race," as he often said, not just for political, but simply for any sort of social activism. As if he had the premonition that life itself was preparing convincing—but for him fatal—proof of this. And here he, an individual who did not believe in politics, a balanced person who was completely devoted to his work—science—became victim of those who hypocritically posed as defenders of science against the intrusion of politics.

Besides, the dilemma which he had to resolve was not a political one, just a human one. He was told to be a flunky, to do whatever he was ordered to do without question, to forget about having human dignity and honor—or leave. He realized that he was not one of those individuals who leave for effect through the front door, knowing that they'll be able to surreptitiously return through the back door. He was not one who in circumstances like these takes the profits and leaves for a practical life; for him, life without science had no raison d'être.[4]

It was practical life that came to the rescue of science. By that time in Russia, the philanthropists who wished to serve science and education had appeared. By the time Moscow University was devastated by the disaster of 1911, Shanyavsky's university and the Ledentsov Society had already been in existence in Moscow for several years. Their full names were the Moscow Public University and the Society for the Advancement of the Experimental Sciences and Their Practical Applications. To their founders they were "The Open University" and "The Society for the Friends of Humanity," which signified public purpose and self-government—freedom from the imperial bureaucracy. The names of the gold-mining tycoon Alfons Shanyavsky (1837–1905) and the merchant Khristofor Ledentsov (1842–1907) might have been no less famous than the names of Nobel and Guggenheim, had it not been for the socialist cataclysm of Russian history.

Shanyavsky's university and Ledentsov's society—in response to the events at Moscow University—helped Lebedev reestablish his laboratory and decided to build a Physical Institute for him. Lebedev worked on the design of the institute but did not live to see it built. We can only speculate what the Advancement of the Experimental Sciences and Their Practical Applications would have achieved under his direction.

Lebedev himself found the first application of his discovery—to explain the movement of comets under the total effect of the sun's gravity and the repulsive pressure of its light. To science visionaries, this hinted at the notion of spaceships—sailboats, driven at high speed by sunlight that required no fuel and would not pollute space with exhaust emissions. Such a practical, and unearthly, application would have been to the liking of Lebedev's and Ledentsov's spirits.

And what would they say about the terrible earthly application that became a reality half a century later? The point is in the brightness of light source. In 1945 a source "brighter than a thousand suns"[5] appeared—the atomic explosion. And ten years later, a fiery ball brighter than a million suns—the thermonuclear bomb—exploded with the aid of this source. Light, whose pressure Lebedev discovered with so much difficulty, became the instrument for creating monstrous energy half a century later. The energy radiating from an atomic explosion compressed matter, harmless in ordinary conditions, to stellar densities, and a thermonuclear star exploded as a result. And both the atomic and thermonuclear explosions obediently followed the same physical law of $E = mc^2$.

The Russian path to the thermonuclear energy began in the building that had been built for Lebedev. It was in this building, the Lebedev Physical Institute of the Academy of Sciences, that the first Soviet hydrogen bomb was designed at the end of the 1940s.

As an irony of history would have it, this was precisely the time when Soviet government patriots were using Pyotr Lebedev's name in their "struggle with cosmopolitanism and kowtowing to the West." Lebedev doesn't need to be excused for this, just as he didn't need to be defended against the self-styled patriots who—forty years before—accused him of being involved in God-knows-what in the house of a suspicious Pole using Jewish funds.[6]

The Russian Intelligentsia

Lebedev's legacy, as is evident from the epigraph to this chapter, includes not only science. For Sakharov, he was one of the moral standards of the Russian intelligentsia. This was a strange social stratum. Let's just begin with the word "intelligentsia" itself, which, despite its Latin appearance, entered European languages from Russia at the beginning of the twentieth century, on the eve of the dramatic events at Moscow University. The Europeans needed the new word in order to name what they lacked. Naturally, there were people engaged in intellectual work in Europe, but they were not united by a feeling of their own moral responsibility for what was taking place in society.

The reason, to be sure, is not in any particular innate moral talents of the Russians but in Russia's social conditions. Lebedev wrote to a European colleague that, "given the prevailing conditions here, which to a European would be completely improbable and incomprehensible, I must abandon my career as a physicist here."[7]

The English language, having borrowed the Russian word *intelligentsia*, somehow left out the Russian word *intelligent* (with a hard G) for an individual who belongs to intelligentsia. It defines those who are both individualistic and socially concerned. The English dictionaries explaining the word "intelligentsia" emphasize the idea of "cultural vanguard or elite." However, let's listen to an undoubted *intelligent*, Russian literary critic and writer Kornei Chukovsky (1882–1969): "The word 'intelligentsia' was created by the Russians (in the 1870s) to designate a purely Russian social stratum, completely unknown to the West, because not every intellectual was called an 'intelligent' in those days of old, only one whose way of life and convictions were imbued with the idea of serving the people."[8]

Pyotr Lebedev never discussed the purely Russian word *intelligentsia*, but in 1911 he considered it his moral duty to abandon his life's work. A European would probably have difficulty understanding the necessity for such a choice.

The Russian intelligentsia took form as Russia began including itself in the life of nineteenth-century Europe. By the beginning of the twentieth century, one could speak about a united European culture with a substantial Russian contribution. The language of Russian music needed no translation and resounded all across Europe. Even books—by Tolstoy, for example—began their European lives just a few years after they were first published in Russia, and sometimes within the same year.

Thus, there was a community of spiritual reference points. It was realized through direct human contact: Russian borders were open to people of the professional class. In Russia, the European *intelligent* saw a society of far sharper social contrasts. Serfdom wasn't abolished here until 1861, a century later than in Western Europe, and naturally, the legacy of unfreedom was much stronger. The anachronistic autocracy made the expression of social views impossible except in literature. In such a society, an intellectual enlightened in the European manner became a Russian *intelligent*, experiencing social contrasts and his moral duty to society more acutely.

Physical objects, placed at the will of an experimenter into extreme conditions—high pressure, temperature, electrical tension—exhibit fundamental properties of their nature. In the same way, perhaps, conscious subjects enlightened in the European manner, who wound up in Russian conditions, exhibited the fundamental characteristics of European intellectual culture and became the Russian intelligentsia.

It appears that the emergence of the intelligentsia presupposes two conditions—sufficient intellectual freedom of a class of educated people, and a nation's social backwardness—the contrasts of social oppression. If this is so, it becomes clear why the intelligentsia became so degraded during Soviet rule, which destroyed the first prerequisite, and why the old-fashioned intelligentsia is doomed to extinction when (let's hope) the second prerequisite disappears in Russia. Then intellectuals will simply take the place of Russian intelligentsia.

During tsarist times, when both prerequisites existed, members of the Russian intelligentsia responded to the powerful tensions of social life in different ways. Their responses depended on their life experience, temperament, and spiritual sensibility. Russia's development was the key issue. Heated discussions took place between Slavophiles and Westernizers.

The year Lebedev was born, the Russian poet Fedor Tyutchev declared famously:

> Russia is not accessible to reason
> and cannot be measured with a yardstick;
> Russia is a special country
> that can only be taken on faith.[9]

This response to the question of Russia's development did not suit people of a scientific orientation, who had resolutely replaced the Russian yardstick (actually, *arshin*) with the European meter.

Another response, a tragic one, rang out like a peal of thunder in 1881 when "the best people in Russia killed the best tsar in Russian history," Alexander II, who had freed the serfs.

Whatever his response, the moral obligation of the Russian *intelligent*—or his social ambition, a skeptical European might have thought—originated from a moral feeling generated by those conditions that were described by Lebedev as unbelievable to a European.

So then, is a Russian *intelligent* a European intellectual who lives in Russia? Yes, but there is more. The Russian word "intelligentsia" came only a few decades before its European version. The word appeared soon after the abolition of serfdom when, due to social reforms, the estate of the intelligentsia began to grow rapidly, and not primarily from the noble aristocracy. First-generation members of the intelligentsia—like Lebedev—easily became persuaded that aristocracy of the spirit was not an inherited trait. What was required was intellect and knowledge, not a pedigree. This time coincided with a powerful leap of European scientific learning. For the Russian intelligentsia, all this combined into a sense of social development—social and scientific progress, a sense of the stagnation of autocratic rule, and a sense of moral responsibility for what was happening around them.

A great deal of what took place in Russia during that time can be understood if you remember that it encompassed the life of Leo Tolstoy. Reflecting on the laws of history and the philosophy of freedom, he used ideas and images from physics in his 1869 novel *War and Peace*. The debate between the Slavophiles and Westernizers affected him, but both these positions were too narrow for him. He decided to renounce his own fiction writings.

Tolstoy, Russia's foremost prerevolutionary writer, renounced his life's work for an entirely different reason, it appeared, than did Lebedev, Russia's foremost physicist: the former undertook to teach all of humanity; the latter wanted to teach only a few of his students in acquiring new scientific knowledge about the world. But on an inner level, both renunciations were motivated by a moral sense, aroused by the prevailing conditions of Russian life. And this, perhaps, speaks more clearly about Russia at that time than a lengthy analysis of its social statistics and social cataclysms.

Count Tolstoy, excommunicated by the state Church in 1901, was not the only one who saw Russian life in somber colors. Another count saw Russia in the same light. Entirely a man of the government, Sergei Witte (1849–1915), the first constitutional prime minister of the Russian empire, tried to combine authoritarian rule with dynamic modernization. In a report to the emperor in

1905, he admitted that the people's uprisings that shook Russia at the time "cannot be explained by the partial imperfections of the present government regime, are not just the activities of extreme parties," that "the roots of these uprisings undoubtedly lie deeper," and "Russia outlived the forms of the existing system and aspires to a legal system based on civil liberty."[10] This was an unpleasant truth for the autocracy. And after recovering from the scare of the revolutionary explosion of 1905, the tsar dismissed the prime minister who said unpleasant things to him.

During the ensuing time, autocratically named the "period of renewal," the new phenomenon of capital punishment appeared. "Never, perhaps since the time of Ivan the Terrible, has Russia seen such a number of executions," says Vladimir Korolenko, in an article in which he describes a new social group "to which prison slang has given the ominous name of 'deathniks,' a group mixing together people from all strata of Russia society, from top to bottom."[11]

Thousands of people were executed in those years—hung with dull efficiency and bureaucratic apathy. "Merely" thousands—the bloodbath that was to wash over Russia in another few years is comparable only in the color of the blood, not in its amount. However, the perspicacious Russian intelligentsia understood what kind of lethal leavening was being added to Russian society.

There were revolutionaries who, along with Gorky's Stormy Petrel, joyfully anticipated, "A storm! A storm is about to break! . . . Let the storm break more powerfully!" Korolenko did not wish this at all, but he understood the nature of the social "stormy petrels."[12] He saw the mood of the nation as uncontrollable anarchy hand in hand with plunder: "We can't go on living this way is what predominates powerfully in contemporary psychology. And since independent attempts to think creatively and actively struggle for a better future are suppressed everywhere, then only naked negation remains unshakeable. And this is what the psychology of anarchy is."

Tolstoy passionately did not wish this for Russia either. After reading Korolenko's article, he wrote the author in the spring of 1910, "I tried, but couldn't hold back . . . sobbing. I cannot find the words to express all my gratitude and love." Tolstoy did not need his eyes opened. He had written the story "The Human and the Divine," published in the collection *Against Capital Punishment* (1907); the lawyer Ivan Sakharov, Andrei Sakharov's grandfather, was one of the compilers of this collection.[13] Equally alien to Tolstoy were theories of the forced restructuring of social life, no matter how "scientifically" based they were, and the stagnant self-preservation of government power. Tolstoy examined the opposition to government evil and sympathized with the moral roots of this opposition, but he also saw its immoral roots. He believed only in the spiritual path to improve society. However, seeing how the first steps toward "a legal system based on civil liberties" were replaced by typical blind violence, originating chiefly in the government, the writer-moralist could not restrain himself. He wrote and illegally published the article "I Cannot Be Silent."

Some of Tolstoy's followers considered that this article, filled with horror and denunciation, was not compatible with his own teachings about nonviolent resistance to evil.

Tolstoy's moral sense turned out to be stronger than his moral philosophy. Similarly, Lebedev's moral sense turned out to be stronger than his political skepticism. The physicist Lebedev's plight was connected to the one of the writer Tolstoy in an even more direct way. Unable to bear the moral discord with his society, with those close to him, and with himself, the 82-year-old Tolstoy left his home, got sick a few weeks later, and died at the Astapovo railroad station at the end of November 1910. The name of this small station became famous in Russia.

Student uprisings followed in the wake of Tolstoy's death. The "indivisibility of thought, feeling and deed" declared by Tolstoy roused others to live with the same brave freedom. The student disturbances gave the police a reason to intrude on the life of Moscow University, which led to the resignation of its finest professors, including Lebedev, in February 1911. This was how the apolitical physicist Lebedev was drawn into politics.

Lenin called Tolstoy "the mirror of the Russian revolution." Some people considered that Tolstoy, like the Russian intelligentsia as a whole, was a concave mirror that focused social heat and finally lit the revolutionary campfire. Even if this is true, it is difficult to answer the question of who was more to blame—those who "chopped the firewood," providing an abundance of fuel in society, or the one who allowed his spiritual fire to burst out into the open.

In any case, the choice made by Lebedev in 1911 was not a sociopolitical one, but rather an individual moral decision.

Lebedev's death reminded the biologist Timiryazev of

> the involuntary cry which had once escaped from Pushkin's chest, racked with accumulated pain—a cry of desperation, a curse at the nation which had birthed him: "Why the devil was I born in Russia with a mind and a heart?" Perhaps with a mind, the cold, sarcastic mind of a dispassionate spectator of history's comedy; perhaps with a heart, too, but not with a mind which recognized the horror of what was occurring around it and could divine what it was preparing in the future. Better still without one or the other.
>
> This unfortunate nation's vast expanse is open to the soulless orgy of irresponsible feeble-mindedness. But the mind and heart cannot coexist . . .
>
> Or will the nation which saw one rebirth live long enough to see a second one, in which the preponderance of moral forces will be on the side of "prisoners of honor," like Lebedev? Then and only then will the opportunity arise for people with "a mind and a heart," to live in Russia, and not just be born in it—to die with a broken heart.[14]

For Lebedev, who died with a broken heart, with his skepticism outside the limits of physics, these exclamations were inordinately bombastic. There is no evidence that he dreamed about "Russia's rebirth." This distinguished him from the majority of his colleagues. And the younger his colleagues, the more revolutionary were their social—socialist—dreams. Student uprisings at Moscow University were a sign of the times Russia was living through. Andrei Sakharov's father graduated from Moscow University in 1912.

The era was having an effect across the entire expanse of the Russian land. It was just then, during those years, far from both Russian capitals, that Igor Tamm was graduating from the gymnasium in Elizavetgrad. By the age of 18 Tamm, who showed evidence of abilities in the exact sciences, had read his fill of socialist literature and aspired to "politics." Aware of their son's passionate temperament, his parents had grounds to fear for him and insisted that he continue his education abroad, far from the turbulent Russian universities. So Sakharov's future teacher spent his first year as a university student in Edinburgh, Scotland. This was the last peaceful year in Russia.

In 1914, World War I broke out. Its bloody experience predetermined the Russian revolution, which grew into a civil war. All this was crammed into the second decade of the twentieth century in Russia. Andrei Sakharov was born in this Russia at the beginning of the next decade, in 1921—born with a mind and a heart.

I

From Tsarist Russia to the Tsardom of Soviet Physics

1

THE EMERGENCE OF SOVIET PHYSICS
AND THE BIRTH OF FIAN

A ndrei Sakharov's life virtually coincided with the Soviet era. The year he was born, 1921, the new regime became definitively established, and the first non-Soviet elections (which brought the humanitarian physicist into the nation's parliament) were held in 1989—the year he died.

During this era the natural sciences became one of the most powerful social forces. The phenomenon of Soviet physics, in particular, cannot be separated from Soviet history. The names of the mathematician Lobachevsky and the chemist Mendeleev had acquired world renown back in the nineteenth century. But the nuclear and space achievements of Russia cannot be explained through its prerevolutionary heritage—indeed, it was during Soviet times that physics blossomed.

But by the end of its civil war in 1921, Russia was in no shape for science.

The Sakharovs

The life of any family grows intertwined with the history of a nation—or breaks up during critical periods. Surviving documents make it possible to see how harshly history intervened in the lives of Andrei Sakharov's parents, Katya and Dmitri, but it did not prevent him from being born into a family atmosphere filled with love and friendship.[1]

The February revolution began on February 23, 1917, with the spontaneous revolt of hungry people. A week later the tsar abdicated the throne and autocracy fell. That same week (February 25, 1917), Katya Sofiano's older sister wrote in her diary: "Today I met some physics teacher, Dmitri, an indescribably unattractive, awkward man. Only his eyes are lovely—nice, kind, and clear. Katya is in love, and so is he, so much so that they cannot and, it seems like they don't want to, hide it. It's a joy and a pleasure to look at them."

Dmitri Sakharov, a 28-year-old physics teacher, was the son of a lawyer and the grandson of a priest from a family of priests. Katya was the 24-year-old daughter of a career military man. The parents of the sweethearts belonged to different strata of educated Russian society. And the point is not that it was purely the noble class on one side and only half-noble on the other—by 1917 this wasn't of any particular importance. More significant was the difference in attitudes toward the ruling power. Andrei Sakharov's ancestors on his mother's side served the government well, whereas those on his father's side implemented the power of the spiritual, until they began to doubt that all power was God-given. Dmitri's parents were under secret police surveillance for nearly a decade, subjected to house searches twice, and even arrested. However, this did not interfere with Ivan Sakharov's (Andrei's grandfather) successful career as a lawyer, and as soon as the light of legal political life glimmered, he took part in the founding of the Constitutional Democratic Party, or the "People's Freedom Party."

By 1917, the autocracy's inability to function became obvious even to the military professionals. The attempt by tsarist authorities to squelch the spontaneous people's rebellions by force was unsuccessful—the military refused to obey. The February democratic revolution was peaceful.

Most of the Russian intelligentsia enthusiastically accepted the establishment of a republic and of democratic freedoms. Even people who were quite removed from politics, and from Marxism all the more, wore red bows. Today, when the word "Reds" can be an intimidating word, this fervor is not easy to understand. It is even more difficult to understand for a Westerner unfamiliar with the Russian language.

Among the differences between the Russian and English languages, one is directly related to political history. The Russian word *krasnyi* (red) is an archaic but still-used synonym for the words *krasivyi, prekrasnyi* (beautiful, handsome), which is the meaning of Red Square in Moscow, whereas *belyi* (white) has a rather more of a negative flavor (*beloruchka* ["little white hand," one who shirks physical work]). The Russian color perceptions may come from Asia—in traditional Chinese opera, for example, the positive characters wear red and the negative ones wear white. In English, the emotional connotation is the opposite: white corresponds to honor and purity, whereas red connotes anger, danger, or financial loss.

"Whites" appeared as a political term only after the "Reds" began to rule. After eight months of highly democratic but ineffective government rule, the Bolsheviks established their dictatorship in October 1917—in the name of the world revolution and universal communist happiness. This government coup was honestly called a coup for nearly a decade. Only after Stalin had established his regime did the phrase "Great October Socialist Revolution" come into use.

The Bolshevik Party was distinguished by sophisticated propaganda coupled with effective militarized organization. In particular, they used the linguistic power of key political words: "reds" (co-opting it from other socialist movements), *bolsheviki* (Bolsheviks; from *bolshinstvo*—majority, while they were a minority), *sovety* (councils, which they used not for meetings but for the implementation of their party's dictatorship). They also used people's feelings skillfully—principally the fatigue from the senseless, bloody war and the aspirations of the peasants to obtain land as private property. As a result, the Bolsheviks secured fairly broad support for themselves among the people and freed their hands to suppress their political opponents.

Soon after the October coup, the government outlawed the Constitutional Democratic Party, and its leading activists, fearing repression, abandoned the Bolshevik part of Russia. Among them was Ivan Sakharov; he left Moscow with his wife, Maria Domukhovskaya, and youngest son in early 1918 for the northern Caucasus, where they owned a house in Kislovodsk.

Ivan and Maria not only had typical Russian names; they were also typical representatives of the Russian intelligentsia. Ivan was born into a family of priests (third generation), but not a single one of his ten brothers and sisters was professionally connected to the church. All received educations and went on to become physicians, teachers, engineers, lawyers, and agronomists.

Maria was descended from an old noble family and educated at the Pavlovsk Institute in St. Petersburg. She learned about the members of the populist organization "People's Will" through a friend and helped them. Ivan's liberal attitudes, along with his legal education and profession, made him an active member of the "People's Freedom Party" (Constitutional Democratic Party) twenty years later.

Both show evidence of free thinking in their family life. Their church wedding, in a country with a state religion, came only after 18 years of conjugal life and the birth of all six of their children. Andrei grew up in a house whose spirit was the grandmother—a person of "exceptional spiritual qualities: mind, kindness and responsiveness, an understanding of the complexities and contradictions of life." His grandmother read his first books aloud to him. She read him the Gospels. He discussed with her "nearly every page" of Tolstoy's books, which he read himself.

Andrei Sakharov's other grandfather, Lieutenant General Aleksei Sofiano, discharged from the army due to his age, also couldn't expect much good from the Bolsheviks. In January 1918 his daughter Anna, Katya's older sister, wrote in her diary: "Mother was here in the evening. They have nothing to live on. They took away Papa's salary and pension. There are four of them. We're inviting them to live here at our apartment, so they can rent theirs out ..." (Anna was married to Aleksandr Goldenveizer, a conservatory professor.) At the end of February: "Today my sister Katya came over. Their material situation is very difficult. It's 8 degrees Celsius in the apartment again, there's nothing to eat, the bread ration is $1/8$ pound per person daily. We've already eaten all the buckwheat groats. There's a bit of rice and some potatoes left. What lies ahead?"

What lay ahead was that, a few months after the seizure of power, the Bolsheviks renamed their party the Communist Party and the second period, called "war communism," began: property was confiscated, money virtually abolished, and forced labor introduced. The government—the Communist Party, to be more precise—became the master of the nation. The security service and the army—newly created by this regime—crushed opponents, the various White armies, and peasant uprisings. The civil war that destroyed the country also deprived many millions of Russians of their lives and deprived Russia of two million emigrés. It's entirely possible that Ivan Sakharov would have become an emigré, too, if he had not died in December 1918 of typhoid fever—the civil war illness.

It seems amazing that it is possible to listen to music, read, get married, and have children in conditions like this. And that, along with the entries quoted from Anna Goldenveizer's diary, there are some entries of an entirely different kind: "And he [her husband] is sitting and playing Grieg . . . and he's playing so well," and "One consolation is Herzen." Her diary from the same year, 1918, on July 7:

> Today at two o'clock in the afternoon Katya married Dmitri Sakharov. The weather was marvelous, the sun was bright, everyone was dressed in white, we all walked to the Church of the Assumption on the Graves; the old priest kept grumbling at them, "Push aside the candle," and harassed Dmitri to death. A beautiful long table, decorated with wildflowers, and pretty darling Katyusha.

The whirlwind of civil war also carried the newlyweds away from Moscow to the south of Russia. "Dmitri is working as a teacher and in addition, he plays at the cinema in the evenings. He earns good money, but it's not enough to buy the basics . . ."

They returned to Moscow in the middle of 1920. And on May 21, 1921:

> Today at five o'clock in the morning Katya gave birth to a son . . . Yesterday at three o'clock in the afternoon she was driven to the clinic . . . Katya is extremely happy, sent her husband such femininely affectionate, happy letters, that I'm amazed that he could read them to us. I suppose because his cup runneth over . . . He is terribly excited and doesn't resemble his everyday self.

Ten days later: "Both of us [husband, who was the child's godfather, Aleksandr Goldenveizer] go over to see little Andryusha [diminutive from Andrei] every day. A very nice boy. Today is the first day I haven't seen him."

The happy mother had a strong character and, from the letter she wrote to her older sister soon after the birth, it is evident that family life was not totally idyllic:

Andryusha gave me such happiness and such a spiritual world that everything confused and violent receded into the distant, distant past, but this
didn't happen right away and also, returning from the clinic, I didn't immediately find the right path. How strange it was to mix up into our relations [with my husband] his love, his cult of the family (as you terribly aptly
observed). I'll tell you a big secret: the Sakharov family as a whole lives
on a very high spiritual level and maybe, a certain bitter contrast caused
my attitude. I'm entirely to blame. Now everything is so clear, simple and
splendid! It's a pity that Dmitri had to spend yesterday so far away and it
was so difficult for him, but I know that he was thinking about us. Lately,
he has demonstrated his feeling with his exceptional concern and fully
deserves boundless feeling in return. The little one is sleeping in my lap
right now . . .

The mother was 28 when she had her first child, and the father 32, rather late
by Russian standards. The father kept a journal in the boy's name, in which
he noted the events of the first months of his life, then later the first words he
spoke.

However, in 1921 much more was required than paternal feelings. The boy's
uncle, Vladimir Sofiano, whose own children died during the civil war, wrote
in a letter: "Katya's husband traveled to Kiev province for provisions, was gone
a long time, but brought back a lot. I'm glad that their little son will be well
provided for and won't have to be as hungry as my children were in 1919 and
1920. Their little son is healthy and a very nice boy."

The power of parental feelings, even in the circumstances of a social whirlwind, are easy to understand. More difficult to understand is how these parents could have given their son a sense of social optimism when both his
grandfathers had suffered at the hands of the Bolsheviks—one was forced to
flee them, and the other was deprived of an honestly earned pension.

Two factors were at work here. One was rather simple—the Bolsheviks had
succeeded in creating order in Russia after an incapable autocracy, the bloody
chaos of the First World War, and the democratic anarchy of the provisional
government. The second factor—a more complex one—lay hidden in the social role of the Russian intelligentsia, to which the Sakharovs belonged. The
writer Pyotr Boborykin, who introduced the word "intelligentsia," was a friend
of the family.[2] Korolenko, the writer who served as the epitome of the Russian
intelligentsia for Andrei Sakharov, corresponded with his grandfather. And
Andrei Sakharov's father worked in Pyotr Lebedev's laboratory right up to the
moment when the latter left the university. They knew about that momentous
event firsthand. And there was the home of the pianist Goldenveizer, which
Andrei's parents (and then he, too) frequented.

Aleksandr Goldenveizer (1875–1961) was a famous musician and professor
at the Moscow Conservatory. He was a close friend of Leo Tolstoy, witnessed
his will, and in 1923 published a book about him.[3] So, when Andrei read Tolstoy
as a child, "discussing almost every page with my grandmother," and saw the

bust of Tolstoy in his grandmother's room, Tolstoy was more than merely a great Russian writer to him.

The Intelligentsia and Soviet Rule

The suppression of the freedom of speech, the brutality of the Red Terror, aroused the resistance of even the left-wing Russian intelligentsia: both Gorky and Korolenko spoke out in defense of the democratic freedoms of specific individuals. However, there were other things besides repression for which the Bolsheviks used their power. For instance, they supported Gorky's idea to create—in the conditions of collapse and hunger—an organization to assist scientists and people who worked in the cultural arena. "The academic ration," which Goldenveizer received, enabled him to feed his relatives, including his godson-nephew. It also made it possible for him to devote himself to his own work: "Yesterday I was at a Bach evening organized at Shura's [Goldenveizer] initiative at the Small Hall of the Conservatory. The best musicians of Moscow participated . . . They played enchantingly. There will be another 12 such concerts." This was in 1921, the time of the civil war.

Already, within its first months of rule, the Soviet government had implemented significant general cultural reforms, associating Russia with Europe—it introduced the metric system of measurement and the Gregorian calendar. Neither the tsarist nor the provisional government had been able to implement these reforms, proposed even before the revolution. The old Russian measurements of the *arshin* (28 inches) and the two-week calendar lag behind Europe were left in the past. The anachronisms of the old orthography, such as the mandatory hard sign after consonants at the end of words, were also dropped.

The new government had slogans that were attractive to the intelligentsia. In a 1919 article, "Successes and Difficulties of the Soviet Power," Lenin, the leader of the nation, made the appeal: "We must take all of the culture which capitalism has left and build socialism out of it. We must take all the science, technology, all the knowledge and art. We cannot build the life of communist society without it."[4]

The slogans were reinforced by some practical actions. In particular, several physics institutes were established in Petrograd in the fall of 1918. And in February 1919 the first congress of physicists was held. This predisposed scientists favorably toward the new regime. In 1919 the 75-year-old biologist Kliment Timiryazev relinquished his title of Honorary Doctor at Cambridge University as a protest against English intervention in northern Russia, and he was elected to the Moscow Soviet of Workers' Deputies in 1920.

People usually discover that the road to hell is paved with good intentions after a major part of this road has already been traversed. It would seem that the reason for restricting intellectual freedom would have disappeared when the civil war ended in 1921. However, it was soon revealed that the Soviet regime did not need all of culture. The regime increased its control over society and in 1922 exiled a large group of "bourgeois intelligentsia," including Professor Nikolai Berdyaev, one of the most preeminent Russian philosophers of that time, and Professor Pitirim Sorokin, who later founded the sociology department at Harvard University. They were indeed not Soviet, but the government pronounced them *anti*-Soviet, and their exile—instead of execution—a prudent humaneness. The ruling "humanists" were operating according to the slogan "Who is not with us is against us."

Timiryazev, whose book *Science and Democracy* elicited Lenin's rapture in 1920, died in April of that year and immediately wound up in the Soviet regime's pantheon. Had he lived to see it, what would he have thought of the 1922 exile? It is hard to believe that he would have given it his blessing—recall what he wrote about the glaringly apolitical Lebedev. The idealistic and religious philosophy in which many of those who were banished placed their faith was undoubtedly foreign to Timiryazev, but his materialism was idealistic enough for him to understand the free thinking of these people who did not wish to convert to the state religion. Moreover, many of the exiled had lived through Marxism in their youth, knew it firsthand, and had outgrown it.

Younger people working in the natural sciences were more inclined to ignore the exile of scholars in humanities—at that time, the natural sciences were not yet subjected to ideological controls. And for the Soviet regime, the words "science and technology" were among the most important in their vocabulary.

The new era promised broad horizons for the creative imagination. The era of social experimentation was in tune with the search for new artistic forms. The film director Sergei Eisenstein went down in history with his "red" film *The Battleship Potemkin* (1925), which rose to the pinnacle of world film art. The artist Marc Chagall was a Red commissar in Vitebsk. It wasn't evident until the end of the 1920s that the ruling regime was more interested in propagandistic support than in artistic exploration. The 1930 suicide of the poet Vladimir Mayakovsky, who placed his poetry at the service of the revolution, signaled this.

The scientific-technological intelligentsia was particularly responsive to the promise of rapid social progress for its nation and for the entire world. Marxism emerged in an age of triumphant achievements in the natural sciences and appropriated the scientific approach to extend its victorious methods to the life of society. Physics was at the forefront of the natural sciences, and it's no accident that Lenin devoted one of his major (and most voluminous) prerevolutionary books to the revolution in physics that started at the beginning of the twentieth century.

Having won power, the Bolsheviks developed a system of public education on a grand scale, with an attitude of great respect toward the natural sciences. All this enabled people who had devoted their lives to science to view everything going on around them with optimism, to forget personal resentments, and to ignore the suffocating reins of the ruling ideology in the humanitarian spheres of social life. The scientific-technological intelligentsia hoped that the openly declared "dictatorship of the proletariat" could lead to social and—principally—scientific and technical progress more rapidly than would a formally democratic power, encumbered by unwieldy democratic procedures. They hoped that the dictatorship would be run by enlightened people. Who could have guessed that without the "unwieldy" democratic organization of political life, what is most certain to arise is a totalitarian society, which is ideally suited for dictatorship by a single individual.

In the meantime, scientists used the resources that the ruling regime—without coordinating with anyone else—generously made available to them. Was this egocentric? More likely, it was science-centric.

The Emergence of Soviet Physics

There were entirely material reasons for the blossoming of Soviet physics, in addition to the state ideology's benevolence. The state's might—the principal concern of a totalitarian regime—requires advanced technology, which is dependent on science.

Soviet physics began at a substantially lower level than chemistry and mathematics. Prerevolutionary physics had no achievements on the level of Lobachevsky's non-Euclidean geometry or Mendeleev's periodic table of the elements. In breaking new ground and using generous funding ("irrigation"), the first harvest was an especially good one, because the "farmers" were genuine scientists, devoted to science. They became leaders due to their organizational abilities, and they put their personal positions at the service of science. It produced results: by the mid-1930s, the number of physicists had increased tenfold in ten years.

The first main events in Soviet physics took place in Petrograd, renamed Leningrad in 1924. The Leningrad Physics and Technical Institute, or PhysTech, became the chief "hotbed of cadres." Its director, Abram Ioffe (1880–1960), a student of Roentgen who blended into Soviet life easily, was skilled at finding a common language with the authorities and did an extraordinary amount for the growth of physics. PhysTech was deservedly called the cradle of Soviet physicists. But it was not the only cradle. Soviet physics was done also by two other institutes in Leningrad—the Optical Institute, directed by Dmitri Rozhdestvensky (1876–1940), and the Radium Institute, headed by Vladimir Vernadsky (1863–1945).

The figure of Vernadsky is particularly interesting. He was considerably older than Ioffe and Rozhdestvensky and wasn't even a physicist, and yet he played an exceedingly important role in the history of the Soviet atomic project. A geochemist, Vernadsky was a pioneer of Russian radiology, as the study of natural radioactivity was then known. He recognized the new phenomenon early on and already wrote in 1910: "Sources of energy have been revealed to us, alongside which the power of steam, electricity and explosive chemical processes pale in power and importance . . . We are looking at a new ally and protector with hope and apprehension."[5]

Radiology grew into radiochemistry and nuclear physics. This is how Vernadsky described the state of affairs to the government in 1922:

> The organization of the State Radium Institute, the result of work which had been conducted at the Russian Academy of Sciences since 1911, cannot be completed without close contact with analogous work in the West and without bringing its equipment up to the level of contemporary knowledge. In the field of radium, the institute cannot be supplemented and organized solely with what is within Russia's borders, as she has been deprived of normal contact with the cultural life of humanity, while enormous successes were made in this field in the period 1914–1921, especially from 1918–1922.
>
> The preservation of the Radium Institute's work is in our time one of those tasks which the State authorities cannot postpone without enormous, perhaps irreparable, damage to the matter. I maintain this because I clearly recognize the inevitable—the revolution in the lives of human beings when the problem of the atomic energy and its practical use is resolved. This is not acknowledged by public opinion yet, but public opinion has no opportunity for its revelation and this has to be taken into consideration given the present situation which has developed.[6]

The tone of the last sentence is quite unusual for a letter to Soviet authorities, but it belongs to someone who, in prerevolutionary times, took part in the public life of the nation along with his professional work in geochemistry. He was one of the most distinguished professors who quit Moscow University along with Lebedev in 1911. He participated in the founding of the Constitutional Democratic Party and was a member of the provisional government. For this reason, he left for the south of Russia in 1918, like Sakharov's grandfather, to be farther away from Petrograd and Moscow (which had become the capital cities of Bolshevism) and, like Sakharov's parents, returned at the end of the civil war.

Although he had no sympathy for Bolshevism, Vernadsky nonetheless saw that a powerful social energy had been awakened in the nation and that the regime was directing a substantial part of it to the development of science. And for Vernadsky, the history of mankind was principally the history of science and technology. This was the basis of his collaboration with the Soviet regime, but it did not obscure his view of the reality surrounding him. Vernadsky did not fear noting his social observations in his diary with the precision of an experimental natural scientist. Later we will come back to his eyewitness accounts of the brutal period and the advent of the nuclear era.

Lebedev's Heirs in Moscow:
Pyotr Lazarev and Arkady Timiryazev

From the time that the new capital of Russia, St. Petersburg, was founded by Peter the Great, relations between the two Russian capitals became an important aspect of the nation's cultural life. The Russian intelligentsia's genealogy is sometimes traced back to Aleksandr Radishchev's *A Journey from St. Petersburg to Moscow* (1790).

The Imperial Academy of Sciences, in which residents of St. Petersburg and humanities predominated, was located in St. Petersburg in the vicinity of the institutions of government power. No position was found at the Academy for the Moscow physicist Pyotr Lebedev. This did not prevent Moscow physics from noticeably outstripping St. Petersburg physics in scientific achievements.

The situation changed after Lebedev left Moscow University. According to Sergei Vavilov, a student at the time and later president of the Academy of Sciences, "for many years before the revolution, Moscow University remained without its own core professorial faculty. Casual outsiders were invited instead of eminent scientists. The University's scientific life stood still and withered during those years."[7]

Lebedev's students quit the university along with him. Pyotr Lazarev (1878–1942), his closest associate, was involved with the laboratory while Lebedev was away. He communicated to Lebedev in one of his 1910 letters:

> Recently a new graduate student, Sakharov, appeared; he completed the work very well. I became acquainted with him earlier, at the beginning of this year when he attended my course, and my conversations with him showed that he has read a fair amount and is bright. It seems to me, for this reason, it would be a good idea to give him a place here with us starting this fall, especially since the topic of diffusion hasn't been given to anyone yet, and it is important for processes inside nerves.[8]

He is referring to Andrei Sakharov's father, Dmitri Sakharov, who began his university studies in the department of medicine and then switched to physics a year later. This could also have additionally disposed Lazarev favorably toward him, because the professor himself was first educated as a physician.

Lazarev worked in biophysics, but that wasn't the only reason he couldn't replace Lebedev. None of Lebedev's first-generation students could replace him. Nonetheless, after Lebedev's untimely death, Lazarev was obliged in 1916 to become director of the just-constructed Physical Institute. In 1917 he was elected to the position of academician—the deciding vote, one might say, was cast by Lebedev from the grave.

The institute was built with funds from a specially created private foundation; however, after the establishment of the Soviet regime, the institute had to find its place in the new government system that had abolished private property. Lazarev found such a place under the aegis of the People's Commissariat of Health, where he became head of the X-ray department. The words "Biological Physics" appeared in the institute's name, and in 1921 Lenin himself (after he was shot and wounded) underwent an X-ray examination there.

The institute, planned for a distinguished prerevolutionary physicist, helped his colleagues survive the difficult postrevolutionary period, but they did not produce any scientific results comparable to Lebedev's or to the work of the Leningrad physicists. Lazarev knew how to find issues related to the national economy, but he lacked the depth and scientific passion to create a first-rate institute.

And what about Moscow University, which Lazarev left along with Lebedev in 1911? Lazarev did not return there after the revolution, but not because he didn't want to teach. Arkady Timiryazev—another of Lebedev's students, who seized the actual directorship of the physics department—did not allow him into the university.

The son of the famous biologist Kliment Timiryazev, Arkady was quite an ordinary scientist with extraordinary social energy. Like his father, he unquestioningly supported the Soviet regime, but his father did not live long enough to develop reservations—he died in 1920. The son, most likely due to his father's name rather than his own merit, was accepted into the party by a special Central Committee decision in 1921. He quickly justified their trust, becoming a member of the editorial committee of the journal *Pod Znamenen Marksizma* [Under Marxism's Banner] in 1922, and earned Lenin's praise for being a "militant materialist."

However, Arkady Timiryazev was not satisfied with a successful Soviet career. He saw a much more distinguished place for himself in science and passionately tried to establish it. He claimed to be Lebedev's principal disciple, but he went down in the history of Soviet physics chiefly as an opponent of the theory of relativity and quantum mechanics. How could this have happened to a student of Lebedev's?

He had graduated from the university in 1904, right when the revolution in physics was beginning—rapid and stormy change in its concepts related to ideas of relativity and the quantum. In order to keep up with events at a time like this, one needs strengths that are inborn and age dependent. Summing up his life experience, Max Planck, the father of the quantum idea, sadly and famously noted that new ideas enter science not because their opponents admit their mistaken beliefs, but it's just that these opponents simply die out, while the up-and-coming generation becomes familiar with new ideas from the very outset.

The behavior of a person from the departing generation depends on his abilities, temperament, and ethical foundations. Some (including Planck) silently experience an internal drama or even tragedy, tormented by the fact that their scientific ideas reveal their limitations. Others, unable to relinquish their habitual ideas, try to dissuade their colleagues from accepting new concepts. Those who are more powerfully creative critically analyze the new physics, making the older physics more classical and elucidating the new—Einstein in the last decades of his life is a good example.

However, attachment to a way of thinking acquired at the beginning of a career can manifest itself entirely differently. Having discovered that their scientific arguments are lacking, and unable to admit their loss of contact with the leading edge of science, learned men sometimes expand their arsenal and arm themselves with their society's nonscientific resources.

There were physicists in Nazi Germany who rejected the theory of relativity as non-Aryan, as a manifestation of the Asiatic spirit. In Soviet Russia, the physicist Timiryazev rejected the theory of relativity as nonmaterialistic and the handiwork of the bourgeois West.

During his student years, Arkady Timiryazev became familiar with the molecular physics of gases. That was what he taught students of the "upcoming generation," solemnly calling it the "kinetic theory of matter" in order to strengthen their materialism.[9] Sakharov attended his "rather boring lectures": "Timiryazev bore a striking resemblance to his father and thus, also to his father's monument which stood at Nikitskie Vorota. We students called Timiryazev 'the son of the monument' behind his back."

The disrespectful students did not suspect how apt this nickname was. Their professor was a foundling. Only "by the Highest Decree of the Sovereign of all Russia, condescending to the humble petition of Professor, Councilor of State Kliment Timiryazev" in 1888 was the eight-year old "Arkady, his ward" allowed "to accept the family name of his guardian, his patronymic after his name and to enjoy the rights of Personal Nobility."[10]

And so the eminent biologist was not biologically responsible for his stepson. What about the "striking" resemblance Sakharov writes about? When you look at the photographs of both Timiryazevs, the only visible resemblance is "the professorial beard." The older Timiryazev's beard did not stand out against the professional backdrop of his time, whereas the younger man's beard stood out greatly and caused him to resemble the statue that stood near the university (which pleased him).

It's easier to engender a sense of resemblance by means of a beard than through scientific stature. Young Timiryazev's ambitions demanded more than teaching the theory of gases. The theory of relativity was the biggest deal in physics at the time. It wasn't taught to him at the university, and he couldn't assimilate it on his own. So he decided to overthrow it.

His ambitions also required holding onto Lebedev's legacy—the physics department at Moscow University. Timiryazev substituted administrative maneuvers for scientific authoritativeness. Only the intervention of the Workers' and Peasants' inspection in 1930 deprived him of power—resulting in the most favorable consequences for the development of physics (more about this coup in chapter 2).

Just a few months after Timiryazev was dismissed from the directorship of physics at the university, the Workers and Peasants' authority suddenly dealt a blow to Lazarev and the institute which he headed. Academician Lazarev was arrested on March 5, 1931. A few weeks later he admitted that he felt guilt for "informing foreigners about several issues related to science," in particular, "about proposed conferences." As a result, Lazarev was accused of espionage and was exiled to Sverdlovsk for three years, where he was allowed to work and teach. The sentence was commuted in February of the next year, and Lazarev returned to Moscow.[11]

No explanation has been found for this arrest. At the beginning of the 1930s, the government's attitude changed toward specialists left over from the old regime. Show trials were held against engineers—the Shakhtinsky case and the Industrial Party [*Prompartiya*] case. However, by his social position and the nature of the charges trumped up against him, Lazarev does not belong in this group.

The fate of his institute—a new "Physical and Chemical Institute for Special Assignments" was moved into his building—suggests a possible reason for what happened to Lazarev. The new institution's director was related to the head of the NKVD (the Soviet secret and overt police, also known as the Cheka, GPU, OGPU, and finally the KGB). But it's a complete mystery what kind of special assignments this splendid building, once intended for Pyotr Lebedev, handled for more than three years.

Physics and historical justice returned there in 1934 when, by government decree, the Academy of Sciences moved from Leningrad to Moscow. The Physical Institute of the Academy of Sciences, recently created in Leningrad, moved into the building in August; in December it was awarded the name Lebedev. This institute is known by its Russian acronym—FIAN.

But how did this institute—in many ways so Muscovite—get started in Leningrad?

The Father and Stepfather
Founders of FIAN in Leningrad:
Georgi Gamow and Sergei Vavilov

Leningrad was the scientific capital of the nation until the Academy of Sciences was moved in 1934. The Soviets made Moscow the country's capital instead of Leningrad in 1918, but they were slow in moving the Academy. They didn't quite understand how to deal with this institution, which was accustomed to a fair amount of autonomy. It took until the early 1930s for the government to gain control over the academy, both by leaning on the support of scientists who genuinely sympathized with socialism and through "classless" careerists. Both the carrot—money for the development of science—and the stick—arrests—were effective.

At that time, practically all of Russian physics lived outside the Academy of Sciences and its feeble Physical and Mathematical Institute (FMI).[12] A new research fellow, Georgi Gamow, appeared in the autumn of 1931 at the FMI. After a three-year stay in physics capitals, he returned to Leningrad with world renown after explaining alpha decay—the first paper on theoretical nuclear physics. In 1928, the main Soviet newspaper *Pravda* printed the proletarian poet Demyan Bedny's praise of Gamow just a few weeks after his work was published:

> *They've Gotten To Those Atoms*
>
> They call the USSR a nation of murderers and cads.
> Small wonder. You take this Gamow, a Soviet lad.
> Whad'ya expect from folks like that?
> Why, that scoundrel! Right to those atoms he has found a path
> Those millions on a pinhead, scores!
> With artful tricks and sly mechanics,
> He penetrated right into the atom's core.
> One, two, three! And smashed its nucleus to bits!
> A Soviet type—now that's a signal to Europe *et al.*—
> He blasphemously solved the toughest riddle of all.
> Now, indeed, is this not direct sabotage
> Against order, the establishment itself?
> Sabotage or not, let us frankly speak,
> Science of the Revolution's eve does reek.

This poetic masterpiece was prefaced by the following note:

> The 24-year old Leningrad University graduate student G. Gamow—who was sent to Copenhagen six months ago to work at the institute founded by Niels Bohr, one of the most pre-eminent physicists of modern times—has made a discovery which has created an enormous impression in international physics. The young scientist solved the problem of the atom's nucleus. The atom's nucleus is known to be important as the field containing giant reserves of energy and the possibility of artificially transforming the elements. Therefore, every new step in the discovery of its structure is of exceptional scientific interest.[13]

The newspaperman conveyed the essence of Gamow's work with about as much precision as the poet. But it genuinely was an enormous success—the first explanation of radioactivity using recently established quantum mechanics.

After spending three years in the best physics locations of Europe, Gamow found that the USSR had a shortage of scientific specialists and paid them very small salaries. He began working at three institutions simultaneously: the FMI, the Radium Institute, and Leningrad University. According to a questionnaire that he filled out, he was fluent in German, English, and Danish and read and

translated ancient Egyptian with the aid of a dictionary. Without his European experience, he would have hardly allowed himself to be so free with the personnel department.

But Gamow had no intention of lingering in Russia for very long. He brought back an invitation to the First International Congress on Nuclear Physics, which was to be held in Rome in mid-October 1931. "Gamow (USSR). Quantum Theory of the Structure of the Nucleus" appeared on the agenda of the Congress, and he saw nothing to prevent him from delivering one of its principal lectures.

But big changes had taken place in his homeland during his three-year absence. The accelerated construction of Stalinism had begun, and centralization had drawn in all the new sectors of public life, including science. It was a complete surprise to Gamow that his trip, which had promised to be so triumphant, got bogged down in bureaucratic alleyways. And this wasn't the only invitation that Gamow was not permitted to accept. Bohr invited him to a conference at his institute. The Poincaré Institute in Paris and the University of Michigan invited him.

Of course, scientific life cannot be reduced to international conferences. A theorist's daily circle of contacts is more significant for his work. His contact with two young theorists from PhysTech—Lev Landau and Matvei Bronstein—was especially close back in his student years. They had trendy student nicknames—Johnny, Dau, and Abbot—and a fashionable name for themselves, the Jazz Band.

The Jazz Band's social activism took various forms but was sustained by a single source. Already wholly independent researchers who needed no scientific guidance, they wanted to devote themselves to physics on an international level. Love of creative freedom plus youth (the oldest of the group, Gamow, was 27) spurred them to actions that made the older generation feel uncomfortable.

One of the Jazz Band's improvisations led to the creation of the Physical Institute of the Academy of Sciences, FIAN. Actually, Gamow and his friends had something else in mind—they wanted to create a small Institute of Theoretical Physics. It required no expenditures; a theorist requires only a pencil and paper to work. There were also walls within which the institute could settle. At the end of 1931, Gamow submitted a memorandum with a proposal to divide the Physical and Mathematical Institute of the Academy of Sciences into two independent institutes: the Mathematical Institute and the Physical Institute, "granting the Physical Institute the role of an all-Union theoretical center, the necessity of which is felt lately." The development of theoretical physics was to be "based on a dialectical-materialist methodology."[14] Gamow wrote the last words over the crossed-out words "in accordance with the modern materialist worldview"—apparently he was informed that this lagged behind the terminology of the day.

Events developed quickly and at the highest academic level. Academician Alexei Krylov, the director of the Physical and Mathematical Institute, sup-

ported Gamow's idea. But opponents turned up as well. The stumbling block was the cornerstone for Gamow and his friends—the proposed institute's concentration on theoretical physics.

As Soviet science centralized, the influence of its leaders grew. Although these leaders in the 1930s were genuine scientists, devoted to science, their inordinate influence sometimes had a negative impact. The leading institutes of physics were run by Academicians Ioffe and Rozhdestvensky, who were both experimenters. But the experiment involving an Institute of Theoretical Physics did not elicit sympathy from either of them. There was much that separated them, but not their views on the relationship between theory and experiment.

Gamow and Landau did not intend to divorce theory from experiment—both were oriented quite strongly toward the experimental basis of physics, oriented but not subordinated. And they knew firsthand about both the Gottingen and the Copenhagen institutes of theoretical physics.

It didn't even help that Gamow was elected to the Academy of Sciences in the heat of this battle. In order to feel the Jazz Band's pressure on the academic "old fogies," let's read the letter Landau wrote to Pyotr Kapitsa in November 1931:[15]

> It's imperative that Johnny Gamow be elected academician. After all, he's indisputably the best theorist in the USSR. In this regard, Abrau (not Dyurso [a famous Russian vineyard], but Ioffe), mildly envious, is trying to put up resistance. The unhinged old boy, with a false idea of his own importance, has to be reined in. Would you be ever so kind as to send a letter addressed to the permanent secretary of the Academy of Sciences, in which you, as a Corresponding member of the Academy, praise Johnny; it's best to send it to me so that I can simultaneously publish said letter in *Pravda* or *Izvestia* along with letters from Bohr and others.

But more than anything, it was another native "old boy," Vladimir Vernadsky, director of the Radium Institute, who helped elect Gamow to the Academy.[16] On February 29, 1932, he was elected a Corresponding Member on the eve of his 28th birthday.

The day before, the General Meeting of the Academy resolved to divide the Physical and Mathematical Institute into two separate institutes: the Physical Institute and the Mathematical Institute. It would seem that the institute that Gamow had planned was born. However, that was only on paper. On another piece of paper, Gamow created the plan of the new institute, according to which, the "Institute of Theoretical Physics is the central institution devoted to the study of the fundamental problems of contemporary theoretical physics based on a materialist worldview" (he forgot the dialectical terminology again!).

The plan was sent for review to the academician physicists. Ioffe categorically objected to the creation of an institute of theoretical physics instead of "a basically experimental institute, linked in a specified relationship to [Ioffe's] Physics-Technical Institute as the indisputably leading institute of the Soviet Union." Rozhdestvensky also considered that "the establishment of a special

theoretical institute in the USSR is harmful, because theorists should work in large institutes of physics (the Physics-Technical, the Optical) side by side with numerous experimenters, promoting their work and receiving stimulus from them for their research . . ."

The meeting held on April 29 was where Rozhdestvensky proposed to call a plenary session of academician physicists and suggested that "Academician S. Vavilov is also interested in the issue about the Physical Institute in connection with his being chosen its director. An increase in the number of staff theorists is not expedient; it is mainly the experimental section which should be strengthened."

Landau attempted to save the cause with the objection that "theoretical physics must play an important role in the proposed institute and not be something supplemental. It also plays an important role in experimental physics: an example can be studying the thin-layer insulation which was conducted without considering theoretical data, as a result of which many resources were expended without producing any results."

Those present preferred not to notice the stone that was cast at Academician-experimenter Ioffe—he had recently suffered a failure in insulation technology that had promised a big leap in building socialism. Rozhdestvensky gave a rebuke to the ill-bred young theorist, declaring that, in the new institute, "theorizing should be done as it is done in other institutes."

Ioffe (with reason to be grateful) joined Rozhdestvensky and suggested that "Academician Vavilov move to Leningrad in order to assume the position of director of the Physical Institute."[17] It was the end of Gamow's FIAN. The beginning of Vavilov's FIAN can be traced to his response of May 7, 1932:

> Without knowing the actual staff, resources and previous work of the Institute, I intended to postpone my evaluation of the Institute's plan of topics which was sent to me before my trip to Leningrad. Unfortunately the trip could not take place for a long time and I must limit my report to the following remarks:
>
> The plan's issues, related to the structure of nuclei, atoms and molecules in ordinary and extraordinary intra-nuclear conditions are, of course, the most significant and interesting issues in contemporary physics and astrophysics and, of course, these same issues are in the pipeline at many European and American Physical Institutes. At the same time, restricting the Institute's work exclusively to such difficult issues, naturally, places the feasibility of such a plan in doubt, especially within one year. Not knowing the Institute's staff and powers, I will not undertake to evaluate the degree of probability of fulfilling the plan, but I think that a certain reduction of the difficult major topics and the corresponding increase in the number of easier and more specific topics would be desirable.

The moderation and good manners are immediately evident. Common sense and a broad understanding of physics by the 40-year-old Moscow University professor, recently elected academician, were also well known. It was easier to deal with him than with the gang of young theorists.

Vavilov soon began taking care of the newborn institute, and he became its director in September.[18] Gamow, the "father of the newborn," also remained a research fellow of FIAN. But it appears that the failure with the Theoretical Physical Institute convinced him that he was in a cage, and not even a golden one. He attempted to escape this cage in the summer of 1932 across the southern border of the USSR by kayak, and in the winter on skis across the northern border, but was unsuccessful.

In October 1933, the Solvay Congress on Nuclear Atomic Physics helped. This time Gamow was permitted to attend, and he never returned to Russia. He didn't slam the door shut, and he was still considered away on a trip abroad for an entire year. Vavilov even mentioned him in an article in *Pravda* on November 5, 1934, about FIAN's move to Moscow and its imminent transformation into an "all-Union scientific center of physics."[19]

There were real preconditions for this move to Moscow, and Vavilov knew them well. The main one was Leonid Mandelshtam's school, which by that time had blossomed at Moscow University.

2

LEONID MANDELSHTAM

The Teacher and His School

The Electrotechnical Consultant
at Moscow University

Moscow physics was in a depressed state when Leonid Mandelshtam, an electro-technical consultant in Leningrad, received a letter in the summer of 1924:

> I've wanted to send this letter off to you for some time, but some uncertainty in the situation restrained me. Things were finally clarified today to the point where I could write to you. It concerns your candidacy for a Chair of Theoretical Physics at Moscow University. In any event, you probably know that after S. Boguslavsky's death, we proposed your candidacy along with those of Epstein and Ehrenfest. However, we were unable to obtain an announcement about the competition until now: the Board, ostensibly for reasons of economy, was refusing to submit a petition about the opening of the competition to the State Scientific Council. Finally, today, an announcement was made that if you, Ehrenfest or Epstein agree to accept this Chair, then the Board will not put forth any objections. It is perfectly clear that neither Epstein nor Ehrenfest will come here. So the whole thing comes down to your acceptance. You will most likely receive an official inquiry regarding this any day now. As for me, I decided to appeal to you with this letter, which expresses my opinion and that of many of my university comrades.
>
> Of course, you are familiar with the situation at Moscow University and know the people who matter there. For this reason, you are fully aware of Moscow's negative side. The other side of the situation concerns the following: many of us hold the firm conviction that you represent the last hope for revitalizing the Physical Institute at Moscow University. Only the appearance of a person like you could initiate the formation of a circle of individuals who are able and wish to work, and will put an end to the endless intrigues which have completely consumed the Institute. There is a small group of students,

hungry for real scientific guidance, who, in spite of their youth, are already disappointed in the present directors of the Institute.

The Department of Theoretical Physics was established as a Theoretical Physics Office and Laboratory—so that the possibility of setting up some experimental work is open to you. There are only two rooms at the disposal of the Theoretical Physics Office at the present time. But if, when you arrived, an increase of space were necessary, I have no doubt that it would be possible. And so, I think that you will find several people in Moscow who fervently await your arrival and that with them you will be able to create a working circle around you.

You are, of course, well aware that one of the negative aspects is low pay. Most likely, you'll also be able to count on other sources of income, specifically State Publishing House. As for an apartment, it seems to me that you might be able to make the provision of an apartment to you a condition, and I think that the University would find it possible to extend the opportunity to you. Pardon me for being so bold as to speak to you about all this: I am quite afraid that you will immediately and categorically refuse.

All the best.

Respectfully yours, Grigory Landsberg[1]

Mandelshtam began working at Moscow University in the fall of 1925. But what was behind this near supplication from the 34-year-old Moscow University senior lecturer to the 45-year-old electrotechnical consultant?[2] Grigory Landsberg was noted for his firm and reserved character. He probably didn't know that Ehrenfest (mentioned in the letter) had written to Mandelshtam in Strasbourg in 1912: "If I wound up in Strasbourg, I would work on some experiment under your supervision." Or that Einstein had sent a postcard to Strasbourg University in 1913: "Dear Mr. Mandelshtam! I just presented to a colloquium your elegant work on surface fluctuations about which Ehrenfest informed me. It's a pity that you are not here."[3]

However, Landsberg, living in the world of physics—at seminars, conferences, in journals—understood without any archival evidence that the electrotechnical consultant was a scientist of European stature. Someone like him was essential to pull physics at Moscow University from the ditch of scientific mediocrity and intrigues. Equally important were other qualities Mandelshtam possessed, which Landsberg described 20 years later:

I was enchanted by his unusual mild manner during my very first meetings with him and felt that, in meeting him, a person who possessed not only a big mind, but a big soul, was entering my life . . . You could be reprimanded by him for harsh and unrestrained language. But he required steadfast behavior at all times and he never advised compromise. Least befitting his image is the word "sternness," as he was always sincerely gentle, humane and kind. Nevertheless, when it came to any sort of equivocation, no one's judgment was more stern. And everyone who came into contact with him felt this.[4]

Sergei Vavilov, Landsberg's contemporary, shared his attitude. Both had graduated from Moscow University (in 1913 and 1914), having had only brief contact with Pyotr Lebedev himself, but they fully sensed the squabbling among his students over his scientific legacy, which destroyed Moscow physics. That is why Landsberg and Vavilov, neither young nor Mandelshtam's students, were so taken by the combination of scientific and moral stature in him that they persistently tried to obtain his invitation to Moscow University and accepted his guidance so readily.[5] This says a lot about them.

At first, only one of Mandelshtam's colleagues at Moscow University had been his student in the ordinary sense of the word—Igor Tamm. His scientific path is especially impressive testimony to yet another of Mandelshtam's remarkable talents—teaching.

Igor Tamm's Path to Science

Recalling his first meeting with Tamm in 1945, Sakharov related: "His study had the same furniture, which I then saw for decades; a desk, strewn with dozens of numbered pages covered with calculations unintelligible to me, dominated everything, and above the desk—a large photograph of Leonid Mandelshtam, who had died in 1944, and whom Igor Tamm considered his teacher in science and life." When Sakharov left, Tamm gave him a book in German about the theory of relativity to study and an unpublished manuscript about quantum mechanics by Mandelshtam.

Tamm's background was so unfavorable to his success in science that only a genuinely remarkable teacher could compensate for his life's circumstances. Suffice to say, Tamm published his first scientific paper at the age of 29, while the aforementioned German book about the theory of relativity was written by a 20-year-old author. For a theoretical physicist, for whom Nobel recognition lay ahead, the former is far more amazing. "Physics is youth's game," the saying goes. But the Russian revolution forced its way into Tamm's youth.

He was already at the mercy of two forces, science and socialism, as an adolescent—passionate, sincere, and not devoid of healthy ambition. Seventeen-year-old Igor Tamm wrote in his diary in 1912: "Science won't satisfy me, personal happiness (in the gross sense—money, drinking bouts) to me are only self-deception; I won't become a petty bourgeois. That leaves only revolution. But will it turn out to be totally engrossing? That's the question . . ."[6]

It's not difficult to understand his parents, who sent him to study at Edinburgh University after he graduated gymnasium—far away from the Russian revolution. He agreed to just one year and combined science, a student socialist circle, illegal Russian publications, and contact with Russian emigrés.

He returned to Russia on the eve of World War I and enrolled in Moscow University. After Lebedev's departure, physics there was in a pitiful state. Maxwell's electrodynamics, which Lebedev's experiments definitively confirmed at the turn of the century, was still too complex for the instructor who took his place.[7] This level of science could not satisfy Tamm's innate temperament and outweigh his sense of social responsibility. The 20-year-old physics student, unaware that he would write the first Russian course on electrodynamics, wrote in his diary:

> The hell with science. It's just hypocrisy . . . Science is the invention of the ruling class—so many good people have been torn away from life by it; if it weren't for science, they would have taken a different path and really messed things up for the masters. Lord, does the "man of science" (what an arrogant word) really live? It's a kind of substitute for living, an absurd, nightmarish substitute. It's better to chop off your arms and legs . . . Well, it's a kind of alcoholism of the mind, even worse: vodka tears you away from life for a time, while ideas strive to captivate and make callous everything else inside you forever. I silenced my fear of the "political" life filled with disasters. Yes, and so I envisioned a future dedicated to science. And I didn't see that it would spell death for my soul. And ultimately it was the same educated philistinism, perhaps the only deep fear I had . . .[8]

This fear sent Tamm off to the front as a nurse in the summer of 1915 and brought him into the Menshevik-Internationalist Party. He spoke at rallies and wrote political pamphlets about the brotherhood of workers of all nations. At the same time, however, he had regrets: "I still can't find work which is (1) interesting, (2) important, (3) absorbs all my time entirely, otherwise it seems to me that I'm being criminally idle."[9]

The 22-year-old Tamm was elected to the Elizavetgrad Soviet, and in June 1917 he was sent to the First All-Russian Congress of Workers' and Soldiers' Soviets. He found out what "the big political life" was, and became more discerning:

> Twice I saw for myself and became convinced that Bolshevism in its mass form exists only as demagogic anarchism and unruliness. Of course, this doesn't apply to its leaders, who are simply blind fanatics, dazzled by that truth, that genuinely big truth which they are defending, but which prevents them from seeing anything else besides it.[10]

It's easy to appreciate how astute the 22-year-old physics student was a few months before the Bolshevik coup. But the success of the coup didn't prevent him from thinking independently: "The 'Great Proletarian Revolution' has taken place, but not only do I not feel any particular enthusiasm; on the contrary, I want to work less than before the 'revolution.' Something is beginning to boil up in me against the Bolsheviks . . . I came into contact with science, and it has attracted me again. Will I remain a politician after everything settles down? Now, at this point, it seems to me more than doubtful."[11]

"The genuinely big truth" of socialism was the goal of both the Bolsheviks and the Mensheviks. The difference was that for the Bolsheviks, this end justified any means. But Tamm and other Mensheviks suspected that the use of some means could destroy the end itself; they were for the evolutionary development of socialism by means of parliamentary democracy and in collaboration with other parties.

The discrepancy between ends and means which Tamm saw in politics ultimately repelled him from it toward his other pursuit, science. But the ideal of socialism that had matured during his youth remained for him "a genuinely big truth" until the end of his life. In theory, he still equated the two realms: "Creativity makes life valuable. Man is the sole creator; he stands out from the swarming masses of petty little folks. It doesn't matter what kind of creativity it is—whether scientific or socio-political—it's of equal value."[12] He wrote this in the spring of 1918 from Kiev, where the whirlwind of civil war had blown him after he graduated from Moscow University.

But in fact he was engaged in physics both in pre-Soviet Kiev and in White Crimea. He did so as circumstances permitted. He studied German physics journals, which turned up in Kiev due to the German occupation. In Crimea he taught practical classes in physics at Tauride University. Teaching certainly aids self-education, but Tamm probably would not have managed to compensate for the lack of scientists at Moscow University and the unscientific circumstances of the civil war had he not set off for Odessa to Mandelshtam upon the recommendation of a Tauride University biology professor: "I found in Mandelshtam a teacher to whom I am indebted for my entire scientific development," he would write two decades later.[13] This was indeed enormously lucky—after Moscow University's neglected state and the homelessness of the civil war, Tamm acquired a first-class teacher who had received a European higher education at Strasbourg University and taught there for a decade.

The Strasbourg school of physics was founded by August Kundt immediately after the Franco-Prussian War in 1872. The first Nobel Prize laureate in physics, Wilhelm Roentgen, was a graduate of this school, as were Lebedev and Mandelshtam's teacher, Karl Ferdinand Braun. Mandelshtam worked as Braun's assistant starting in 1903, became a professor in 1913, and returned to Russia on the eve of the war, in 1914.

The 40-year-old European professor, at the height of his creative powers, could not find a place to apply these powers. Not in St. Petersburg, Tiflis, Odessa. "No instruments, no books, no journals, no frame of mind . . . Publications are out of the question," he wrote to mathematician Richard von Mises, a friend he made in Strasbourg. If it weren't for Mises's food parcels, hunger and typhus fever might have added Professor Mandelshtam to the list of civil war victims. No less important were the parcels with the books and journals—after all, the fascinating revolution in physics continued. How would a musician, deprived of his instrument and even the possibility of listening to music, have felt? This is what lies behind the desperate sentence in his letter: "All my efforts are now directed toward a single goal—to be engaged in science in Germany again."[14]

But there is no bad without good, after all. Tamm was fortunate to have come to Mandelshtam precisely the right time. Exhausted by his yearning for scientific contact, Mandelshtam directed all his scientific ardor, knowledge, and pedagogical gift toward the young physicist. Two years of contact between them enabled the undereducated graduate of Moscow University to attain a European level of science and to subsequently write first-class papers, including the theory of the Cherenkov radiation that garnered him the Nobel Prize.

He did this work at the same time as he helped his teacher raise the new generation of physicists to the European level—that "group of students, hungry for genuine scientific guidance" in Landsberg's letter: Aleksandr Andronov, Aleksandr Vitt, Mikhail Leontovich, Semyon Khaikin, and Semyon Shubin.

Mandelshtam's School
of Physics and Life

Mandelshtam began his path in science with radio physics at a time when this field was just emerging, and worked under the supervision of Braun, whose achievements in this field were celebrated by a Nobel Prize in 1909. At the time, the radio was the cutting edge of both science and technology. Maxwell's electrodynamics was in the ascendancy, and electromagnetism was considered the only force that explained the properties of matter and light. The latest word in science was being rapidly implemented in the highly scientific technology of the radio. Participating in this implementation, Mandelshtam mastered the theory of oscillations that served as the "international language," in his words, for the most diverse fields of physics. The works of the English physicist John Rayleigh on the theory of oscillations and waves played an important role. Mandelshtam never met him in person but considered him teacher on a level with Braun.

As befits a good student, Mandelshtam was not constrained by the authority of his teachers. Theoretical analysis led him to an important discovery in radio technology—he found that a so-called "weak connection" between the receiver and antenna is needed to improve radio reception, not a strong one, as everyone including Braun thought.

In his first paper on optics (1907), Mandelshtam called into question Rayleigh's famous work on the blue color of the sky.[15] In 1871, Rayleigh explained a picture familiar to all—different rays of sunlight, depending on their color, scatter differently on atmospheric molecules. Mandelshtam was doubtful: was it on separate molecules or on their microscopic accumulations—fluctuations? However, in 1907 the concept of fluctuations had not been properly developed yet. This occurred several years later, with Einstein playing a particularly important role. Even though fluctuations arise randomly, they obey some laws and are responsible, if not for the blueness of the heavens, then for many other less graphic, but no less important, phenomena of nature.

In 1913, Einstein gave a report on Mandelshtam's work "On the Roughness of the Free Surface of Liquids"[16] at his seminar. Its title is reminiscent of the choppy surface of the ocean in windy weather, but the topic under discussion is the ideally smooth surface of liquid in a laboratory container. The roughness of such a surface is visible only to the mind's eye and physical instruments, but it is a manifestation of the same random and law-abiding fluctuations.

The theory of oscillations and its most important applications—radio physics and optics—remained at the center of Mandelshtam's interest, but the field of his interest itself "continually expanded and deepened," according to N. Papalexi, who remained his associate and friend from their Strasbourg days to the end of his life.[17] Two revolutionary ideas entered this field that transformed physics: quanta and relativity. Both arose in the bosom of electromagnetism: the first article on the theory of relativity was titled "On the Electrodynamics of Moving Bodies," and the first quantum idea was advanced in order to explain thermal electromagnetic radiation.

Mandelshtam essentially did not distinguish between pure science and applied science: "Mathematics, physics and technology are so closely interconnected that there is neither the need nor the possibility of dismembering a living whole into separate parts."[18] Fundamental questions of theory concerned him as much as concrete radio physics. In the 1930s the discussion of the nature of quantum theory was at its height. One of its founders, Albert Einstein, aspired to classical clarity and posed difficult questions to his colleagues about the shortcomings of the theory. It concerned tricky thought experiments, a cat [Schrodinger's] that was neither dead nor alive, and the electron's free will, but essentially it concerned the nature of scientific knowledge.

According to Tamm, Mandelshtam, who was a contemporary of Einstein's,

> would immediately analyze and find an argument to refute every successive critical article by Einstein. When we asked him to publish his thoughts, he always refused on the grounds, he'd say, that Einstein is such a great man, that he probably knows something that he, Mandelshtam, doesn't know. Several months would pass, a new article responding to Einstein by Niels Bohr would appear, and it always turned out that its conclusions coincided with those of Mandelshtam.[19]

The term "universality" usually refers to a physicist's ability to work in various fields in his science; however, in the twentieth century, it is limited to either theory or experimentation. Mandelshtam was one of the rare exceptions. He was fluent in both aspects of the science. Questions about epistemology, raised by twentieth-century physics, were just as organically part of his thinking. He included these questions in his physics lectures without worrying about whether his views fit into regimented state philosophy. Mandelshtam considered all physics his native land. This "scientific cosmopolitanism" as well as his gift as a teacher accounts for the diversity of his students—from radio engineers to elementary particle theorists.

An example of his vision of science is his speech at the general meeting of the Academy in 1938. The topic didn't sound at all intriguing: "An Interference Method of Studying Electromagnetic Wave Propagation." But this is what Vladimir Vernadsky, a geochemist by profession, wrote in his diary: "An interesting and brilliant lecture by Mandelshtam yesterday at the Academy. I listened to it as I rarely have occasion to. For some reason I remembered Hertz's lecture which I heard in my youth about his major discovery" (the experimental discovery of electromagnetic waves).

In his lecture Mandelshtam didn't merely sum up some research; he presented it as an organic part of a developing science. Without artful contrivances and external effects, he brought together radio technique and philosophical lessons of quantum physics, the historical development of pure science, and prospects for practical applications. It was a picture that conveyed the spirit of living physics to nonphysicists and deepened the understanding of his colleagues.

The diversely gifted men who blossomed quickly and robustly under Mandelshtam's influence shared a similar attitude toward their teacher. Their feelings of love and respect seem at times exaggerated and incomprehensible. There were no signs of a generation gap. Mandelshtam's "scientific sons" speak about him in such exalted tones that it's easy to attribute it to the overly formal style of socialist realism.

However, Mandelshtam also had a major weakness—a strong lack of ambition that, as is commonly the case, arose from his virtues and that helps us understand the qualities and the tone of his admirers' pronouncements. He lacked even the kind of healthy ambition, without which it doesn't seem possible to be a creative individual. By stating something new—in science, in art, or in any field—an individual is in effect saying that he considers he has a right to say this new thing before others do, and that he is prepared, despite their silence, to consider himself "smarter and bolder" than they are.

Mandelshtam had enough boldness to take on problems that the greatest theorists, Einstein and Bohr, racked their brains over, and to offer his solution to these problems to a circle of research fellows and students, but he lacked the ambition to rush to publish his solution and stake his claim to being first. A few traits of his scientific biography stem from this. He didn't hurry, to put it mildly, to publish his work, and would check and recheck his findings. His fellow researchers and colleagues had to persuade him to send his work off to a journal. But it was clear to those in daily contact with him that this came from a sense of responsibility toward scientific knowledge, a moral sense.

Delaying publication about the discovery he had made along with Grigory Landsberg cost them the 1930 Nobel Prize. They discovered a new type of interaction between light and matter but did not hurry to publish their finding (the Indian physicist Chandrasekhara Raman, who did hurry and beat them by several weeks, received the prize). Landsberg said that Mandelshtam's scientific restraint "did not at all spring from his underestimating the significance of his findings; on the contrary, he understood them very well and, for this reason, considered that he was especially obligated not to come forth with important claims without the most painstaking verification."[20]

In this case, his country's history also contributed to the delay in publication. Traces of that time remain in the 1928 letters written to family friend Richard von Mises after a hiatus in their correspondence of more than a year: "You haven't heard from us for a long time because we had various troubles with our relatives and we didn't feel much like writing . . . We have not been having very joyful days of late. Lots of family concerns and the like, they aren't completely over yet."[21]

The "troubles" and "family concerns" were the arrests. Evgeny Feinberg's memory preserved a scene from that time: "Mandelshtam walks into the room with a wet photograph in his hands. He examines the spectrum and pensively says: 'Nobel Prizes are given for things like this . . .' To which his wife hotly exclaims: 'How can you possibly think about such things, when Uncle Leva is in prison!?'"[22]

Landsberg, who didn't receive the Nobel Prize because of Mandelshtam, received something more important:

> I was no longer a boy when I first met Leonid Mandelshtam. Now I'm already elderly. But I'm not ashamed to admit that during the two decades of my closeness to him, when making some responsible decision or evaluating my behavior and intentions, I asked myself the question—what would he think about this? I could disagree with him, especially if it concerned some practical steps, but I never had any doubts about his moral opinions about people and behavior. And I hope my memories of him accompany me during the years that fall to my lot and will serve as a source of moral strength, as those meetings and conversations with him served in the preceding fortunate years.[23]

Landsberg said this in 1944 at a meeting devoted to the memory of Mandelshtam. It was unthinkable then that within a few years accusations against Mandelshtam—from idealism to cosmopolitanism to espionage—would start resounding loudly and threateningly. We will learn more about this somber time and will see again that Mandelshtam truly left a powerful source of moral strength for those disciples who defended him. Perhaps not realizing it, Andrei Sakharov, who came to Mandelshtam's school a few weeks after his death, also drew upon the same source.

Everyone who knew him noted the organic union of science and morality in Mandelshtam. And it was precisely this union that created the atmosphere of his scientific milieu. As Ilya Frank, who coauthored a Nobel Prize–winning paper with Tamm, wrote: "Scientific generosity was one of the characteristic features of the Moscow School of Physics, whose foundation was laid long ago by Lebedev and developed by Mandelshtam in my time."[24] Coming into the sphere of action of Mandelshtam's school, into the field of continuous scientific discussion, Frank did not immediately realize that "new ideas were expounded during these talks long before publication and naturally, without fear that someone else would publish them."[25]

An important characteristic of Mandelshtam as a teacher was that he did not choose his students—they chose him. He was prepared to teach anyone who genuinely wanted to study. In 1918 he began the introductory lecture of his physics course this way: "Studying physics, delving deeply into its fundamentals and those broad ideas on which it is based and, especially, independent scientific work bring enormous mental satisfaction. I don't want to persuade you of this. I even doubt whether it is possible. Each individual must convince himself. But I wanted you to know that if any of you should feel this aspiration, that it will always be a great pleasure for me to assist you with everything I can to realize it."[26]

Mandelshtam mentored solely by his own personal example, his way of life. Regarding the dishonorable behavior of a certain physicist, he said: "You don't re-educate adults. You either have dealings with them or you don't." Not everyone who passed through Mandelshtam's school withstood the temptations of ambition or overcame fear of the powers-that-be, but an amazingly large proportion of his students combined scientific and moral qualifications.

Tamm spoke about the rare combination in Mandelshtam of two ordinarily mutually exclusive qualities: "Indescribable kindness and sensitivity, loving gentleness in dealing with people were combined in him with an unbending firmness in all issues which he considered of fundamental importance, along with total irreconcilability with compromises and appeasement."[27]

With such a spiritual bent and without any sort of political inclinations, it's not easy to create a scientific school even in civilized conditions, much less in Stalinist Russia. Considering his personality, we should marvel at the recognition of his scientific contributions during his lifetime. He was elected to the Academy of Sciences in 1928 and awarded a Lenin Prize in 1931 and a Stalin Prize in 1942.

But the most important thing to be amazed at is that Mandelshtam succeeded in so fully actualizing himself. The main reason was that his personality also attracted people of a practical cast, ready to provide the walls and a roof for his school in the real world of Soviet life.

For six years, from 1930–1936, this was Boris Gessen's main mission.

The Roof and Walls of
Mandelshtam's School

Who was Boris Gessen? Was he the professional physicist whose paper on Isaac Newton at the 1931 International Congress of the History of Science in London "was one of the most influential reports ever presented at a meeting of historians of science," as Loren Graham, a prominent Western authority on the history of Russian science, wrote?[28]

Or maybe Gessen was "the 'red director' of the Physics Institute of Moscow University," whose

> job was to see that the "scientific director" (the well-known physicist Professor L. Mandelshtam) and the staff did not deviate from the straight path of dialectical materialism into idealistic swamps. A former schoolteacher, Comrade Gessen knew some physics, but was mostly interested in photography, and he shot some very good portraits of pretty coeds.[29]

This is from Gamow's autobiography, written in America in the 1960s.

Western historians of science may consider Gessen one of its founders, readers of Gamow's scientific adventure story may be amused by the Marxist schoolteacher's pretensions, but Gessen's role in Russian history was entirely different. He was not a professional physicist, nor was he a schoolteacher. And he was wrongfully accused of being keenly interested in photographing girls—a different professor at Moscow University, one from Timiryazev's camp, had a passionate interest in it.[30]

Gessen's main mission began in September 1930 when he—a communist involved in the philosophy of science—was appointed director of Moscow University's Physics Institute. The flourishing of the Mandelshtam school began with his appointment.

The eve of the 1930s in Soviet history is referred to, thanks to Stalin, as the time of the "Great Turn." The nation had not yet felt his heavy hand to the fullest extent. Stalinism was only just turning into a totalitarian system. The leader had managed for the time being to deal politically with his rivals in the highest leadership, but there remained members of the revolutionary generation at other levels who were perhaps dazzled by the socialist idea but not dispirited by fear. The tragedy of the peasantry and the Great Terror of 1937 still lay ahead.

Stubborn facts are an impediment to finding a single formula for Soviet history, and one of them is the fact that, at the beginning of the 1930s, the state regime had not yet trampled down the life of science. Evidence of this is, for instance, the fact that in 1931 the highest prize of the nation—the Lenin Prize—was awarded to Leonid Mandelshtam and Aleksandr Friedmann. The latter award was even more amazing because Friedmann had died (from typhoid fever) in 1925, soon after he glorified his name with the discovery—from the tip of his pen—of the expansion of the universe. In fact, he understood Einstein's theory of gravitation better than Einstein, who didn't immediately acknowledge the correctness of the Russian mathematician.

It's true that Friedmann got the award for atmospheric dynamics, not for his cosmology. But cosmology had been attacked earlier and more fiercely than other physics theories by the Soviet "militant materialists." Involvement with the "pro-theist" theory could outweigh any scientific contribution.[31] The fact that it didn't says a great deal about the time.

Another remarkable fact of that time was that the newly appointed red director of the Physics Institute of Moscow University, Boris Gessen, took such good care of physics, right up to his arrest in 1936. In order to understand why Gessen took this concern upon himself, we must first know that he was a friend of Tamm's from their gymnasium years, that they also shared socialist ideals, studied physics and mathematics at Edinburgh University, and returned to Russia together.

Then for several years their life paths parted—Gessen had to leave for Petrograd. Like Tamm, he wanted to continue his studies, but the quota system for Jews blocked his path to Russian universities. There was no such obstacle in Petrograd to the Polytechnical Institute, which was created at the turn on the century at the initiative of Sergei Witte, the broadly thinking tsarist minister.

Gessen studied economics at the Polytechnical Institute and simultaneously audited physics and mathematics at Petrograd University. His inclination toward basic sciences clashed with a Marxist understanding of history, according to which economics is the foundation of social life. Calling economics a "basis" and science a "superstructure," Marxism paid little attention to the issue of personality and could hardly appreciate the contribution that the quota system of prerevolutionary Russia's basis made in Gessen's revolutionary Marxist superstructure.

He joined the Communist Party in 1919, worked as an "instructor of political education" in the Red Army, and then from 1921 on was a manager at the Sverdlov Communist University (which took over Shanyavsky University's building).[32] Boris Gessen was in this capacity when Igor Tamm saw him, having returned to Moscow in the fall of 1922. As Tamm then explained to his wife:

What is Sverdlov University? Party youth from all across Russia are dispatched there for a three-year course that prepares them for socio-political work (a variety of several specialties). The formal educational requirements are few, the determining factor is general maturity. The natural sciences are taught since they are essential for the creation of a scientific worldview (Boris's formulation).[33]

Boris immediately offered his friend a position in the Scientific Association at that (red) university, which meant an academic food ration, a room, a salary, working in your own science, and no more than four hours of lectures a week. But friendship and worldviews don't mix.

"It's obvious," Tamm wrote to his wife,

that I'll have to turn this down, because it includes a materialistic worldview in philosophy, science and social issues as a pre-requisite. Meanwhile I can say this, and with certain reservations at that—only with respect to social issues—on the whole, I have no firmly established views in philosophy, and

I don't understand at all what materialism there is in the exact sciences—
there is science, and that's all.[34]

Conversations between friends lasting long hours led not only to clarifica-
tions of the philosophical terminology but also, most important to Gessen, coming to understand better what was science and what was philosophy. When
Gessen enrolled at the Institute of Red Professors (a kind of graduate study in
Marxism), he requested that Mandelshtam be his adviser and chose the serious (and not at all red) subject of the foundations of statistical physics—the
same physics that causes the roughness of the free surface of liquids and the
blueness of the sky. Gessen published his results in 1929 in a physics journal.[35]

At that same time, the "Moscow Worker" publishing house came out with
Gessen's popular book *The Main Ideas of the Theory of Relativity*. Stating these
ideas competently and in an accessible form, he made an effort to convince the
reader that the theory of relativity is the concrete manifestation of Marxism's
teachings about space and time.

In a society where state ideology played such a militant role, science was
not left on the sidelines. The first to start military operations were the opponents of the theory of relativity headed by Arkady Timiryazev. Having exhausted scientific arguments, they began to accuse the theory of relativity of
being incompatible with Marxism. If one considers that administrative authority was in Timiryazev's hands, the defenders of the newest physics had to take
up the same—dialectical—weapon, because it was double-edged and could be
bent whichever way you pleased.

Gamow imagined, from his Leningrad or, rather, American distance, that
the "Red Director" Gessen was supervising Mandelshtam and his fellow researchers, but in fact Gessen was more likely hanging on their very words. With
their help he was finding out what the new physics was and searching for a
proper—respected and safe—place for it in the Marxist worldview. In addition,
he was protecting this place against the "militant materialists" with Timiryazev
at their head. This was Gessen's Marxism in science.

But along with it an entirely different form was in effect—the most materialistic form of Marxism. Aleksandr Maksimov embodied it. Getting acquainted with
his life, it's not difficult to understand that the "basis" of his Marxism was simply
a powerful survival instinct. Its flexibility was limited only by his spine's flexibility. He had graduated from Kazan University with a specialty in physical chemistry, but only one of his publications, his first, is not focused on Marxism. All
the others are a dreary tedium of endless quotes. He wrote with pride that in 1918
he worked in the cultural education section of the Kazan Soviet, "was imprisoned by the White Guards and lived through an execution attempt during the
retreat." From 1920 he was in Moscow as deputy director of a section of the Department of Workers' Faculty in the People's Commissariat of Education. The
party cell of this department admitted Timiryazev into the party, and he in turn
admitted Maksimov into the physics department of Moscow University.

Nonetheless, it was Maksimov who played a not inconsiderable role in Timiryazev's dismissal from power at Moscow University. His Bolshevism came down to a striving to align himself with those who had the most influence. By the end of the 1920s, under the influence of the successes of the new physics, Timiryazev's position had visibly weakened. Maksimov oriented himself in the direction the wind was blowing and in the autumn of 1929 wrote a report to the Central Committee, "On the Political Situation at the Physics Institute."[36]

A few months later a Workers and Peasants' Inspection brigade arrived at Moscow University. The inspectors interviewed the research fellows and graduate students and examined documents and laboratories. In its report, the commission came to the conclusion about the low level of the institute's professionalism, where some research fellows had not published a single paper in five years. The only exception was Mandelshtam's group, which had completed nearly 50 papers. However, the commission found a different shortcoming: "a complete indifference to questions of dialectical materialism."

The report described the atmosphere of confrontation in which Mandelshtam had to work: "The predominant group felt a great danger to its monopolistic position. Since it was impossible to directly oppose Mandelshtam's work, passive resistance to it had to be organized. In the words of A. Andronov, a former graduate student: 'Professor Mandelshtam was absolutely ill-treated.'" The Workers and Peasants' inspectors saw that this was simply a disguised battle for material resources, the disguise being "Timiryazev's assertion that there was a battle being waged between the Soviet and anti-Soviet professors."

In the end, the commission proposed that the institute director be dismissed and "a reliable person, not necessarily a physicist" be appointed. In September 1930, Gessen became the institute director.

A victorious career as a professional party overseer of science lay ahead of Maksimov, but his 1929 report to the Central Committee described the situation surprisingly accurately. The communist Timiryazev, for instance, is accused of including the "reactionary" professor Kasterin in the list as prominent only because he refuted the theory of relativity.

And this is what Maksimov said about physicists close to Mandelshtam:

Vavilov: "A well-known physicist . . . Oriented to the right, but aspires to work with us lately."

Tamm: "A fine young physicist. Completely loyal. Recently some vacillations noticeable."

Landsberg: "A good physicist. Oriented to the right."

About Mandelshtam himself: "A major physicist with a European name. Splendid teacher. Loyal"—to the Soviet regime, one has to suppose. Apparently this is how Maksimov perceived the European professor's politeness.

Tamm saw his teacher somewhat differently:

> Incidentally, Mandelshtam's aversion to Bolshevism—although things are
> good for him—has become painful, to the extent that having to sit at a table
> (at opposite ends and not speaking) at dinner with a communist—despite
> the fact that, according to him, this communist behaved himself quite
> decently—gives him an excruciating migraine all night.[37]

This was in December 1922, just a few months after the deportation of a large
group of non-Soviet writers and scholars from the country. Ten years later,
Mandelshtam knew at least one communist—Gessen—who induced not mi-
graines but appreciation for his efforts for the benefit of science.

Let Maksimov's biographers clarify exactly what the circumstances were
that forced him to promote science, too. Circumstances were changing at a
breakneck pace. He was twisting himself into full agreement with the "general
party line." And in 1937 when this line pierced through Gessen, Maksimov
added his kick at his former colleague without delay.[38]

But in 1931 they were side by side, working on *The Great Soviet Encyclopedia*, for which Maksimov chaired the section on the natural sciences and Gessen
was responsible for physics. This was at a point when Gessen had written his
article on "Ether" for the encyclopedia.[39] This article angered the young Leningrad theoretical physicists, and they responded with a malicious telegram. This
is how Gamow informed Stalin about the event in a letter in January 1932:[40]

> Dear Comrade!
>
> I am obliged to address a letter to you about the condition in which theo-
> retical physics here in the USSR finds itself. In the last few years theoretical
> physics has been unceasingly attacked by philosophers who proclaim them-
> selves materialists, but in fact continually fall into the most revolting vari-
> eties of idealism.

Having included Timiryazev and Gessen along with such philosophers,
Gamow informed Stalin about an "outrageous fact" of this sort—about Gessen's
article, "anti-scientific nonsense which comprises Marxism," published in a
book "which was intended for the education of the broad masses and cost the
government a great deal of money":

> When I and several of my theorist friends, who work at the Leningrad Phys-
> ics and Technology Institute, saw this ludicrous article which asserted that
> "physics is only now beginning to study ether," we sent Gessen an ironic
> telegram: "Having read your article, we are enthusiastically beginning to
> study ether. Awaiting instructions about phlogiston and thermogen" (the
> theory of phlogiston and thermogen are theories refuted over a hundred
> years ago and which became synonymous with scientific rubbish). By send-
> ing this telegram, we intended to start a campaign against the falsification

of scientific materialism. But Gessen did not lose his head: he complained to the Presidium of the Communist Academy.

As a result, an investigation was conducted during a general meeting at the Physics and Technology Institute. "The most abominable, slanderous accusations" were showered upon those who had signed the telegram, that "'had they been experimenters, not theorists, they would have thrown bombs at the leaders of the revolution; that they weren't throwing bombs because they didn't know how.' The intimidated assembly voted on a resolution which confirmed that the motives for sending the telegram to Gessen were counterrevolutionary."

The letter ends with an appeal for help: "Considering that the events which took place at the Physics and Technology Institute have great importance as an example of the scandalous distortion of the party's policy in the field of science, I expect that you will take measures toward the liquidation of the disgraceful persecution of theoretical physics which has arisen."

Alas, Gamow was treading on a slippery path not only in the area of party policy but also in the field of science. In the ardor of his self-defense, he does not distinguish between Gessen's ether or Timiryazev's ether, although the difference was enormous. To see the difference, we must also discuss the third ether, which Einstein wrote about in his 1930 article "The Problems of Space, Ether and the Field in Physics."

Stalin hardly had time to sort out how Einstein's ether differed from Timiryazev's or Gessen's, but we should do it to understand Boris Gessen's actual role in the history of science. Simply put, Timiryazev inherited his ether from nineteenth-century physics; he studied it before the appearance of the theory of relativity. And he carried it through his entire life. Firmly held notions are fine in morality, but not in science, especially when science is going through revolutionary changes.

It was Einstein who changed concepts in physics. His 1905 theory of relativity made the ether of previous centuries just as unnecessary as phlogiston. But ten years later he created the theory of gravitation, connecting the curvature of space-time with the distribution of matter. Space-time is no less universal and ubiquitous than ancient ether; therefore, it's possible to keep the term, if one doesn't try to hang onto its old meaning. It was in this spirit that Einstein talked about ether in 1930—pouring entirely new wine into old skins. Sometimes social legacy also intrudes on the life of words—the concept "ether" had, after all, disappeared from physics.

Of course, to physicists who firmly hold the real content of their concepts in their hands, the loss or substitution of a term is not a major event, whereas philosophers devote too much attention to words. Encyclopedias usually sum up the current state of affairs and do not speculate about the future. If one strays from the constraints of the encyclopedic genre, one can even find in Gessen's article the problem of the quantum generalization of gravitation, which still faces physics today. But if not, then one must admit that Gamow and his friends had reason to become angry.

Gessen's qualifications were sufficient to make his lecture at the London Congress on the History of Science an event. However, when Gessen attempted to interpret the physics of the day, his limits in physics became evident. In Gessen's articles you will not find crippling blows at his ideological opponents. No wonder his Marxist comrades criticized him in 1931:

> Included in Comrade Gessen's works are "the probabalistic basis of the ergodic hypothesis" . . . and others—these articles are remote from the Party's actual concerns . . . There isn't even a whiff of Bolshevik spirit in these articles . . . We see a single thread through all of Comrade Gessen's work—worshipping bourgeois scientists like icons . . . The general basis of his errors is reverence for modern theories without analyzing or critiquing them.[41]

It's clear who helped Gessen chose objects to "worship." As director of the Physics Institute of Moscow University, Gessen punctually attended Mandelshtam's lectures and seminars where genuine physics was alive. That was where he found out which physics was correct. All he needed to do was try to find the appropriate Marxist formulations, supported by Lenin's words, "Marxism is not a dogma, but a guide to action."

In 1934 when FIAN moved to Moscow, Director Sergei Vavilov invited Gessen to be his deputy. Some research fellows came from Leningrad, but Vavilov saw Mandelshtam's school as the foundation of the institute's scientific potential.

After Gessen's arrest in August 1936, Mandelshtam's people were forced out of the university. Fortunately, they had somewhere to go—FIAN. At that time Vavilov publicly declared his attitude toward Mandelshtam:

> I made many compromises during the establishment of the Institute in Moscow, as I wanted Leonid Mandelshtam to concentrate his work here. Leonid Mandelshtam is not on our staff. He has the right to this kind of existence—it is guaranteed him by the Academy of Sciences. Naturally, this method of working may perhaps seem strange—a person receiving his fellow researchers at his apartment. I think that, in time, the situation will change, but in any case right now Leonid Mandelshtam is of great use. We would like him to take even larger part in the life of the Institute, critique their work, give instructions. He is a man of uncommonly high scientific standard.[42]

Academician Alexei Krylov, from whom Vavilov received the FIAN directorship, said: "Sergei Vavilov is a remarkable man—he created an institute and was not afraid to invite physicists to it who were stronger than he himself was."[43] Vavilov consciously sought out such people.

3
THE YEAR 1937

In Russian, "37" isn't simply a numeral, it's a noun of double mourning: the first began with the death of a single man, the second with the deaths of millions.

At the end of January 1837, Russia's greatest poet, Alexander Pushkin, was fatally wounded in a duel. Pushkin's role in the life of Russia is so great that his death is a historical event for every educated Russian. There is no comparable figure in the English-speaking world. Nor is there a parallel to the plague that befell Russia one hundred years later. In the West, it is called the Great Terror; in Russian, it is simply "thirty-seven," although it refers to a period of nearly two years beginning in mid-1936. At that time Sakharov was embarking upon adult life; he entered the university in 1938.

"Thirty-seven" was not the first wave of the Stalinist terror or the last one. This wave crashed mainly down on the educated stratum of the population and is marked by an incomprehensible irrationality. The state terror machine devoured those in the state and party elite whom Stalin had designated as "enemies of the people" and, along with them, millions of people uninvolved in politics—engineers and scientists, poets and actors, workers and peasants. The public saw only show trials of highly placed enemies of the people; they were pilloried at mass rallies and slandered in newspapers.

The centenary of Pushkin's death was just as prominent in the newspapers and a genuine subject of cultural life at the time. Publication of the complete sixteen-volume collected works of Pushkin began in 1937, and a plethora of materials about the poet's life was published. There were official ceremonial gatherings as well as school evenings at which Sakharov's contemporaries read Pushkin's poems and staged his plays.

Sixteen-year-old Andrei Sakharov listened to "splendid broadcasts about the Pushkin festivities" on the radio; fourteen-year-old Elena Bonner, Sakharov's future wife, cut out the serialization of Veresayev's documentary book about Pushkin's life from the newspaper. In Sakharov's words:

> It was then, in 1937, that Pushkin was officially proclaimed a great national poet. Imperceptibly, ideology drew closer to the famous trinity which characterized Nicholas I's era—Orthodoxy, Autocracy, Nationality. Moreover, this national character was personified by Pushkin, communist orthodoxy by Lenin's body in the mausoleum, and autocracy by a hale and hardy Stalin.

How was all of this linked together? Why did the first Soviet human rights activists meet for a demonstration at Pushkin's memorial statue thirty years later, and why did Sakharov declaim Pushkin's lines inscribed on it?

> Long will I be by my people beloved,
> For awakening kind feelings with my lyre,
> For celebrating freedom in this cruel age
> And calling for mercy for those who have fallen.[1]

And then ten years later, the disgraced physicist and his wife reread Pushkin, read everything they could find about him during their Gorky exile. And it was with a line from Pushkin—a prearranged secret signal—that Elena Bonner asked him to stop his hunger strike.

A Feast in the Time of the Plague?

Until Sakharov's autobiography appeared, no one knew about his attachment to Pushkin. Nor did Sakharov's university comrades suspect it, with one exception. Mikhail Levin chose *Walks with Pushkin* as the title of his memoirs about Sakharov: "Sometimes I got the feeling that, in addition to the real space-time in which we lived, Andrei had at hand another one, displaced in time by 150 years, right where Pushkin dwelt with his circle. And I was lucky that while we were still young, Andrei allowed me into this world, hidden from outsiders."[2]

In his memoirs, Sakharov begins the list of his childhood books with Pushkin's "The Tale of Tsar Sultan," and then, after many famous Western names (including Jules Verne, Charles Dickens, Mark Twain, Mayne Reid, Jonathan Swift, Jack London, H. G. Wells), he adds, "a little later—nearly all of Pushkin."

The power of Sakharov's attachment to Pushkin is unusual, but the poet's position in Russian culture is even more unusual. A simple experiment provides persuasive evidence. In a word association game, a person of Russian culture will name with astonishing frequency: nose for "part of the face," apple for "fruit," and Pushkin for "poet." The third association is more ambiguous for Americans, not because there are no remarkable English poets, but simply because none of them is of such significance as Pushkin for Russians—in any case, not for contemporary culture. One can speak at length and in various ways about Pushkin's preeminence in the world of Russian culture, but the very fact that Pushkin's name was known to all across the nation was sufficient for the leader of the Soviet people to grant him status as the official great poet. The dictator Stalin's genius manifested itself both in the ability to eliminate all his potential enemies and in his skill in exploiting popular feelings. Of course, Stalin's propagandists, manipulating quotes and facts, created a frame befitting the poet's official portrait, profiting from the fact that his life was rich in events and ended long enough ago. There was even the book *Pushkin's Legacy and Communism,* which came out at the time of the anniversary.[3]

And yet, when you open a volume of Pushkin, you can't envy the party's Pushkin scholars. This is from one of his last poems:

> On the king or populace depend?—
> What's the difference? Be gone!
> Shut them out and bolt the door.
> To serve and please yourself alone:
> Neither thoughts, neck nor conscience bend
> For power—its symbols or its ends,
> But wander here and there, fully at your ease,
> Marveling at nature's divine beauty as you please.[4]

Now, how can this hymn to individual freedom be reconciled with communism? It's not likely that Andrei Sakharov concerned himself with this question. He got his Pushkin from his home library, not secondhand from the propagandists. Pushkin's books in the homes of the Russian intelligentsia were more essential than the Bible.[5]

How could a poet of the previous century who lived by his passions attract a levelheaded youth infatuated with physics? This wasn't simply a deep feeling for poetry—for Sakharov there was no one comparable with Pushkin. Both his friend Mikhail Levin and his wife Elena Bonner attempted to share the riches of Russian poetry with him but didn't succeed.

Andrei Sakharov didn't feel cramped in Pushkin's world, possibly because it is an enormous universe that the poet fearlessly and indomitably explores. He explores all its realms—love and death, power and liberty, faith and doubt. He explores the very freedom of his exploration. The circumstances of Pushkin's life furnished him with rich material. He was friend with the Decembrists, the participants of the enlightened nobility's unsuccessful uprising against the autocracy. He spent over six years in exile for his freedom-loving verses, which spared him from participating in the uprising and saved him for Russia. He never spent time abroad but traveled freely in time and space using his artistic imagination. His itineraries led him to ancient Russia, medieval Spain, plague-ridden London in 1665.

With age, his youthful struggle against tyranny was enlightened by a sober understanding of national psychology, while the imaginative power of writing along with philosophical dispassion helped get to the essence of the elements which rule the world.

Although the poet was removed from science in itself, somehow he managed to divine its essence as well:

> How many discoveries full of wonder
> Does enlightenment for us prepare,
> As does experiment through painful error,
> And genius, who is a friend to paradox,
> And chance, oh God, the Divine inventor . . .[6]

Perhaps he divined science because various forms of human creativity only appear superficially different and are, in their depths, rooted in a single realm. What was happening in the depths of the young physicist's soul resonated with the great lyrical poet's spirit.

But how did this creative resonance sound against the backdrop of the year 1937? The answer is hinted at in one of Pushkin's "Little Tragedies"—*The Feast During the Time of the Plague*.[7] On the stage there is a real feast and the real plague, and "a wagon, filled with corpses." And a hymn in honor of the plague resounds,

> Everything that threatens destruction holds inexplicable bliss for the mortal heart. At the edge of an abyss, amid stormy waves and tumultuous night, and in the breath of the Plague . . . Fortunate is he who amid the strife can feel the pleasure. So, hail to thee, Plague! We do not fear the blackness of the grave, nor do we fear your call.

This hymn is a reminder that even a plague does not abolish the ability to create. In contrast to Pushkin's plague, in 1937 there was no diagnosis—it simply wasn't there, not for the young Sakharov and his generation, not for many of those who wound up in "plague"-infested barracks of those times. There were wagons filled with corpses onstage in Russia of 1937, but they were invisible—only the creaking of their wheels could be heard. The nation lived in a thick fog of ignorance and fear. People were dragged one by one onto the plague-corpse wagons. Relatives of the lone individual had no idea that the sentence "ten years without the right of correspondence" meant execution right there in prison. No one, except for those who served the repressive machine, knew that even "more lenient" sentences virtually meant death in distant camps as well, maybe put off for a few months. Total control of the press and orchestrated lies helped people to not see the dismal abyss at the edge of which they lived, helped them invent explanations for what was happening right next door: "a misunderstanding," "a judicial error," "they'll sort it out and let them go."

People who were sensitive to poetry could think, "Why, you can hear Pushkin's poems all around and remarkable new ones are being written!" as salvation. There were also other, more prosaic, things with which to shield oneself from the abyss—broad access to education and the flourishing of children's literature—and it's difficult to say today whether they arose because of or in spite of Soviet rule. Finally—and even more removed from lyrical poetry and closer to Andrei Sakharov's calling—the powerful ascent of Soviet physics: most of the Soviet Nobel Prizes in Physics were awarded for work done in the 1930s.

The universal force of the calling to science arose many centuries before Soviet rule, and Stalin's plague—in contrast to Hitler's—did not directly offset the impact of this force. It turned out that it is possible to live and create at the edge of an abyss—if there is no other choice.

Boris Gessen, the "Enemy of the People"

On the wave of the 1937 centenary of Pushkin's death, a poem addressed to the poet appeared:

> Long did you await happiness.
> Now Party activists—
> focused and hushed—
> Listen to your poems
> At regional meetings until dusk.[8]

A quarter of a century later Sakharov remembered this verse: "The invaluable testimony of a contemporary, as Pushkin would have said. You see, such Party activist meetings really were held everywhere that terrible year. They were unusual—after them all the participants went home."[9]

A meeting was held at the Lebedev Physical Institute (FIAN) in April 1937 at which Pushkin's verses were not heard. Not long before, a Plenary Session of the Central Committee had expelled the last two of Stalin's former Lenin Politburo comrades from the party. Naturally, the concluding resolution of the FIAN meeting hailed "the Plenary's decision to expel Bukharin and Rykov from the Party as Trotsky's allies, Japanese and German agents, and to transfer their cases to the organs of the NKVD."[10]

Much more attention was paid to their own "enemy of the people"—Boris Gessen, the deputy director—who had been arrested back in August 1936. But this issue did not predominate—three-quarters of those who spoke didn't mention Gessen's name at all. Thus, to narrow down the state of science in 1937 to repression only is to paint too dismal a picture by a factor of four.

Reading the 1937 shorthand transcripts half a century later, knowing what was unknown to those speaking, it is difficult to understand how they could have spoken about their scientific concerns. However, none of them knew that Gessen had been executed back in December, that his "investigation" was a pile of hastily written, absurd papers, that according to the "court's" decision, delivered on the day of the execution, the corresponding member of the Academy of Sciences and historian of science took part in the "counter-revolutionary Trotskyist and Zinovievan terrorist organization which carried out Kirov's villainous murder and which prepared a series of terrorist acts against the leaders of the Bolshevik Party and the Soviet government with the aid of the Fascist Gestapo in 1934–36."[11] None of this was known beyond the walls of the NKVD. Gessen simply vanished without a trace—"ten years without the right of correspondence." It wasn't until his "rehabilitation" in 1955 that the fact of his death was officially admitted.

And his colleagues, what did they say about Gessen in April 1937? Vavilov, the director of FIAN, spoke first, respectfully calling the suspect by his first name and patronymic and assuming responsibility for having invited him to FIAN. The institute was transformed after its move to Moscow. Several dozen research fellows arrived from Leningrad, but Mandelshtam's school and his closest co-researchers, Tamm and Landsberg, became the institute's scientific foundation. They came from Moscow University, and Gessen came along with them.

Bentsion Vul from Leningrad, distancing himself from those touched by the plague and holding firmly to the general party line, recalled:

Who invited Gessen to the Institute, who tried to secure his appointment and who promoted him to deputy director? Here Sergei Vavilov must say that for a long time we fought against attracting Gessen to our Institute. We were against him, not because we knew that he was a provocateur and spy. We didn't know this. We were against Gessen purely due to work considerations. And the people who promoted Gessen to deputy were not guided by these considerations. The staff who came to us from Moscow University had group interests which were more important to them than the interests of the State. They exerted pressure on S. Vavilov, on the Academy of Sciences Presidium and, in the summertime when we were all away from the Institute on vacation, secured the position of deputy director for Gessen; he become the head of two Institutes and could do harm at both. I repeat, responsibility for Gessen lies with the Moscow University group which promoted him, and with the board of directors, those who succumbed to the influence of this group.

In response to the accusation that he was dodging Gessen's exposure, Landsberg briefly and firmly replied:

If I knew what there was which could serve as the basis for Gessen's activities as a saboteur, I probably would not be silent about it. This is my last statement on this issue: I categorically declare that it is a lie that I am trying to suppress any type of accusation. I cannot bring any evidence in this regard, naturally, because it's impossible to prove that you don't know something.

Landsberg demonstrated ever more obstinacy at the university raking-the-coals sessions. When the whistleblowers uncovered evidence of Gessen's wrecking in "defects" in the educational program, Landsberg declared that he had put the program together himself.

Three weeks after Gessen's arrest, a denunciation was sent from the university:

Information was received from the Party organizer Umansky that on September 9, Gessen's wife came, very insistently requesting to look for professor Landsberg. She didn't find him in the department; when she met professor Tamm, she told him that he had to go somewhere, and they left. In addition, we have information that Tamm was Gessen's childhood friend, that they studied together at Edinburgh University in Scotland. According to unverified information, Tamm was a Menshevik in the past, and had ostensibly participated in the Second Congress of Soviets.

September 10, 1936. Dep. Secretary of the Moscow University Party Committee.

At the FIAN meeting, Tamm had to answer not only for his friend Gessen but also for Tamm's younger brother, arrested in the fall of 1936:[12]

You see, comrades, of course there can be evidence, although not entirely serious enough to warrant taking it to higher levels of authority, but serious enough to arouse a person's mistrust and suspicion. And so I maintain that I didn't have even that kind of evidence. If I had, I wouldn't have trusted either my brother or Gessen. Whereas I trusted my brother until the very moment of his testimony at the trial, and trusted Gessen until his arrest. If the question is that either you report something else which no one knows or you lose society's trust—then, unfortunately, I am doomed to society's mistrust, because there is nothing more that I can say. But I personally consider it impermissible to pose the question in this way. No one has evidence to suspect me of knowing anything counterrevolutionary, that I have somehow had something to do with diversions and things of that nature.

D. Blokhintsev (1908–1979), who spoke next, testified:

I have known Igor Tamm for a long time, since 1929, met with him very often; moreover I have had occasion to have conversations with him on the most diverse subjects, not only scientific but political ones too. And I must say that, not only have I never heard any remarks, but couldn't even detect a suggestion in his words that might be called un-Soviet. Meanwhile, it must be said that I couldn't declare this about the entire staff, because people are often inclined to be clever for the sake of making a clever remark. And I had a very definite attitude toward Igor Tamm: I could vouch for him as a completely Soviet scientist. The question really did arise for me: how could a person who came into contact with a brother not be able to detect at least some sort or another of anti-Soviet nuances? This remains a riddle for me. But I do not think that Igor Tamm could know everything, because one could consider his behavior crazy when he, after the list of witnesses in which his brother appeared was published, could go and say: "I vouch for my brother"!

Tamm also took responsibility for Gessen:

> I was an advocate for the assignment of Boris Gessen as Deputy Director of the Institute, because I considered that Gessen brought great benefit during his first few years (I emphasize this) of directorship at the university, and I was guided precisely by these considerations, and not at all by group ones. I must say that it's true, that the last year-and-a-half to two years his directorships were different: while the first period of his work was considered useful, then it later turned into a complete lack of activity, particularly with respect to the university and, apparently, the same with respect to this institute. And I must admit that I didn't pay enough necessary attention to this last period of Gessen's work, to the period of wrecking or lack of activity. I attributed it to an intense neurasthenic condition. Thus, being an advocate of his work here, I was not operating from group interests, but nonetheless, responsibility in this regard lies with me.

Others also spoke about Gessen's inactivity during "the last year-and-a-half to two years." Subtracted from the time of his arrest, it would begin around December 1, 1934, when Sergei Kirov, the party boss of Leningrad, was killed. The events following Kirov's murder could have plunged party member Gessen into neurasthenic apathy—Stalin moved from political battle for power to the physical annihilation of his rivals. In contrast to his physicist friends, who were immersed in boiling science, it was easier for Gessen to discern the harbingers of 1937.

In 1955 references from people who knew Gessen were required for his "rehabilitation" (the removal of accusations from his name), according to the customs of the time. Tamm, by that time an academic who had received the award of Hero of Socialist Labor for his contribution to the creation of the hydrogen bomb, wrote such a reference.

Why didn't Tamm disappear into the maws of the Great Terror in 1937, following his younger brother and childhood friend? After all, he didn't disown them at all; on the contrary, he vouched for them. We are still to discuss why an individual with such a record was allowed to head up a new group in the Soviet thermonuclear project, but for now, Tamm's own response requires a leap into another civilization to read it:

> To the Office of the General Prosecutor of the USSR:

> I would like to report the following in connection with the matter of Professor Boris Gessen's rehabilitation, which is under consideration at present:

I was a friend of B. Gessen's since childhood. We were in the same class from the day we entered gymnasium until we graduated in 1913, after which we studied together at Edinburgh University in England in 1913–14. Although we continued our educations in different cities upon our return to Russia in 1914, we met often, then from around the end of 1922 we started meeting in the same city—Moscow. A close friendship always linked us. Moreover, from about 1928 to B. Gessen's arrest in 1936, we worked at the same institution—the Physics Department at Moscow University, where I was a professor and B. Gessen was the Dean of the Physics Department, then Director of the Research Institute of that Department.

In my opinion, from a scientific point of view, B. Gessen was the most prominent of all the Marxist philosophers I have known who worked on issues of contemporary physics, and stood out clearly among them with his combination of deep erudition and clarity of thought both in philosophy and in physics.

From the political point of view, B. Gessen was one of the most logical and deeply staunch communists with whom I have had occasion in my life to have close contact. Communist studies and Marxist philosophy defined not only his political and philosophical convictions, but his entire life and work as well. From the very moment he joined the Party in 1918 or 1919, the communist worldview—not in words but in deeds—determined the entire path of his life, his attitude toward his surroundings, all the serious decisions that a person must make in life.

I would especially like to emphasize that, during the course of his lengthy political activity, B. Gessen was always a firmly convinced advocate of the general Party line and opponent of all oppositions. Always during our talks as friends when, in the turbulent 1920's and early part of the 1930's I had any kind of doubts in political issues, B. Gessen knew how to dispel these doubts in me with the uncommon clarity and logic of his thinking and to persuade me of the correctness and historical necessity of the line being pursued by the Party. Our close ties of friendship, of course, absolutely ruled out any insincerity on his part.

This is why I am convinced that B. Gessen not only was not guilty of any sort of crime, but that his life and work can serve as a model of the life of a true communist.

Hero of Socialist Labor, Academician I. Tamm
20 October 1955

This was who Boris Gessen was in the eyes of his friend Igor Tamm in 1955. And this was Igor Tamm in 1955: Andrei Sakharov's mentor in science and in life. Although not a party member, for Tamm the expression "true communist" sounded then like unconditional praise. For a long time he was unaware of how his understanding of this expression differed from the party's. Traces of Soviet vocabulary are evident in the quoted reference, but this is not weak-kneed opportunism. Tamm did not adapt himself even in 1937, when—like a madman—he refused to condemn those among his associates whom the plague of terror had singled out.

Chaos and the Logic of the Plague

At the same FIAN meeting in 1937, Yuri Rumer (1901–1985) explained why he "felt like he was under the reliable protection of our Party organizations" and then "by direct invitation of a Party organization" spoke out about his contacts that aroused suspicion.

> In January I went to Kharkov where Landau worked. There was a critical situation there. Landau was under suspicion then, and I considered it my duty to openly speak out in defense of my friend Landau. And I now declare: "If Landau turns out to be a saboteur, I too undoubtedly will be held accountable," but that now when this, my statement, is entered in the minutes, I still vouch for him, as my best friend. I do not vouch for anyone else—neither Gessen, G. Landsberg, nor for I. Tamm, because I do not know them well, but I am always prepared to vouch for Landau.
>
> I have a brother who is 17 years older than I am. When he was arrested, I came to the university and told my comrades about it, including the Party organizer. My brother was arrested by the NKVD and exiled for three years. Twenty-nine months later, he had another seven months of exile remaining. I never concealed this and maintained that my brother was not a saboteur or a Trotskyite.
>
> If an individual feels that he is politically clean, like I do, then he can boldly say to everyone: "Investigate my contacts, my work!" I declare that not a single person among my acquaintances was arrested. It's true that my brother was, but that's another matter: my brother was 17 years older than me. Besides which, I choose my friends, but not my brothers.

Rumer had reason to be concerned about his friend Landau. "The critical situation" was, in fact, the beginning of the destruction of the Kharkov Physics and Technology Institute. At that point, in February of 1937, Landau left Kharkov for Moscow and managed to avoid danger. However, a year later, on April 28, 1938, Rumer's prognosis was justified—he was arrested on the same day as Landau.

Landau's case was a rare exception—in contrast to the millions of other victims, there was a legal basis for Landau's arrest. He took part in preparing a leaflet that contained the diagnosis: "The Stalinist clique carried out a Fascist coup . . . Socialism remains only on the pages of completely falsified and deceptive newspapers."

To understand how such a suicidal document could have appeared, one has to carefully look into the circumstances of Landau's life.[13] But to understand the time of the Great Terror, a more important fact is that the actual anti-Stalin leaflet was but a minor charge brought against Landau—the major ones were the falsified accusations of wrecking at the Kharkov Institute. Even more telling was that despite material evidence—the leaflet—a year later Landau was free. Meanwhile, millions of his contemporaries—with no evidence except statements made under torture and entirely spun out of thin air—perished in the gulag or spent many years there, as did Rumer (who knew nothing about the leaflet).

Millions of people died merely to "legalize" Stalin's reprisals against his enemies at the highest levels of the nation's leadership, in order to firmly establish his dictatorship. There were only, perhaps, a few dozen of such enemies—potential or imaginary. But in order to proclaim each of them an enemy of the people, the members of each enemy's group had to be selected, those one step below or to the side among his fellow workers and relatives. Each of their accomplices then had to be found, and so on. The pyramid created by the NKVD required more and more victims. Investigators feverishly sought out new criminals and fabricated more lies and more absurd crimes.

This is how the burial mounds of 1937 appeared. Only when Stalin was satisfied with the results of the purge at the top did he stop the sacrifices and, by the summer of 1938, sent off those who executed his will—the leadership of the NKVD—to the same graves.

A thick fog of ignorance and lies covered the general picture for those living in that time. I had the occasion to speak with three participants of the 1937 FIAN meeting and found that they did not remember it, although they had made speeches at it. When I reminded Ilya Frank about this meeting, after a pause the Nobel laureate asked me whether he had said anything "horrible" at the meeting. No, not a word about politics, only about the scientific matters of his nuclear laboratory, which was taking its first steps then, and about the great help they were receiving from Igor Tamm. Perhaps no politics and some gratitude to one of the main "defendants" could be considered political—moral politics.

But how could anyone forget the horrible speeches heard at FIAN in April 1937? How could Tamm and Frank in that terrible year of 1937 devise the theory of radiation emitted by superluminal electrons, for which they were awarded the Nobel Prize twenty years later, the first Soviet Nobel Prize in Physics?

The archival shorthand report elicited bitter bewilderment from the eyewitnesses. In their memories, the prewar FIAN was filled with "the atmosphere of immersion in science, mutual good-will, combined with tactful rigor, so unlike what one had to deal with in other places at the time."[14] And all of them cherished grateful memories of Sergei Vavilov, whose efforts created this atmosphere.

Of course, psychological self-defense was in effect. Scientific creativity was sustained both by the contemporary state of physics of the time and by youthful enthusiasm and, possibly, the aspiration to seek escape from the life around them, from the irrational cruelty of what was occurring. Besides, FIAN seemed an oasis then, in contrast to what was happening in Moscow University.

Yes, the people of FIAN did not comprehend what was occurring in the nation, and they were unaware that the crowning contribution of Soviet physics to world science attained its maximum during the second half of the 1930s. The physicists of 1937 would not have been surprised by the steep rise to the peak, while our contemporaries are not surprised by the decline, just by its slight slope. It is no accident that the chart peaks near the year 1937—the growth curve sagged under the blow of the Great Terror and the weight of the central-ized organization of science formed by the end of the 1930s. The nuclear and space successes of Soviet physics are essentially a side product of the 1930s, when their creators entered science "amid stormy waves and tumultuous night, and in the breath of the Plague."

At the 1937 FIAN meeting, Tamm talked about his participation in the First Congress of Soviets in June 1917:

> There were three resolutions that were moved on there: one granting gen-erals the right of capital punishment at the front, the second against grant-ing it, and the third resolution was not to grant generals the right of capital punishment at the front, not because it was impossible, but because it was possible only at the hands of the proletariat. Five people voted for this reso-lution, and I was one of them.

The right of capital punishment, which Tamm in 1917 considered possible at the hands of the proletariat, continued to tear the nation apart. A week after the FIAN meeting, Tamm found out that Semyon Shubin (1908–1938), his favorite student, who was head of the theoretical department of the Ural Physics and Technology Institute, had been arrested in Sverdlovsk. Aleksandr Vitt (1902–1938), professor at Moscow University, a brilliant member of the Mandelshtam school, was arrested in May, and in August so was Matvei Bronstein (1906–1938), a remarkable Leningrad theorist for whom Tamm acted as official reviewer when Bronstein defended his doctoral dissertation. They were given different sentences: eight years, five years, and death by execution—but all three died in 1938.

Three decades later, Tamm summed up the development of theoretical physics in a commemorative tome dedicated to the 50th anniversary of Soviet Russia. In his summation, he pointed out the untimely death of these three "ex-ceptionally brilliant and promising" physicists from the first generation edu-cated in Soviet times.

In his *Memoirs*, Sakharov also talks about these deceased 30-year-old physi-cists; he found out about them from his teacher. In the 1960s, Sakharov came to deal with the theory of gravitation and cosmology, in which Matvei Bronstein did his major work, and shortly after that he met Bronstein's widow, Lydia Chukov-skaya. The human rights movement brought them together, and Sakharov ex-plained the meaning and significance of her husband's scientific papers to her.

Lydia Chukovskaya is remembered for her eyewitness accounts of the 1937 era, which she wrote at that time:

> My records from the era of terror are noteworthy, incidentally, in that only dreams are reproduced in them. Reality did not lend itself to my description; moreover, I didn't even make an attempt to describe it in my diary. It couldn't be captured by means of a diary; besides, was it thinkable to keep a real diary at that time? The contents of our conversations, whispers, conjectures, failures to speak at that time in these entries is carefully absent.[15]

Reality was captured by her books, which were based on what she had lived through and written without any hope of publication in her lifetime.

Summing up this reality half a century later, Andrei Sakharov wrote:

> Speaking of the nation's spiritual atmosphere, the general fear that gripped virtually the entire population of large cities and left its mark on the rest of the population and continues to exist secretly to this day, nearly two generations later—it was generated mainly by this era. Fear was induced by the brutality and mass character of the repressions, as well as their irrationality and daily occurrence, when it's impossible to understand who is being sent to prison and why.

The irrationality of what occurred in 1937 even astonishes now, when, for instance, you read in the investigation of someone who was executed that he was recruited into "a Fascist terrorist organization" by a person whom the terror bypassed, or you find out from carefully filed papers that a Jewish physicist from Holland recruited a Soviet Jewish physicist to work for Nazi intelligence.

Irrationality was unbearable for people who were under the power of rational "scientific" social ideology. So they tried to find a reason for the arrest of an intimate person or an acquaintance. And if you seek you will always find. The reason for an "investigative error" could be the fact that a suspect was acquainted with a known member of the opposition at one time, that he spent time abroad, that he spoke out too harshly about the flaws of life around him and was misunderstood or, finally, that saboteurs themselves had infiltrated the NKVD and were imprisoning honest citizens loyal to the Soviet regime.

What, for example, could Tamm think about the arrest of his younger brother?

> When he was arrested, I was in torment, trying to understand what he could have been guilty of. I couldn't tolerate the thought that they could imprison an innocent person. I was tormented until I found a satisfactory, as it seemed to me, explanation. I thought: Lenya never could have committed anything bad. But maybe he knew something about other people's crimes and didn't inform on them. He was an honorable person; he never would have informed on anyone. And at that time not informing was punished by the law, and rather strictly. And so they arrested him. When I made all this up, I felt a lot better. And only much later did I realize that he had been completely innocent.[16]

Andrei Sakharov on the Threshold
of Adulthood

Andrei Sakharov was only sixteen years old, and life spared him so that he did not see 1937 in its full flush. The terror did not touch his parents, and they tried to protect their son from brutal reality, in particular by affording him the opportunity to study at home right up to the seventh grade. This was unusual in those times and an expensive undertaking, prompted most likely not only by a distrust of schooling but also by apprehensions about the standards of Soviet education. Their boy was too special—their first child, and a rather late one. Andrei was a "little prince" to his mother and a grateful student to his father, a born teacher.

Lengthy homeschooling, in Andrei Sakharov's own opinion, intensified the "lack of communicability from which [he] suffered in school, at the university, and in life in general as well." But at the same time, the walls of his home, love and a lengthier-than-usual solicitousness protected his inner world, and this, perhaps, contributed to his characteristic feeling of self-esteem, mild manner, and moral firmness. When he was taking exams before entering the seventh grade, his teachers were particularly surprised by his "free, simple and unaffected manner."

Along with an individual education, freedom at home shaped his inner world. His father, a physics teacher and author of textbooks and popular science books, taught him physics and mathematics:

> We did the simplest experiments which he made me carefully write down and illustrate in a notebook. It seems to me that I grasped everything without explanation. I was very excited by the possibility of reducing the entire diversity of nature's phenomena to the comparatively simple laws of atoms interacting, as described by mathematical formulas. I didn't entirely understand what differential equations were yet, but I already sensed something and was delighted with their omnipotence. It's possible that this excitement inspired my desire to become a physicist. I was immensely lucky to have a teacher like my father.

It's not difficult to understand this father who, while revealing the elegant laws of nature, kept to himself his thoughts about the chaotic lawlessness outside the window: "While I was still young, my father was afraid that, if I knew too much, I wouldn't be able to get on in this world. And maybe this typical hiding of thoughts from one's son best characterizes the horror of the era."

The world of Andrei's youth was at the mercy of two totalitarian forces simultaneously, and the deadly threat of Hitlerism made people forgive Stalinism far too much. At the neighbors in his communal apartment, Andrei listened to a radio set for the first time in his life and heard "Hitler's Nuremberg Congress, the senseless and frightening chanting of its participants' 'Heil! Heil!

Heil!', Stalin's speech at the Congress in 1936," about the new Socialist constitution. He also listened to the broadcasts about the Pushkin anniversary in 1937
on the same radio. And so, Hitler's and Stalin's speeches and Pushkin's poetry
resounded over the airwaves and shaped the mentality of Andrei Sakharov's
generation.

I was able to locate only one of his classmates who said that he realized the
scale of government deception as early as the end of the 1930s. The bright boy
accidentally read *Ten Days That Shook the World*, a book about the October
Revolution by American John Reed, with an introduction by Lenin, and found
that the revolution was made by the "enemies of the people" and not Stalin,
who was hardly mentioned in the book. His father never discussed the Great
October Revolution with the boy but explained in great detail how the French
Revolution devoured its children.[17]

Perhaps particularly revealing is the mindset of another classmate who was
born in the United States and spent the first thirteen years of his life there. His
father brought his family to the USSR in 1931 to build the new world of socialism. The father was arrested in 1937. The 19-year-old youth, who was sure that
it was a misunderstanding, headed over to the NKVD. They did not arrest him
or take away his belief that he was living in a nation where the radiant future
of all of humanity was being created, the only nation which was capable of
resisting fascism.[18]

This belief united him with Igor Tamm—whose brother was arrested, with
Elena Bonner—whose parents disappeared in the abyss of 1937, and with many,
many others. They all knew the quote from Marx that was displayed everywhere throughout the land: "Religion is the opium of the people." And they
all were under the power of an equally potent narcotic—the atheistic religion
of socialism, which promised the scientific road to creating heaven on earth
and provided the hellish patience to walk this road.

It was after 1937, under the influence of this narcotic, that Alexander
Solzhenitsyn "tried to carefully investigate the wisdom of *Kapital* by Marx," and
David Samoilov and his poet friends tried to design a life platform of "candid
Marxism."[19] And these were people of strong, independent, and honest spirit
and humanitarian orientation. For a person of science, the air castles of scien-
tific socialism were an even greater temptation.

The optimists of that time have forgotten how the socialist anesthesia, along
with psychological self-preservation, buoyed social optimism. When you are
in an unbearable situation, there is nothing to do but bear it. No one has yet
succeeded in postponing one's life until more auspicious times.

In any case, in addition to the general background of life, there is also its
concrete daily content, especially important in youth. By the summer of 1938,
the plague of the terror had sated itself. And young people, who had entered
university along with Andrei Sakharov, sank their teeth into science, read
poetry, fell in love, and quarreled just like other generations of students. The
proportions varied in each individual. One of Sakharov's female classmates

recalls how "a tall fellow, skinny as a rail" strolled along the halls during the breaks between lectures, "his head lifted high, staring into the ceiling, preoccupied with his own thoughts. He wasn't friends with anyone, he was on his own. Of course, we were interested in boys, and boys were interested in us, but not Andrei."[20] Many years later, this classmate found out from Elena Bonner that Andrei was not indifferent to her during those student years. Andrei Sakharov's classmates found out with no less surprise—after his death—that Pushkin was an important subject of his association with one of them during those prewar years.

Sakharov's inner world was pretty closed not because of arrogance—he readily helped others, explaining difficult questions. His class wasn't lucky with their university teachers. Gessen's arrest, called "the defeat of the Trotskyites at the Physics Department," engendered the removal of the "Mandelshtam group," who had combined first-rate research with teaching. The remaining teachers were more ambitious than discriminating. It is true that mathematics was taught at a traditionally high level. And students also had the excellent university library at their disposal.

There was a physics club with approximately a dozen students. A photograph of one of its sessions has been preserved. According to the reminiscences of the circle's monitor, Sakharov gave a talk at that session on the topic of Fermat's Principle, which governs the propagation of light. Despite the principle's old age, it is not easy to elaborate at the second-year level, and the speaker was unable to achieve transparency in his first scientific report.[21] It's difficult from the photograph to confirm or deny this by the appearance of those listening. It is even more difficult to imagine that these young people, so peacefully soaring in the clouds of theoretical physics, had been living in 1937 only two years before.

But they were allotted only three years of peaceful study. The war began. Andrei, along with other students, served on duty during air raids, extinguished incendiary bombs, and unloaded railroad cars. During these same days he started working on his first scientific inventions. A magnetic device was needed to detect shrapnel in wounded horses.

Physics students were accepted into the Air Force Academy, but the medical commission did not pass Sakharov: "I was disappointed then, but later considered that I had been lucky—the Air Force students spent nearly the entire war at their studies, while I worked at a cartridge plant for two-and-a-half years, and made a timely, albeit small, contribution."

This is what he wrote about his attitude to the front and to life:

> Some people like me, who were not drafted—young women in particular—enlisted in the army as volunteers (during this time my future wife Lusya volunteered). I don't recall thinking about it. I did not want to hurry fate, I wanted to leave things to flow naturally, without rushing ahead or trying to be clever in order to stay out of danger. This seemed proper to me (and still does now). I can honestly say that I never had the desire or attempted

to scheme either with respect to the Army or other things. As it turned out, I never served in the Army, as did the majority of my generation, and I remained alive while many others died. That is how life turned out.

The way life turned out was that, in October 1941, the remaining university students were evacuated to Central Asia by freight train. The journey took a month. One sentence tells everything about that time: "Once I saw a spice cake which someone had dropped in the snow and ate it immediately." The period of study at the university was reduced by a year, "while the program, out of date as it was, was sharply cut back. This was one of the reasons why there remained yawning gaps in my education as a theoretical physicist for the rest of my life."

In the summer of 1942, having passed his state exam in the physics of metals, Sakharov graduated from Moscow University with honors. He was offered the opportunity to continue in graduate school, but he refused and was assigned to a military plant: "It seemed to me that continuing my education during the war, when I already felt myself capable of doing something (even though I didn't know what), was wrong."

We can find parallels for the 21-year-old physicist's feeling in the Russian tradition. At about the same age Pushkin wrote:

> To greatness I aspire,
> I love my Russia's creed.
> My promises inspire—
> God, help me to succeed![22]

II

Intra-atomic, Nuclear, and Thermonuclear

4

THE MORAL UNDERPINNINGS OF THE SOVIET ATOMIC PROJECT

Behind all the discussions of Soviet nuclear history, there lies a simple and harsh question: how could they have made a lethal nuclear weapon for a dictator who was lethal enough on his own?

"They" were the Russian scientists of whom Soviet Russia was proud. Many of them are the pride of world science, and a few of them are worthy of the respect of the humanity—if not for the superbomb in the hands of Stalin. Let us not ask if the wheel of history could have turned differently. Better to ask how the Russian scientists who were moving the wheel of the nuclear age felt about their work.

The history of nuclear weapons is an unprecedented blend of pure science, dirty technology, and state politics both dirty and clean. Does morality have anything to do with this history, especially in the Soviet Union, where spiritual life had been totally suppressed? Were Soviet scientists all equally immoral, agreeing not only to live in the country but to work for the "evil empire?"

Only by understanding the moral underpinnings of their agreement can we truly understand the Soviet atomic project and, thereby, world history, because Soviet nuclear might in many ways determined the life of the world for four decades.

The roots of the Soviet atomic project go back to the prewar years, when three main institutes worked in nuclear physics, competing for resources that came from only one source—the Soviet government: the Radium Institute, headed by Vladimir Vernadsky (1863–1945); the Leningrad Physics and Technical Institute (PhysTech) of Abram Ioffe (1880–1960); and the young Physical Institute of the Academy of Sciences (FIAN), which was headed by Sergei Vavilov (1891–1951) and based on the school of Leonid Mandelshtam (1879–1944).

These scientists were not specialists in nuclear physics, but their students were responsible for the main achievements of the Soviet atomic project.

Vernadsky's disciples were responsible for the radiochemical component of the project or, simply put, the manufacture of nuclear fuel. Graduates of Ioffe's school helped to create the nuclear reactor and the atomic bomb. And finally, Mandelshtam's school opened the Russian path to thermonuclear energy in uncontrolled and controlled versions.

These three "scientific families" were united in a common cause and by a common social stratum. However, there were stylistic differences stemming from the founders, who embodied three philosophies of science. Put schematically, Vernadsky considered the power of science to be greater than all other social forces, Ioffe saw the embodiment of science in state Soviet ideology, and for Mandelshtam science and social ideology were two different worlds.

Ioffe's Pragmatic Philosophy

For Soviet historians of the era of "socialist realism," Ioffe was the only founder and father of Soviet nuclear physics, and that must be because his philosophy was in accord with those in power.[1] The Stalinist doctrine of socialist realism—unlike ordinary realism—demanded that life be depicted the way it should be according to the will of the highest Ruler. This doctrine, invented for the arts, reigned over all of Soviet life. Ioffe was very attentive to the voice of Soviet leadership, and his philosophy could be called pragmatism, even though he proclaimed his adherence to dialectical materialism, the official state philosophy of science, dubbed DiaMat. The physicist-experimenter-administrator Ioffe readily used DiaMat vocabulary to secure what he considered true science. Bits of official Soviet wisdom and praise for Soviet leaders tripped off his tongue.

Here is but one example of his rhetoric from the post-Stalin period (1955), when no one forced him to use the terminology:

> Our Party has equipped the young scientist was a trusty compass—the philosophy of dialectical materialism . . . Our country, moving ahead of progressive humanity, is embodying the highest form of social life—communism. Every thinking person, and the scientist must be a thinker, must strive to have his work contribute to the construction of communism.[2]

A gift for this sort of political poetry helped Ioffe build his PhysTech. He managed to obtain ample funding and to take talented young people into the institute. In 1933, summing the first fifteen years of his institute, Ioffe called one of his principles "attracting young people to responsible creative work" and added, not without satisfaction, "at one time they referred to us sarcastically as an orphanage."[3] The nickname "Papa Ioffe" stuck to him, but the only thing he could teach his "children" was a general awe of the primacy of science and a kind of sportive scientific derring-do. Young talents, as a rule, do not feel that they need mentors; they easily underestimated the father's care and were clear-eyed about some of Ioffe's errors that resulted from his overenthusiasm. This meant that Ioffe's school also taught them an honest attitude toward science.

That said, Ioffe did more than any other scientist for the growth of Soviet physics. Compared to the other sciences, only perhaps Nikolai Vavilov, brother of Sergei Vavilov, did as much for Soviet biology, even though he differed radically in every other way.

Let's believe the president of the Royal Society of London, Sir Henry Dale, when he wrote of Nikolai Vavilov in 1948:

> In 1942, the Royal Society of London elected Nikolai Ivanovich Vavilov to be one of its fifty foreign members. At first director of the Lenin Academy of Genetics, Vavilov was reputed to have been of great assistance to agriculture in the Soviet Union. We desired to honor his work as a great contribution to science for the whole world. However, it had been reported in Britain already in 1942 that N. Vavilov had somehow fallen from favor with those who came after Lenin, though the cause of his trouble was still unknown; we might have supposed it to be political or otherwise irrelevant to his scientific achievements.[4]

Laureate of the first Lenin Prize (1926), first president of the Academy of Agricultural Sciences, and director of the Institute of Genetics of the Academy of Sciences, Nikolai Vavilov was arrested in August 1940. He was replaced by Trofim Lysenko, who several months later became a laureate of the Stalin Prize.

In contrast, sixty-year-old Ioffe applied for membership in the Communist Party in 1940. It is difficult to separate Ioffe's concern for Soviet physics from his concern about his own position in it; however, the latter was perfectly obvious to his colleagues. His social conformity was based both on his socialist sympathies and his pragmatic materialism. His Leningrad PhysTech deserved its reputation as the cradle of Soviet physics, and it can certainly be considered the cradle of the Soviet atomic bomb.

For Ioffe, nuclear physics was simply one of the hot scientific directions, on a par with, say, semiconductors. The problem of nuclear energy served primarily as an instrument for developing his science and getting additional financial support from the government. There were also other instruments, such as the project for creating a battery, on the basis of thin-layered insulation, and the round house project, which conserved heat. This philosophy helped Ioffe influence the policies of Soviet physics and, in particular, the selection of the scientific leader for the atomic project.

Two other people with completely different worldviews remained outside the "socialist-realist" history of Soviet nuclear power.

Vernadsky's Noospheric Philosophy

Vernadsky's specialty in the narrow sense was geochemistry. However, one could never call his perspective narrow. Even by Russian standards, he was a broad thinker.

Central to his philosophy was the concept of the noosphere, meaning the part of the biosphere that was under the influence of mankind. In Western culture, the term (from the Greek *nyo[s]*, meaning mind) was introduced in the 1940s by Teilhard de Chardin (1881–1955), the French paleontologist and religious philosopher. He combined a scientific view of the evolution of the world with a Christian worldview, which brought harsh criticism from science and the church.[5]

For Vernadsky, religion was one of the forms of spiritual search, and he was close to religious people and clergy, but he considered the scientific approach more profound. Soon after his first public discussion of the concept of the noosphere, he wrote in his diary (March 12, 1938): "Read Tolstoy. Now (and for a long time) the concrete ideas of Christian (and any other) religion and philosophy seem so petty before the inner I in its scientific revelation!"[6] Vernadsky's noosphere is a concept of natural sciences, of geology in the most general sense. It is the stage in the life of humanity when science and technology become a force of geological scale:

> Science is a natural phenomenon, an active expression of humanity's geological manifestation, turning the biosphere into the noosphere. It expresses the real attitude between the human living substance—the sum of all living people—and the environment, primarily the noosphere. Man and human communities can be removed from here only mentally. Man and noosphere are inalienable. (May 28, 1938)

From the early years of studying radioactivity, Vernadsky saw the power of the noosphere in this new form of energy. He applied his own "noospheric" energy far beyond geology, becoming head in 1922 of the Radium Institute with physics, radiochemical, and geochemical departments.

Vernadsky's school reflected more than any other a continuity with pre-revolutionary scientific life. In February 1922 Vernadsky wrote a friend in Paris:

> Managed to restore work on radium that I had begun in 1910. Work begun in 1916 by my colleagues on obtaining radium from Russian ore was completed in December 1921; we found new methods of processing and had obtained the first Russian radium from Russian ore. Now we have to defend ourselves to keep it in the hands of science. The salvation and restoration of this work is the great deed on the part of young scientists; one of them died, unable to survive the barbaric living conditions.[7]

The tradition of serving science was combined for him with the tradition of defending science's interests from the state power—imperial or Soviet, it did not matter. He could be called a constitutional democrat in science—he respected the independence of experts and knew how to provide them with creative space.

Vernadsky's efforts determined to a great degree the starting position of the Soviet atomic project. The first European cyclotron, the apparatus of the nuclear age, was built at his institute. It was possible only because in 1931, when the first cyclotron was built in America, Vernadsky defended the right of his institute's existence. He wrote to Stalin then: "The study of cosmic rays and nuclei of atoms will lead us to the discovery of new, powerful sources of energy. A state that looks forward instead of back cannot live without attention to the inevitable great discoveries. We are facing the future sovereignty of radioactive energy, more powerful than electricity."[8] And when the discoveries were made, on Vernadsky's initiative, the Soviet government was informed (in July 1940) of the importance of "the practical use of intra-atomic energy."

Now, under "practical use" nuclear weapons first come to mind. But Vernadsky was thinking of the general—"geological"—effect. In October 1941 he recorded in his diary a conversation with an academician economist, who

> feels that a new form of energy—atomic—will not change the economic structure of society, will not make that revolution that I imagined when I talked and thought about it. I think there are no "laws" of economics that will not change radically once man gets such a concentrated energy, and 5 kilos of it will equal 200,000 tons needed now for the same effect. (October 12, 1941)

But what social philosophy stood behind his attitude toward the Soviet regime, to which he offered this concentrated energy? For Vernadsky the history of mankind was primarily the history of science and technology, and he was prepared to collaborate with any "noospheric force." In March 1943 he sent a telegram to Stalin: "Please use 100,000 rubles of my prize named for you for the needs of defense, as you see fit. Our cause is righteous and now spontaneously coincides with the coming of the noosphere—the new state of the sphere of life—when man's mind becomes an enormous geological planetary force." "Now spontaneously coincides" means it might not coincide.

Twenty-one years previously, Vernadsky wrote from Petrograd to his friend in Paris:

> The more you think about your surroundings, the more you become convinced that the present great current happening in mankind—at this historic moment—is the current of scientific thought. It must prevail on its own, and before it all political, social, national, and even religious strivings of life are petty. In the final analysis, it is creating the future.[9]

Perhaps this science worshipper was merely hiding from reality in vague illusions to create spiritual peace for himself and therefore was no less a conformist than Ioffe, just adapting differently? This hypothesis does not fit the real life of the noospheric geologist.

Vernadsky had a clear understanding of the people who stood at the head of the Soviet state and to whom he wrote many a letter to save the lives of his colleagues and friends. For instance, he believed it quite possible that Hitler could be behind Soviet nuclear efforts. He wrote about it in his diary of 1940:

> Hitler proposed to Stalin and Molotov an exchange of scientific achievements between Germany and the Soviet Union. It turns out the achievements are not that great—a commission from the NKVD headed by Beria or some other important official was sent. Apparently, it hasn't reached a tragedy yet. Perhaps the Central Committee's resolution on uranium is related to Hitler's proposal, too?

Hitler and Stalin's secret police, supervising science? That had to add a tragic tone to Vernadsky's noospheric visions, but as he wrote in his diary, "of course, both Hitlerism and Stalinism are a transitional stage, and life is unlikely to go without explosions. Which?"

Anyone who could write that in December 1938 was no conformist. He knew that no social status and no contributions were a defense against the plague of terror. Therefore, when he quotes the dangerous thoughts of others, he does not give their full names. For instance, an entry of December 4, 1938:

> Many regard the near and distant future grimly. L. (academician) "Man is headed toward savagery." I'm of a completely different opinion—Man's headed for the noosphere. But now it's becoming clear that we will have to live through a collision, and the next few years are very unclear—War? I do not believe in the power of German fascism—but the Western democracies fear collision more than fascism, this is a dangerous situation . . . The transition to the noosphere will apparently take place in paroxysms.

So Vernadsky's collaboration with the state was the result of his noospheric views. He attributed social evolution not to class struggle but to an objective process: "I've seen that the basic traits of democracy are a noospheric phenomenon" (November 14, 1938).

And how did the Soviet authorities regard Vernadsky? He was one of the founders of the Constitutional Democratic Party, had been a member of the Provisional Government, and, despite that, in his diary—in the twentieth year of the Soviet regime!—referred to the newspaper *Pravda* [which means truth] as *Falseda* and was not afraid to write about the party congress, "The newspapers are filled with ridiculous babble. It was a gathering of bureaucrats afraid to speak the truth. Not a single almost-living thought. They do not touch upon the course of real life. Life goes on—as much as is possible in a dictatorship—without them" (February 20, 1941).

The authorities had to know Vernadsky's anti-Soviet views. But they also knew that (according to the *Great Soviet Encyclopedia*) he was "a Soviet natural scientist, mineralogist, and crystallographer, one of the founders of geochemistry," or more simply put, a specialist in useful minerals. And that may be the answer to the puzzle why Vernadsky lived to die a natural death. Minerals are too useful a thing to lose when building socialism in a single country. And if the geologists of the country consider Vernadsky their teacher, you can shut your eyes to his dark past, his inappropriate appeals to free enemies of the people, and his anti-Soviet feelings.

Mandelshtam's Old-Fashioned Morality

The moral position of the third key figure, Mandelshtam, was equally distant from both conformity and any sort of global social philosophy. It was an old-fashioned idealistic morality of the prerevolutionary era, rooted in the spiritual world of the Russian intelligentsia.

Unlike Ioffe and Vernadsky, Mandelshtam did not leave a single philosophical publication; there are only a few remarks of a philosophical nature in his physics lectures. However, Mandelshtam's moral philosophy is evident in his philosophy of science.

Mandelshtam was to become—posthumously—the most prominent target of the "militant materialists" in the late 1940s, and in February 1953 a special meeting of the Scientific Council of FIAN condemned Mandelshtam's "philosophical errors," that is, his "subjective idealism," eight years after his death.

But perhaps the philosophy watchdogs made up their charges and Mandelshtam was a "spontaneous materialist," in the standard Soviet terminology of the day? Gessen and Vavilov both practiced Marxist philosophy. If these Marxist administrators defended Mandelshtam, how could his philosophy have not been Marxist?

Mandelshtam helps us answer that question. His philosophy did not notice Marxism. We do not need to analyze oblique semiphilosophical phrases in his physics lectures, for he left an entire manuscript on epistemology in physics. And this manuscript on the philosophy of science speaks eloquently of his life philosophy.

The manuscript was written in Borovoe during the war. Elderly and physically weak academicians were evacuated to this resort in Kazakhstan at the start of the war. Mandelshtam developed very warm ties there with Vernadsky and Alexei Krylov. Those two Russian scientists were contemporaries but otherwise very different, with a relationship that was completely respectful but distant. Mandelshtam, sixteen years younger, attracted both, even though their subjects of conversation were poles apart.

The mathematician, shipbuilder, translator of Newton, and tsarist general Krylov talked with Mandelshtam most about science and life—he was completing his memoirs in Borovoe. Geochemist and thinker Vernadsky, occupied in Borovoe primarily with his noospheric writing, talked with Mandelshtam about physics and geology but also about the philosophical ideas of thinkers as varied as Goethe, Einstein, and even Jaspers. The name of the German religious philosopher, far from natural science, in the conversation of a Russian physicist and geologist, at the height of a world war, characterizes the breadth of their philosophical foundations.

There are no "isms" in Mandelshtam's philosophical manuscript, written in Borovoe, and only one brief quote from Austrian philosopher Wittgenstein: "*Zu einer Antwort, die man nicht ausspretchen kann, kann man auch die Frage nicht aussprechen,*" which freely translated means, "If a question cannot be answered, that means there is something wrong with the question itself." On the whole, the manuscript is an incredibly free and pedagogically clear exposition of philosophical positivism. To use the labels of the period, it was Machism. In the Soviet era the name of Ernst Mach, Austrian physicist and philosopher (1838–1916), was one of the most reviled. The official state philosophy is glaringly absent from Mandelshtam's writing. It is clearly addressed not to philosophers but his young colleagues, students who were being intensively brainwashed in philosophy.

Deprived of his usual milieu, lab and students bubbling with ideas and questions, Mandelshtam probably tried to satisfy his pedagogic need by preparing a lecture on epistemology for his physics students—the very first lecture. For physicists of his generation—which is the generation of Einstein—positivism was the most fruitful philosophical viewpoint. Richard von Mises (1883–1953), a friend from Mandelshtam's Strasbourg days who managed to combine both mathematics and flying (during World War I), was also interested in the philosophical side of life. After the Nazis came to power, he left Germany and spent several years in Turkey. Far from his usual scientific surroundings, he decided to sum up his views on science and life. In a letter dated March 15, 1937, Mandelshtam asked him, "When will your 'Short Textbook of Positivism' be published? I'm awaiting it with great curiosity."[10]

Similarly, Mandelshtam, cut off from his physics, decided to lay out his philosophical approach. The ruling ideology gave students very short rations in philosophy. So positivism, branded "subjective idealism," deserved special care.[11] A diary entry of that period (April 5, 1942) by Vernadsky: "Yesterday in conversation with Mandelshtam—a very interesting and logical mind—he correctly said that today a physicist cannot work in science without philosophy, and the flourishing of modern physics is contingent upon it."

The very act of writing a philosophical manuscript was a manifestation of Mandelshtam's "subjective idealism" in the nonphilosophical sense, or his astonishing independence from totalitarian materialistic circumstances. This independence may have attracted Vernadsky. His final diary entry (December 24, 1944) is about Mandelshtam, who had died a few weeks earlier. Noting that

Mandelshtam was "one of the most interesting scientists with ideas that I have had met in recent years," Vernadsky added, "Mandelshtam told me that he had been offered the chance to convert to Christianity and remain in Germany, but he preferred to return to Moscow."

It may seem strange that Vernadsky chose to note this not most important fact of Mandelshtam's biography on that sad occasion. It might have been an implicit comparison with another physicist academician, about whom Vernadsky felt quite differently. It was well known that before the revolution Ioffe converted to Christianity. There was no reason to suspect the thirty-year-old physicist had had a religious revelation, and it was easier to see his baptism as readiness to make too-great concessions to the powers-that-be in order to achieve a practical goal—to get a good job in tsarist Russia. That same ability to make concessions could be seen in Ioffe's demonstrative loyalty toward the Soviet regime. In the pages of his diary, noting Ioffe's achievements and his talent, Vernadsky had written, "ambitious and dishonest because of it, morally—I know him from the Radium Institute—false. He cannot be trusted."

Mandelshtam and Ioffe were the same age; both had gotten doctoral degrees in Germany and began their work in physics in the bosom of European science, but with that their similarities ended.

Igor Kurchatov, a Special Physicist

Ioffe, Vernadsky, and Mandelshtam characterize the actual multiplicity of moral positions in Soviet physics. These positions were not automatically recreated in the schools they headed and by their people in the Soviet atomic project, but these three men did serve as role models.

The influence of the three leaders on the course of events in the project is indisputable. During the war years that Vernadsky and Mandelshtam spent in Borovoe in far-off Kazakhstan, Ioffe was in the center of events and applied considerable efforts to be there.

In January 1942, as his two-year candidacy expired, the sixty-two-year-old academician became a member of the Communist Party and, in May, the highest-ranking physicist of the land. He was elected—or rather selected—to be head of the physical-mathematical division and vice president of the Academy of Sciences, since those who were to be elected to such high positions had been first approved by the Soviet government.

And in 1942, when the situation at the front gave no cause for optimism and the Battle of Stalingrad was still in the future, the government had to make a decision about the creation of the Soviet atomic project. Intelligence reports showed that in England scientists and the government were taking the development of the atomic bomb quite seriously, and there was information about similar efforts in Germany and the United States.

On Ioffe's recommendation, Igor Kurchatov (1903–1960) was appointed director of the Soviet project in February 1943. Ioffe knew him very well, and since 1932, when experimental discoveries made nuclear physics a hot topic, Kurchatov had been Ioffe's "nuclear" right hand. Now, looking back and knowing the goals of the project and the circumstances in which it had to develop, it is easy to agree that Kurchatov had been the best choice to lead the project. The best for everyone—for Stalin and Beria, for Soviet and world science, and for international security. But at the time, Vernadsky was not pleased by the appointment. He did not know that the goal of "intra-atomic" research for many years to come would be weapons or that this research would be supervised, inside and out, by the KGB. He foresaw that the end of "the stage of Hitlerism-Stalinism was unlikely to pass without explosions" and that "the transition to the noosphere would take place in paroxysms," but he did not imagine that the first explosions would be nuclear, and the first paroxysms the agony of the victims of Hiroshima.

Vernadsky knew that Kurchatov was a good physicist who had gone through the school of nuclear experimentation at his Radium Institute, where he helped build the country's first cyclotron. But he did not see Kurchatov's rare talent as scientific manager. Distance interfered—the remoteness of the Radium Institute from Moscow, where Vernadsky had moved in 1935.

But Vernadsky knew Vitaly Khlopin (1890–1950), the founder of Russian radiochemistry, with whom he had collaborated for many years, and to whom he passed on the directorship of the Radium Institute in 1939. On his initiative, Khlopin became head in 1940 of the Uranium Commission of the Academy of Sciences, created for broad research into intra-atomic energy.

The war interrupted this work, and when it was resumed the authorities did not even ask the opinion of Vernadsky or Khlopin. In the era of socialist realism, a myth was generated that they had been called to Moscow by Stalin for consultation,[12] but Vernadsky's diaries show that he never left Borovoe and was definitely against Kurchatov.

The government point of view is understandable. Dealing with old-fashioned academicians with their old-fashioned sense of dignity would have been burdensome. The government needed someone more manageable than Khlopin. But the uranium project could not proceed without radiochemist Khlopin. For his development of the technology of nuclear fuel, he would eventually be given the star of Hero of Socialist Labor and the Stalin Prize—and radiation sickness. Sharing Vernadsky's views, Khlopin regarded nuclear technology with caution. His associate, Mikhail Meshcheryakov, who was present at the first Soviet atomic bomb test and who later visited Khlopin when he was ill, recalled:

> After looking at me for a few seconds, Khlopin asked, "I heard you were at the test site?"
> "Yes."
> "Did the military take part?"

"Yes, in the most direct way."

Khlopin shut his eyes, as if he did not expect anything new from life or from people.[13]

Kurchatov worked in the atomic project from the start in close contact with Beria's agency. The People's Commissar of Internal Affairs was involved in all the work of state security, which in a totalitarian state meant in everything.

After his appointment as head of Laboratory No. 2, the scientific headquarters of the Soviet atomic project, Kurchatov immediately received an assignment from Mikhail Pervukhin, deputy chairman of the Soviet government: to evaluate the intelligence materials on the uranium issue. Kurchatov prepared on March 7, 1943, a fourteen-page handwritten analysis that he summed up this way: the intelligence materials suggest a way to solve the problem much more quickly than thought possible by "our scientists who are not familiar with the course of work on this problem abroad." "Naturally, the question arose" whether the material was accurate or just "an invention intended too disorient our science." His opinion was that "it reflects the true position of things."[14]

Kurchatov ended his analysis with the phrase, "This letter will be brought to you by your assistant Comrade A. Vasin, who has the rough drafts for destruction. The content of the letter must not be made known to anyone besides him for the time being." Clear, businesslike, no emotion—it was as if the physicist felt quite at home in the top secret atmosphere and was used to working with the intelligence services. It would be hard to picture Khlopin in that situation.

In putting Kurchatov in charge of the atomic project, the government decided to make him an academician, too, in 1943. However, the academicians, for their own reasons, did not accept the will of the Central Committee, and on September 27 elected another—A. Alikhanov. The very next day, the government permitted an additional vacancy in "the special division of physics," and on September 29 Kurchatov became an academician.[15] And apparently, he didn't suffer doubts over the government's interference in the affairs of the Academy of Sciences. Being much younger than Khlopin, Kurchatov had spent his formative years under Soviet rule, so his experience of Soviet "serfdom" could have also played a role in his attitude.

Kurchatov's administrative gifts could be considered genius if only because his name still elicits kind thoughts in those who knew him—even after glasnost opened people's eyes to the country's true history. Only one of the dozens of people who knew Kurchatov and left reminiscences of him spoke in a less than delighted tone—that was Meshcheryakov, one of his deputies and the first director of the Dubna accelerator. When asked, "Do you think that chairman of the uranium commission Khlopin could have handled Kurchatov's work?" he replied grimly, "Kurchatov was manageable, while Vernadsky and Khlopin would not have allowed anyone to control them."[16]

"Manageability" was neither the only nor the most important quality in Kurchatov. He was a true scientist, dedicated to science and appreciating such dedication in others. He was a doer who found satisfaction in the success of work. He used his influence to support science beyond armaments, honoring the logic of the development of science. And he honored Beria's direct orders. His main instrument in his relations with scientists was the ability to infect them with enthusiasm and instill a sense of being protected; dealing with the government, he knew how to inspire trust. In both directions, his charm and personality were at play.

Within his professional competence, he was capable of taking bold steps and even working against the system, but he knew his limits. His strokes and early death (at 57) bespeak the difficulty of being the buffer between the world of Soviet autocracy and naturally democratic science.

In keeping with the spirit of his time, he saw science as the main force of world progress, but as a son of his Soviet times and graduate of Ioffe's school, he did not need the concept of the noosphere, trusting the concept of communism that was not his invention. And he trusted the leader of Soviet communism, Stalin, to set the general course. The country that was building communism got a guidebook from their leader, Stalin's *Short Course of the History of the Communist Party*, and it was studied zealously in all auditoriums.

But Vernadsky did not trust the *Short Course* or the general course taken by Stalin, and on November 16, 1941, he wrote clearly in his diary:

> These facts strike me, sharply contradicting the words and ideas of the communists: 1) the dual government—the Central Committee [of the Communist Party] and the Soviet of People's Commissars. The real power is Central Committee and even the dictatorship of Stalin. That is what ties our organization with Hitler and Mussolini. 2) The state within the state: the real power of the GPU [aka KGB] and its further permutations. This is a cancer, gangrene, corroding the Party—but without it, the Party cannot survive in real life. As a result, millions of imprisoned slaves include, along with the criminal element, the flower of the country and the flower of the Party.

5

ANDREI SAKHAROV, TAMM'S GRADUATE STUDENT

From a Cartridge Factory
into Theoretical Physics

The moral and political complexities of nuclear physics were unknown to the twenty-three-year-old engineer at the Ulyanovsk Cartridge Factory, Andrei Sakharov, when in July 1944 he sent a letter to the director of FIAN:

> Please allow me to take admissions examinations to the graduate school of the Physical Institute in Theoretical Physics, which I consider my calling. Since I am working in the People's Commissariat of Armaments system, I need an invitation to take the examinations. My address: Ulyanovsk. Zavolzhye. General Delivery.[1]

The two years in the factory had not been in vain: attached to the admissions request was a patent for an invention and manuscripts of three works. In another hand, there is a notation, "Handed to Prof. I. Tamm." It is apparently the handwriting of Andrei's father (then a docent at the Pedagogical Institute), who passed the manuscripts to Tamm. In the required autobiography, Sakharov described his studies during "my engineering period" and proudly wrote about the economic effect of his invention—a magnetic method of testing bullets.[2] He had already made a contribution to the war effort when he heard the call of science.

Thoughts about his "cartridge" inventions led him to some problems of theoretical physics, but it was a calling to which he would have been led in any case. What is more surprising is that he decided not to go on to graduate school when that was offered to him in 1942. A talent for science includes the ability to concentrate on it even in the most unsuitable conditions.

A classmate of Sakharov's cannot forget him, sitting on a backpack, reading a scientific journal in the fall of 1941, when the physics students were waiting for a train that would evacuate the university to Ashkhabad. "I looked over his shoulder. It was a review on colorimetry [methods of measuring colors, not too fascinating]. And I asked in shock, "What are you reading that for?" Andrei replied with exhaustive clarity: "It's interesting."[3]

Andrei himself recalled something else he read while waiting for the train. "Those days turned out to be fruitful scientifically—reading Frenkel's books on quantum mechanics and theory of relativity, I somehow understood a lot right away."

The train set off at last, and life settled down on it.

In every car with two rows of two-tier wooden beds and a stove in the middle, there were forty people. The trip took a month, and over that time each car developed its own life, with its leaders, chatterers and quiet ones, panic-mongers, go-getters, gluttons, indolent and hard workers. I was among the quiet ones, reading Frenkel, but listening and observing everything going on around me, inside and beyond the car, to the war-torn life of the country through which we were traveling.

The physics that attracted Andrei Sakharov then was not nuclear. He knew about the fission of uranium and chain reactions before the war, but as he wrote, "to my shame," he did not appreciate the importance of the discovery and "until 1945, simply forgot that the problem existed." Why to his shame? Physics has so many fascinating problems. In the evacuation in Ashkhabad, despite the hardship and his final compressed year of school, Andrei and his comrades organized a study group on gravitation theory.[4] It is hard to find a field more remote from "defense physics of metals," which was the degree they received at university.

But Sakharov, apparently, had mastered his major, since they would not let him leave the defense plant even when the Academy of Sciences invited him. It was a military plant and it was still the war autumn of 1944. In December Vavilov, the director of FIAN, had to make a special appeal to release Sakharov from the plant to the institute. They found a way around the obstacle. Sakharov was accepted in extramural graduate school "without leaving his main work." It was only on February 1, 1945, that he became officially a full-time graduate student. He had three years to pass exams and to write a dissertation to get the Candidate of Sciences degree (equivalent to Ph.D., with the next Soviet scientific degree being Doctor of Sciences).

During the first few months at FIAN, the new graduate student heard rumors about Laboratory No. 2, which was allegedly "the center of physics," but the whole world of science had opened before him and he was too busy to focus on just one center. He had the opportunity to do real physics next to the masters of the craft. The graduate student was working on higher science—the theory of relativity and quantum theory—from German books that Tamm gave him, "almost with no break I worked through books by Pauli, and they changed my world."

He tore himself away for a small project that gave him pleasure and that he remembered long after. The talk the new student gave in March was titled "On the Reasons for Anomalous Absorption of Sound in Water with the Presence of 'Bubbles.'" The FIAN acoustic researchers had a question about the absorption of sound in foam that occurs when water is shaken, and they came with the question to the theoretical department.[5] According to Tamm, "Sakharov immediately found a good explanation and came back a week later with a ready theory."

In May 1945, there was his lecture on electrical breakdown in dielectrics. And in May, "an unforgettable event—Victory over fascism." August follows victorious May in Sakharov's memoirs:

> On the morning of 7 August I left the house for the bakery and stopped by the newspaper displayed on the newspaper stand. I was struck by the report of Truman's announcement: on 6 August 1945 at 8 a.m. an atomic bomb of enormous destructive power equivalent to 20 thousand tons of TNT was dropped on Hiroshima. My knees buckled. I realized that my life and the life of very many people, maybe all of them, had suddenly changed. Something new and terrible had entered our lives, and it had come from the side of the Grand science—the one that I worshipped.

Soon afterward the newspaper *Britansky soyuznik* [British Ally], published in Russian by the British Embassy, started printing the official American Smyth Report on the creation of the atomic bomb.

> Impatiently, I grabbed and studied every new issue. My interest was purely scientific. But I wanted to invent, too—of course, I was inventing things that were either long known (three years) . . . or impractical . . . My university chum Akiva Yaglom used to say then—Andrei has at least two methods of separating isotopes every week. When the publications ended in *Britansky soyuznik*, I cooled toward these things and almost never thought of them for two and half years.

He thought of "grand science." The 1945 report from the theoretical department on "the basic problem of elementary particles" said, "I. Tamm proposed a new hypothesis on the character of interaction between the proton and neutron" and "with the assistance of graduate student A. Sakharov began the necessary calculations for a quantitative check of the proposed theory. The corresponding calculations are very labor-intensive." Sakharov's November lecture on the same subject was called "A Rational Scheme for Calculating Traces."

This is nuclear physics, and Tamm was one of its pioneers. And grand science for him and his new graduate student was understanding the law that regulated the life of the atomic nucleus.

A Little Bit of Nuclear Physics

Nuclear physics determined the life of this graduate student and the life "of many people, perhaps all of them." Therefore, for curious nonphysicists, here is an attempt at a simple answer to the question of what Tamm was so interested in.

Now, every nonphysicist knows that all the objects around us consist of atoms. Fewer than a hundred elements, the standard units of the universe, occur naturally, very few compared with the variety of objects composed of them. In the nineteenth century, the chemist Dmitri Mendeleev discovered an order in the qualities of chemical elements. He laid out all the then-known elements on shelves and predicted new ones, for which he assigned empty spaces on those shelves. The predicted elements were discovered and put in their prepared places, and the entire "cupboard" was called Mendeleev's periodic table.

But it remained completely unclear why the elements had such varying qualities—why some were shiny metals and others transparent gases. It was just as unclear what an atom was and how it was constructed. The mysteries of the atomic world grew until history got sick of it. And then, as often happens in the history of science, one answer fit a hundred enigmas. One day, Andrei Sakharov was demonstrating his ability to do mirror writing and used that saying as an example. It is the general formula for scientific triumph. One such triumph was when the atom was discovered or, rather, uncovered.

In 1911 Rutherford discovered through experimentation, studying radioactivity, that atoms are basically emptiness. Only the very center of the atom, taking up one million-billionth of the whole, is filled. That is the nucleus, around which electrons move at enormous distances from it. If the nucleus were to be enlarged to the size of an apple, the electrons would be proportionately dozens of kilometers away from the apple. That meant that the problem of radioactivity was hidden in the nucleus and that this radioactive, or "intra-atomic," energy

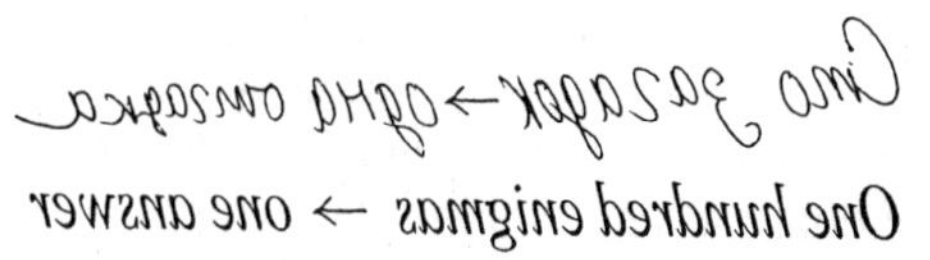

One hundred enigmas → one answer

Figure 5.1 Sakharov demonstrated his ability to do mirror writing using this "law of scientific progress" as an example.

should be called nuclear. When the nucleus changes its state, the excess energy is carried away by particles or light rays—alpha and beta particles and gamma rays. The first three letters of the Greek alphabet are the first letters of the nuclear alphabet, too.

Two years later, Niels Bohr unlocked the laws governing the movement of the electrons in the atom—the quantum laws of atomic physics. And on that basis, he then explained the way the Mendeleev chart was set up. It became clear that an atom's qualities are determined by its nucleus.

The nuclei differ in charge and mass. Nuclei differing in charge are chemically different elements. Nuclei equal in charge but different in mass are chemically indistinguishable isotopes of a single element. Hydrogen's nucleus is the lightest nucleus—just one particle, and it was called the proton. Uranium has the heaviest nucleus, made of more than two hundred particles.

This is the second general rule of scientific progress: "At the heart of the answer lie new enigmas." The organization of the nucleus turned out to be a much more difficult puzzle than the organization of the atom. Electrons and the nucleus have opposite electrical charges, and so they are tied by a force of electrical attraction known since the days of Maxwell and long domesticated. Then what holds the same-charge particles in the nucleus? What overcomes the enormous force of electric repulsion? All these forces are billions of times stronger than atomic ones because the nucleus is a hundred thousand times smaller than the atom.

This nuclear question has still not been answered completely, but one of the first steps toward its solution was made by Igor Tamm in 1934. Not long before that, a new particle—without electric charge, neutral—had been discovered and named the neutron. In every other respect, the neutron resembled the proton. They were considered equal components of the nucleus and united under a single term, nucleon. This solved a few nuclear brain twisters, but the question of the force holding the nucleus's particles together was still open.

Tamm hypothesized that the proton and neutron could be tied by an exchange of known light particles (the best known is the electron), as if the nucleons were tossing balls between them. This was a new idea—new and, alas, wrong. Tamm did the calculations, realized that the force was too small, and published his negative results. But they were negative only at first glance.

The path Tamm indicated was followed in 1935 by the Japanese theorist Hideki Yukawa, who did not specify the particle whose exchange tied the nucleons in the nucleus. He also got a negative result—the appropriate particle would have to have a mass 200 times that of an electron, and since no one had ever observed such a particle, he noted sadly, his "theory seems to be on a wrong line."[6]

The line was correct. Two years later, in 1937, a particle with that mass was found. They called it "meson," from the Greek word for "middle," since it was between the electron and proton in mass. They found a new particle, but not the law of nuclear interaction. The line was correct, but twisting. Physicists did not suspect then that the particle was not the one predicted by Yukawa. They would come to that conclusion only ten years later, when they discovered the right particle and transferred to it the name "meson."

In the meantime, for the subsequent ten years, the problem of nuclear forces was the hot issue for physics, and Tamm faced it.

Igor Tamm, Unemployed Fundamental Theorist

This decade was the darkest of Tamm's life. In 1937 he lost three of the people closest to him: his younger brother, a friend, and his favorite student. It is hard to understand why he himself was not branded "an enemy of the people," but in the chaos of the Great Terror there were many mysterious things. It is clear that the title of Corresponding Member of the Academy of Sciences was no protection, and nuclear physics was far from being a strategic profession then.

The losses of 1937 led to "organizational conclusions," and the university rector "recommended" that Tamm submit his resignation as head of the theoretical physics department. After the arrest in 1938 of Yuri Rumer, a member of FIAN, "measures were taken" at the Academy of Sciences as well. Because of "a lack of leadership from the head of the department [Tamm], inadequate work in preparing cadres," the theoretical department was formally shut and its staff relocated in other laboratories.[7]

Then came the grim years of the war and the institute's evacuation to Kazan until the fall of 1943. Only upon FIAN's return to Moscow was the theoretical department restored, so that Tamm could take his rightful place.

In the ten years after 1937, Tamm did not solve any problem that was comparable to his previous achievements. Wartime conditions explain a great part of this. But these explanations meant little for Tamm's passionate nature. What affected him was the fruitless effort to build a theory of nuclear forces.

Tamm's innate enthusiasm was compounded by the fact that he came to physics at a revolutionary time, when the basic concepts of the science were radically changing: space, time, causality. The alchemists' dream was coming true—the nuclear "alphysicists" had learned to turn one element into another. Outstanding physicists, beginning with Bohr, were quite seriously chasing another wild goose—perpetual motion. Tamm managed to contribute to the understanding of the nonelementary nature of elementary particles.

Now it is clear that the revolutionary period in fundamental physics ended in the early 1930s. The generation that saw this revolution take place retained revolutionary daring for many years. Especially Tamm, who was a risk taker by nature. He had seven first-class results to his credit, including the theory of Vavilov-Cherenkov radiation for which he would receive the Nobel Prize.[8] But what he valued most was his 1934 idea—incorrect in the narrow sense of the word—on the mechanism of nuclear forces. Then he was dealing with the leading edge of physical knowledge, and his idea was a step over the edge.[9]

He was very emotional about physics. "You can fall in love with a beautiful theory as if with a beautiful woman," he used to say.[10] And when the affair turned out to nothing more than temporary intoxication, exhausted and miserable, he would beg his young colleagues "to throw me a problem of some kind," calling it a hangover treatment.[11] He gave the last fifteen years of his life to unrequited love for an enchantingly beautiful and bold idea that seemed to promise him a fundamental movement into the depths of the microworld.

When the new graduate student Sakharov came to FIAN, there were seven laboratories (atomic nucleus, oscillations, optics, luminescence, spectral analysis, dielectrics, and acoustics) and a theoretical department.[12]

Tamm was head of the theoretical department, with Vitaly Ginzburg as his deputy; the senior scholars were D. Blokhintsev, M. Markov, E. Feinberg, and Academician V. Fock (part-time), and there were eight doctoral and graduate students.[13]

The Theoretical Colloquium, led by Tamm, had lectures by FIAN members and other leading physicists, including Landau and Pomeranchuk, for audiences of three dozen. The topics covered the whole range of physics, from quantum theory to the expanding universe, from nuclear physics to the propagation of radio waves. But inside the theoretical department, according to the 1945 annual report, "attention was concentrated on the problem of elementary particles and their interactions." This problem was to remain central:

> In the coming five years, the theoretical department proposes to concentrate its work even more than before on the basic problems of contemporary physics: a) general questions of relativistic quantum theory and possible ways of overcoming very profound obstacles; b) theoretical analysis of the possible number and characteristics of elementary particles found in nature; c) mechanism of nuclear forces.[14]

It is hard to read this without compassion, knowing how much was still unknown in 1945 about elementary particles. Fortunately for Sakharov, neither he nor his teachers suspected. Therefore, the graduate student successfully mastered the craft of theoretical physics, making complex calculations and preparing his first publication, which appeared in 1947.[15] He saw then that the problem of nuclear forces was in the area of national security: "The journal's editors changed the title from 'Generation of Mesons' to the inaccurate 'Generation of Hard Components of Cosmic Rays.' Tamm explained the change to me this way: 'Even Beria knows what mesons are.'"

However, at that time even the chief of state security could not have really known what mesons are. In the summer of 1947 FIAN's theorists published a collection of articles titled *Meson*, edited by Tamm, devoted to the state of the theory of nuclear forces—an unsatisfactory state.[16]

The introduction said, "All the authors of this collection are first and second-generation students of the unforgettable Leonid Mandelshtam, to whose memory they are dedicating this book." This is a sad book not only because the authors felt orphaned, but because it demonstrates very clearly the difference between theoretical physics and chopping wood, to use Einstein's comparison: the results are not immediately apparent. The futurist poet Vladimir Mayakovsky once borrowed a metaphor from science:

> Poetry is also radium extraction.
> A gram of extraction, a year of labor.
> For the sake of a single word
> You consume thousands of tons of verbal ore.[17]

This applies even more to theoretical physics. *Meson* was all verbal ore. The only grain of optimism in the book lies in the research of A. Alikhanov and A. Alikhanyan, who announced the discovery in cosmic rays of a whole family of new particles, which they named varitrons. FIAN's experimenters were very skeptical about these discoveries, but the theorists needed a breakthrough so badly that they joined the authors in taking the wish for the reality.

They were mistaken, and the varitrons, awarded the Stalin Prize in 1948, turned out to be a mirage. But they were correct in their expectations: experiments of the same kind, at approximately the same time but in another country, England, discovered real mesons. However, even this discovery, which won the Nobel Prize, did not solve the problem of nuclear forces in the way that Tamm had dreamed. As Sakharov would write forty years later: "This very clever mechanism of nuclear interactions is still not cleared up, even though every passing decade brings astonishing experimental revelations and profound theoretical ideas."

The research dead end in which Tamm found himself may have been beneficial for his graduate student, just as Mandelshtam's scientific unemployment in the early 1920s in Odessa helped Tamm: the teacher could give the student more attention. It was more than personal contact. The graduate student had to report at seminars on the latest articles in scholarly journals, especially from the most important one, *Physical Review*. That meant picking up English on the run and mastering the language of living theoretical physics, which is harder because that language changes with every new step in science.

Tamm's other requirement was that his graduate students teach. Sakharov taught courses on electricity, theory of relativity, and nuclear physics at the Power Engineering Institute. In preparing his lectures, he learned his subjects systematically—filling in the blanks in his wartime education—and later he expressed regret that he only had those eighteen months of teaching and did not have the opportunity to go through the other fields of theoretical physics in the same manner.

He considered "the understanding of science that I acquired in my first FIAN years under Igor Tamm" the foundation for his scientific life.

Transitions of the 0 → 0 Type

That was the less-than-inspiring subject of the dissertation Andrei Sakharov chose after having found that Tamm's nuclear hypothesis of 1945 would not work. It was still about the life of the nucleus—no longer about "the meaning of its life" but about those infrequent incidents when the nucleus changes its state without radiation for some reason. Sakharov explained the reason in his dissertation. It was a topical question. Three years later the British theoretical physicist Richard Dalitz wrote his dissertation at Cambridge University on the same issue.[18]

Was this the main issue in physics then? No. But it was a vital one, a mystery of nature, to use an old-fashioned phrase. The history of physics shows that the question considered to be most important with time can lose meaning completely and be washed away by the flow of knowledge, while a question that seemed very particular opens up a new channel for that flow. And no one can guess what is to be expected from a given "mystery." It seems that nature does not divide its mysteries into big and small ones. And that was Sakharov's approach to science. Landsberg, chairman of the examination commission, basically said this at Sakharov's defense of his dissertation on November 3, 1947:

> Young theorists working in such areas as cosmic radiation, or the nucleus, often have a certain disdain for more classical sectors far from this sphere of questions. But Sakharov has shown a complete understanding of all questions with which he dealt. He had no difficulty with the damned, treacherous questions that trip up most graduate students. This is a manifestation of the fact that this man at a young age has a broad scientific view not only in special areas but in all areas of theoretical physics.[19]

Coming from Landsberg, this was high praise indeed. An integral comprehension of the building of physics—from foundation to rooftop—was characteristic of his teacher, Mandelshtam. It was under Landsberg's direction that the first edition of *Elementary Physics Textbook* (1948), the best physics course for high school, which went through ten editions, was prepared. Andrei Sakharov's father, who initiated him in physics, took part in the writing of this textbook.

A broad view is not the same thing as deep and solid knowledge. Landsberg was not present at the examination in quantum mechanics (June 1946), which Tamm described at the defense, explaining why there was one "B" among Sakharov's "A" grades. Tamm thought some of Sakharov's explanations on the exam were incorrect. "I argued with him for a long time, because I thought they were wrong. And I gave him a 'B.' A day later he came to my house and persuaded me that I was wrong."[20]

Tamm concluded with

> Andrei has a very rare combination of what theorists in particular need—two basic things—the ability to create a clear picture for themselves of a

phenomenon and a mastery of the mathematical apparatus needed to find the solution. The independence and originality evident in his dissertation and in many of his talks, which play a great role for our theorists and our laboratories, show that Andrei has many gifts and that we can expect much of him. And I am very happy to think that our theoretical department of FIAN could be enriched in the near future by such a colleague.

It is no surprise that after such reviews the scientific council voted unanimously to award Andrei Sakharov the degree of Candidate of Sciences.

The Scientific Council did not pay attention to the fact that Sakharov had failed the exam on Marxist-Leninist philosophy the first time he took it.

> I was asked whether I had read any philosophical works by Chernyshevsky —by that time the fashion for purely Russian scholars and philosophers had begun, without a whiff of the West. Out of excessive frankness I replied that I had not, but knew what it was about—and got a "D." A week later I had read all the required material and took the exam again—with an "A."

His memory fails him a bit—he got a "B."[21] And the second round of questions were more interesting. First, he had to talk about the views of Herzen, a Russian Westernizer, whose profound essays are incomparably deeper than Chernyshevsky's Utopias. Another question, on the role of intelligentsia in Soviet society, can be regarded as harbinger for Sakharov's future. The most difficult was the third question—on Lenin's struggle "against physical idealism." Sakharov was the same kind of idealist as his mentor Tamm, who wrote twenty years earlier, "what materialism is in the exact sciences I simply cannot understand—there is science and that is it."[22]

After being tested in dialectical and historical materialism, Sakharov had the opportunity to experience the materiality of the world in an empirical way. It was the end of June, vacation time, and his defense was postponed to the fall. Supporting a family on a graduate student's fellowship, when his wife did not work but took care of their two-year-old daughter, was very difficult. Their vegetable garden, where the young theorist planted potatoes side by side with other FIAN members, helped a bit.

The greatest problem was lack of housing. They had to rent rooms—cold, damp, without privacy, or in cellars—for brief periods. While they looked for the next one, they stayed with Andrei's parents—five in a room. Sakharov's mother never accepted his wife, from the moment she arrived in Moscow with the baby (their daughter had been born in Ulyanovsk a few weeks after Andrei's departure). His mother thought her beloved son deserved more than Klava Vikhireva, a provincial girl he met at the Ulyanovsk cartridge factory. Andrei felt that "everyone was equally guilty—or not guilty" in that triangle—he, his wife, and his mother.

Andrei earned a little on the side teaching at the Power Engineering Institute and a night school, but they were always short on rent money. His parents helped out, and Tamm gave him loans. It was only after he defended his dissertation that FIAN arranged for them to have a small room.

Understanding their acute need for housing, we can appreciate his refusal to leave Tamm. He got an enticing proposition twice—a year before his dissertation defense and then on the eve of it. The first came from someone calling himself "General Zverev"; the second, from Kurchatov. The first offer was to "come work in our system to participate in executing important state assignments"; the second, to move to the Kurchatov Institute and work on theoretical nuclear physics. A. Migdal and I. Pomeranchuk—his dissertation opponents, theorists well known to him—worked there. He was offered a high salary and, most important, an apartment.

But he refused: "I told them briefly that now I wanted to continue my pure theoretical work in Tamm's department." The dissertation was only a small part of that work, and its contents were related to other issues the young theorist was pondering. One touched on the fundamental problem of quantum theory; others were more specific, but Sakharov posed them all to himself and to nature. The creative freedom was worth more to him than the perks of "special physics." As Pushkin put it:

> Neither thoughts, neck nor conscience bend
> For power—its symbols or its ends,
> But wander here and there, fully at your ease,
> Marveling at nature's divine beauty as you please.[23]

But Sakharov also knew that a higher degree at FIAN would reduce his financial strain. He had to be aware of the how much his mentors appreciated him. Two days before his defense, Vavilov, the director of FIAN, appointed him to be a scientific researcher.[24] And Vavilov was not just the director of FIAN, but the new president of the Academy of Sciences of the USSR.

6

SERGEI VAVILOV

The President of the Academy of Sciences

The Choice of Vavilov

How did it come to pass that Sergei Vavilov assumed the country's top scientific position? A preselection at the highest level of Soviet authority preceded the election, held at the Academy of Sciences on July 17, 1945. A KGB document dated July 8 helps us understand the circumstances surrounding this decision. This document, "regarding the scientific and social activities" of twenty-three academicians, was prepared from "agent information."[1]

The document begins with V. Komarov, the 76-year-old former president, who was so ill that he "couldn't get around without assistance." Intelligence on eight academicians follows in alphabetical order—the most probable candidates, it would appear; the rest are not in alphabetical order.

Vavilov wound up in the first eight. The document says about him:

> The author of widely known scientific papers on fluorescence (created the theory), on the study of the nature of light. The author of many books and translations (Newton's "Works").
>
> A participant of international congresses. Politically loyal. A representative of the State Committee on Defense in the Optical Industry during the War.
>
> Vavilov possesses organizational abilities, has good relations with the majority of the USSR Academy of Sciences scientists, and commands authority among them. Simple in manner, modest in lifestyle.
>
> Vavilov is now in the prime of his creative powers and personally conducts scientific research. Has major students and followers. Well-known in the USSR and abroad.
>
> Sergei Vavilov's brother, Nikolai Vavilov, a geneticist, was arrested in 1940 and sentenced to 15 years for sabotage in agriculture. He died in 1943 in the Saratov jail.

If the last paragraph of this description is discounted, then 54-year-old Sergei Vavilov was the best candidate.[2] If it is taken into account, then he was just the only possible candidate. In fact, Stalin made this choice a few days before the Potsdam Conference began, before the first atomic explosion, before the beginning of the Cold War. The country's image in the Western world was then an object of his serious attention, and science, international in its nature, was very appropriate as window dressing. It was no accident that the Academy of Sciences' 220th Jubilee Anniversary was magnificently celebrated with invited foreign scientists—even though the date was not a round figure and the celebration was a year late—against the backdrop of a country destroyed by war.

An ominous shadow was cast over the Soviet academic window dressing by the death of world-renowned geneticist Nikolai Vavilov. His name appears in the KGB document again—in the character reference of Lysenko: "Not regarded as an authority by the biologists of the USSR Academy of Sciences, including [President] V. Komarov and [Vice President] L. Orbeli; moreover, they attribute N. Vavilov's arrest to him."

Lysenko had clout with Stalin. When Stalin decided to get rid of his protégé's chief scientific opponent Nikolai Vavilov in 1940, he could ignore how England and the United States would react. The friendship treaty concluded in 1939 between the USSR and Germany hermetically walled off Soviet affairs from the Anglo-American world for almost two years.

The situation for Stalin was entirely different in July 1945, on the eve of the Potsdam meeting with the leaders of his military allies, the United States and England. There had been no official information about Nikolai Vavilov's lot for a long time. When the Royal Society of London elected him a foreign member in 1942, the British Embassy was not afforded the opportunity to personally deliver the diploma. That was when his Western colleagues suspected foul play; however, all their inquiries remained unanswered. It wasn't until 1945, apparently right at the time of the Academy anniversary, that they were able to find out that Nikolai Vavilov "was dismissed from his position, disappeared along with several of his colleagues in genetics and died sometime between 1941 to 1943."[3] In this situation, the election of his brother as head of the Academy of Sciences turned the secret crime into a mysterious Russian riddle, to paraphrase Churchill.

What can be said about Sergei Vavilov's decision to accept the position of the nation's highest scientific administrator? People who knew him well confirm that he accepted it with a heavy heart, seeing Andrei Vyshinsky as an ominous alternative.[4]

Before becoming an academician in 1939 (when Stalin was "elected" an academician as well), Vyshinsky had served as Procurator General of the USSR throughout the duration of the Great Terror. He served as prosecutor at the show trials. In 1939 he became the Deputy Head of the government—the Deputy Chairman of the Council of Peoples Commissars of the USSR—and directed the Academy's affairs; it was to him that the president of the Academy of

Sciences addressed a request for an additional position for Kurchatov in 1943.[5] The fact that there is no mention of Vyshinsky in the KGB "pre-election" document didn't at all rule out his candidacy. Stalin had no need for the description of such a close assistant.

The documents do not allow us to say whether Stalin actually held Vyshinsky's candidacy in reserve or whether he used his name as an additional argument to "assist" Vavilov in coming to a decision. More important, had Vavilov rejected the ruler's offer, he would have jeopardized too much besides his own security—principally FIAN, his brainchild, into which he had put so much work and soul. To Vavilov, FIAN was his associates, part of world science and a vital component of his homeland's culture: "He felt that he was an heir of its past, personally and deeply responsible for its future."[6]

A sense of responsibility forced Vavilov to take on a doubly heavy burden— to become part of the government and help cover up the death his brother had suffered from the very same government. He bore this burden only five years— his sudden death in January 1951 liberated him.

Although a sense of responsibility characterized both brothers, they differed in temperament and moral resilience. "We'll go into the bonfire, we'll burn at the stake, but we won't renounce our convictions" is a sentence typical of Nikolai Vavilov.[7] Sergei was ready—and able—to carry the large burden of shame in order to defend his life's work.[8] He could not, however, foresee the heavy weight of this burden in July 1945, while the country was still in the grip of euphoria of victory and durable peace after the difficult war. In Soviet Russia there was widespread expectation of greater liberty at home for the country that had liberated the world from the Nazi plague.

In July 1945, only dispassionate political analysts could have surmised the looming Cold War and the even stronger grip of Stalin's plague within the country and beyond.

What the President of the
Academy of Sciences Could Do

The situation changed radically just a few weeks after the election of the Academy president, when the shadow of the nuclear mushroom cloud fell across the entire world. The first atomic explosion in the United States occurred on July 16, the day before the Academy election. It was only on July 24, a week after the Potsdam Conference began, that Truman mentioned to Stalin in passing that the United States had a new weapon of extraordinarily destructive power.[9] The American president did not say that it was the atomic bomb, but Stalin understood everything—first and foremost, that the intelligence agents and physicists were not trying to pull the wool over his eyes, and that his atomic project had to acquire government importance.

The first consequences of this realization benefited Russian science far beyond the atomic project, and Sergei Vavilov played an important role in this. The president of the Academy of Sciences was not included in the top management of the atomic project, but he used his new personal position for public ends, taking advantage of the new social authority that physics acquired after Hiroshima.

Vavilov met with Stalin on January 25, 1946.[10] Kurchatov had been in Stalin's study just before him, and during the fifty-minute audience he had received instructions to conduct the work "with Russian sweep"—this is from the record that Kurchatov made under the impact of that conversation.[11]

Stalin gave Vavilov 15 minutes more. Vavilov left no records about the conversation, but the fact that Vavilov's heart stopped precisely on the fifth anniversary of that meeting might have bearing on the impression that Stalin made on him. However, we can surmise from the events subsequent to this meeting that Vavilov made an impression on the ruler, too—that the skillful popularizer of science and expert in its history was able to explain to Stalin the significance of broad scientific research. A new, highly scientific weapon was of greatest importance to the dictator, but under the influence of Hiroshima and of the new president of the Academy of Sciences, it appears that he thought, "There's no telling, those physicists—today they created a super-bomb from imperceptible atomic processes, and tomorrow, from something else?"

In March the salaries of scientific workers were raised by several notches and the country's scientific budget by a factor of three (aside from secret funds).[12] And the FIAN physicists received tangible reasons to think that "the party and the government," as the expression of the time went, were taking care of the broad development of science. In the conditions of postwar ruin, when the system of ration cards was still in effect, a station to study cosmic rays was built in the Pamirs for FIAN, and an unprecedented large-scale expedition was organized to Brazil to observe the solar eclipse. Both purely scientific events required decisions at the very top.

FIAN's Pamirs station had been launched in 1944, but the physicists lived there in a barn that, in the spirit of wartime, was called "a common grave."[13] The main station building was constructed by the party-cum-government in 1946–1947 by the labor of inmates from a labor camp.[14]

Even more impressive was the astrophysics expedition to South America, whose aim was the first study of a solar eclipse by means of radio waves. The physics of cosmic rays was at least scientifically connected to nuclear physics and, thus, with "defense topics," but the research of radio waves emitted by the sun was entirely pure science. To be sure, this pure science itself took advantage of the fruits of defense research: the development of radar during the war years powerfully advanced the methods of receiving radio signals.

The radio observation of the solar eclipse was Nikolai Papalexi's idea. He was the head of the Oscillations Laboratory at FIAN and a longtime friend and associate of Mandelshtam's, and this idea was a natural development of their research in radio physics. Papalexi saw "the basis for a new science—radio astronomy" in the radiation of the sun and outer space.[15] The solar eclipse which was to be observed in Brazil on May 20, 1947, provided a good opportunity for researchers. Of course, the director of FIAN was well informed about all this when he went to the meeting with Stalin on January 25, 1946.[16]

Apparently, Vavilov was able to explain to the ruler the law of history that made pure and applied science serve one another by turns, since Stalin decided to make a display of supporting science. Evidently, the decision to support the expedition was made during the January 1946 audience, although preparations for it—not a simple matter in a country devastated by war—took nearly a year.

Yakov Alpert, one of Mandelshtam's and Papalexi's associates in research before the war, worked on the logistics of the expedition. The expedition, which included nearly 30 physicists and astronomers, set sail toward Brazil's shores on April 13, 1947, and returned on July 27. This was a unique event in the scientific life of the USSR, and Vavilov, president of the Academy, personally took it under his wing.[17]

Alpert had his personal reasons to notice a change on Stalin's part toward science after Hiroshima: there had been a strange incident on the eve of the war with a Stalin Prize that was *not* awarded. For his 60th birthday in 1939, Stalin was elected an Honorary Member of the Academy of Sciences, which in so doing had called him the "Coryphaeus of science." Whether he believed this or simply decided to personally take up science and make the Academy into something like a ministry of science, he established the highest Stalin Prizes for achievements in science. In 1940 the Academy of Sciences, preparing for the first award of Stalin Prizes, proposed that the first prize for physics go to "The Propagation of Radio Waves across the Earth's Surface," whose authors included Alpert as well as Mandelshtam and Papalexi; next in order was the work "The Spontaneous Fission of Uranium" by G. Flerov and K. Petrzhak. The Academic Secretary of the Physical and Mathematical Sciences published an article about the upcoming awards in *Pravda*. Alpert was offered an opportunity to speak on the radio, and the program was heard across the entire country. Flerov even organized a banquet in honor of the inevitable—it would seem—award.[18]

However, the "Coryphaeus of science" decided otherwise; he ignored the opinion of the Academy of Sciences and did not award the prize named after him to the two aforementioned studies. It is now clear that at that time, on the eve of the war, the physicists were lucky—Comrade Stalin did not reconsider the Academy's rejection of several papers "competing for the Stalin Prize," such as Comrade I. Ogurtsov's "Design for a Perpetual Engine."[19] In biology things fared much worse—the ruler awarded the first Stalin Prize to his protégé Lysenko and his bio-alchemy.

Five years later Stalin came to see that the atomic bomb and radar—the most important scientific weapons of the recently ended war—had emerged from physics, which he had considered of little interest on the eve of the war. This embarrassment might have helped the president of the Academy of Sciences secure Stalin's trust.

What the President of the
Academy of Sciences Could Not Do

The president of the Academy of Sciences could do a lot, but far from everything. Although the party government had built a high-mountain station and outfitted a transoceanic astronomy expedition, at the same time it tightened the reins of governing science. Vavilov felt this clearly in the autumn of 1946 during the first elections to the Academy during his presidency.

The candidacy of Igor Tamm, FIAN's main theorist, a Corresponding Member of the Academy of Sciences since 1933, was proposed, among others. His election to academician seemed entirely inevitable due to his position in science as well as because the new president of the Academy knew him so well. But the party Central Committee, where all the candidacies were reviewed before the election, knew him even better and Tamm was not approved.[20]

Vavilov had to nominate another candidate, another FIAN theorist— Mikhail Leontovich. More than anyone, Leontovich himself did not agree with this, and this was the only case in the history of the Academy when a candidate so opposed his own election. He wrote a letter to Vavilov on November 24, 1946: "I'm addressing you with the request to use your position in the Academy and your authority to take measures which would guarantee that I not be elected Full Member of the Academy of Sciences."[21] And in a letter read at his request at a meeting of the department of physics and mathematics, Leontovich explained: "There are already two candidates, the theoretical physicists who undoubtedly deserve full membership—I. Tamm and L. Landau. For this reason, not wishing to compete with these candidates, I consider that my candidacy should be removed."[22]

Before the secret vote, the will of the party was delivered to the party member academicians, who were to make sure that the party recommendations were made known to the other academicians. In 1943 during Kurchatov's election, academicians did not initially accede to the party's will. In November 1946, they turned out to be more obedient—they kept Leontovich on the candidate list, elected him by a majority of votes (13–2), and rejected Tamm (4–11).[23]

Most likely, the academicians were made more obedient by the stifling decree on literature by the Central Committee, which had thundered across the country on August 16. It was delivered to the people by Andrei Zhdanov, the chief party ideologue, who had trampled on Anna Akhmatova's poetry and Mikhail Zoshchenko's prose. Having learned that it was the same Zhdanov who had spoken out against Tamm's candidacy, the academicians could have entertained doubts about the secrecy of their ballots. Credit is due to the four who disobeyed and voted for Tamm. One of them could easily have been Vavilov himself.

Vavilov felt the government's tightening of the reins not only within the Academy but also within his FIAN. In April 1947 the institute pass of graduate student Leon Bell was taken away without any explanation. This meant that he was deprived of "access." At that time, there appeared a new division between physicists, in addition to the distinction between theorists and experimenters—those with access and those without. The ever more powerful security service considered that Bell could not be at FIAN with a resume like his. The questionnaires for applications, introduced in 1946, became twice as extensive (and included, for instance, information about a spouse's parents). They interpreted Bell's American origin as a crime; he had come to the USSR with his parents in 1931 at the age of 13. Vavilov could do nothing about this. However, he helped Bell get work at the Institute of Plant Physiology and made it possible for him to defend his dissertation at FIAN in 1948.[24]

What Sakharov, the Graduate Student, Didn't Notice

However, what do all these stories have to do with Andrei Sakharov's life, if there isn't a word about them in his memoirs? His silence is, indeed, surprising.

The graduate student from whom the custodians of state security had taken away the FIAN pass (thereby taking away his opportunity to do physics) was a classmate of Sakharov's, the initiator of the circle where Andrei had presented his first scientific report. They had been evacuated along with a few other classmates to Ashkhabad, where they graduated from the university.

Sakharov did not talk about the candidacies of Tamm and Leontovich in his memoirs, which seems strange. Tamm's role in his life is commensurate only with the role of his father. He also inherited his father's attitude toward Leontovich: "In Father's remarks about Leontovich, I remember great respect, even admiration, combined with a certain warmth, which also predetermined my assessment of him then (which I subsequently retained)."

And, finally, it's difficult to imagine that Sakharov had not heard about the expedition to Brazil, had not heard about it from Vitaly Ginzburg, who was, in Sakharov's words, "one of the most talented and favorite students" of Tamm. A year after the Brazilian expedition in which Ginzburg had participated, they began to work shoulder to shoulder on the hydrogen bomb. An open and emotional person, Ginzburg could not help sharing his impressions about the unusual journey—his first trip abroad and so unusual for the circumstances of that time, after the Iron Curtain had dropped. Politics aside, there had been the journey to South America, Rio de Janeiro, the visit to Holland along the way, meetings with Western physicists, the unique solar eclipse . . . how could one remain indifferent to this?

Nonetheless, Sakharov doesn't even mention these striking events. His silence is remarkable and thus eloquent. What does it speak about? Anything but a weak memory—he fully recollects the events of 1947 related to his scientific work, about how he "caught hold of the anomaly in the hydrogen atom and continued on with it persistently," how he understood "that the meaning of this idea far exceeded the limits of the particular assignment," how he "was very excited," having seen the path beyond the limits, and how he didn't have enough confidence to take this path without the blessing of his beloved teacher.

Brazil faded against the backdrop of this dramatic story, to which Sakharov devoted several pages in his *Memoirs*. The superficial exotica of transoceanic voyages couldn't compare with the expedition into the unknown depths of the structure of matter. And the horizon's breadth was sacrificed for scrutiny's depths. The capacity of the young theorist to concentrate—the ability to remain undistracted by the events of unscientific life—was the key to his achievements in science. But this same concentration kept Sakharov from noticing what had happened to Bell and the Academy disturbances around Tamm and Leontovich. No more than "kept"—both events were less conspicuous than it might appear now. And this very inconspicuousness speaks to the reality surrounding Sakharov.

Bell was also very focused on his science: he worked in an entirely different laboratory and had no scientific contact with Sakharov. He was dismissed a few months before completing his graduate term, and most important, he himself wasn't certain of the meaning of what happened to him. The security organs didn't go to the trouble of explaining, and discussing government actions was unsafe. Sakharov might have thought, "Leon isn't to be seen at the FIAN? Perhaps he's already completed his graduate study."

Tamm and Leontovich, whom Sakharov saw on a daily basis, were connected both by personal friendship and by belonging to Mandelshtam's school and its scientific and moral traditions. The strength of this connection did not suffer in the least from the election or nonelection to the Academy. Therefore, anomalies in the hydrogen atom occupied Tamm's graduate student far more than the anomalies in the awarding of highest academic ranks. As for Leontovich's stand, which he had expressed in that situation, there was nothing particularly anomalous about it for Sakharov, who had observed his teacher and teacher's friend on a daily basis at FIAN.

One can envy the graduate student who was surrounded by such normal people, despite the madness in society. And one can understand his lack of desire to leave FIAN for a career in the atomic empire—even if it were richer—under the command of Lavrenty Beria, Marshal of State Security.

7

NUCLEAR PHYSICS UNDER
BERIA'S COMMAND

Two weeks after Hiroshima, which finally convinced the Soviet leadership of the new weapon's feasibility, Stalin gave the atomic project governmental status and appointed Marshal Beria, the People's Commissar of Internal Affairs—its highest-level director—chairman of the Special Committee. The sinister head of the Stalinist gendarmerie towered over Soviet nuclear physics for nearly eight years. At first he even planned to have the nuclear physicists wear the NKVD uniforms to which his eyes were accustomed.

What did the physicists *then* know about him? When Beria was appointed head of the NKVD in 1938, the bloody maelstrom subsided. One could have thought that it was Beria who had stopped the bacchanalia of lawlessness, all the more so because a certain number of the arrested were freed. That thousands of Polish officers were executed in 1940 was not then known. The inhuman deportation of "guilty" peoples during the war—first the Germans, then the Kalmyks, and the peoples of the Northern Caucasus and the Crimea—carried out by the NKVD seemed based on a kind of military logic against the background of the inhuman war.

In actual fact, the state machine of repression remained the same. It was merely used more in the interests of "the economy"—for the building of socialism by means of slave labor. It was precisely during Beria's time that scientific research slaveholdings—*sharashki*—flourished.

It was a matter of a certain reorientation—the principle itself had long been in operation. Vernadsky had recorded this in his diary back in 1932, citing the testimony of a geologist acquaintance about the extraction of radium—"the GPU extracts it." "GPU" was the name at that time of both the secret and overt police, also known as the Cheka, NKVD, and, finally and mainly, the KGB. Vernadsky sometimes used the old Russian term *gendarmerie*. Impressed by a meeting of the Council on the Study of National Production Capacities, he wrote in his diary:

A most interesting phenomenon has come to light. A remarkable anachronism which I would have considered impossible before. Practical-scientific interest and the gendarmerie. Could this be possible for the future? But right now the scientists are working in slave conditions. They are trying not to think about it. This amorality is felt, it seems to me, everywhere: moral sense does not reconcile itself to this. Are they closing their eyes?[1]

People differ greatly in their ability to close their eyes and in the nature of their moral sense.

Kapitsa's Mutiny

The Special Atomic Committee established by Stalin in August 1945 included only two physicists—Igor Kurchatov and Pyotr Kapitsa. Very soon Kapitsa found that he could not work under Beria's command. He wrote to Stalin on October 3: "Comrade Beria has little concern for the reputation of our scientists (your work, he says, is to invent, research, what need have you of a reputation). Now, having clashed with Comrade Beria of the Special Committee, I especially clearly felt the unacceptability of his attitude toward scientists."[2]

It appears that Kurchatov did not feel this. By that time—in two and a half years—he had already received thousands of pages of atomic intelligence materials from Beria's department, about which Kapitsa knew nothing. To work with this enormous volume of material, Beria created a special department in the KGB headed by General Sudoplatov in September 1945.[3] Yakov Terletsky, a physicist from Moscow University, was also enlisted into service. And within a month, Beria, already in his new capacity uniting the atomic project and state security, decided to send Terletsky to Denmark to Niels Bohr for the missing atomic information. The very special physicist was quickly prepared for the trip, educated in atomic matters, and furnished with questions for Bohr.[4]

Terletsky brought Bohr a letter from Kapitsa. For Beria, it was a letter of recommendation to obtain Bohr's trust; for Kapitsa, it was the only opportunity to communicate with Western colleagues. A friendship of long standing linked Kapitsa with Bohr: they were bonded by their mutual love for their teacher, Ernest Rutherford. An understanding of the problems of the newly born nuclear age also united them.

We read in Kapitsa's letter (written in English and dated October 22, 1945):

The recent discoveries in the nuclear physics, the famous atomic bomb I think proves once more that science is no more the hobby of university professors but is one of the factors which may influence the world politics. Nowadays it is dangerous that scientific discoveries, if kept secret, will serve not broadly humanities but be used for selfish interests of particular people or national groups. Sometimes I wonder what must be the right attitude of scientists in these cases. I should very much like at the first opportunity to discuss these problems with you personally and I think it would be wise as soon as possible to bring them up to a discussion at some international gathering of scientists. Maybe it will be worth while to think over that [*sic*] measures should be included into the status of the "United Nations" which will guarantee a free and fruitful progress of science.

I should be glad to hear from you what is the general attitude on these questions of the leading scientists abroad. Any suggestions about means to discuss these questions from you I shall welcome mostly. I can indeed inform you what can be done in this line in Russia.[5]

And Bohr, in a letter to Kapitsa written only a day earlier, is concerned about the same thing:

I need not say that in connection with the immense implications of the development of nuclear physics the memories of Rutherford have constantly been in my thoughts and that I, like his other friends and pupils, have deplored that he should not himself be able to see the fruits of his great discoveries. His wisdom and authority will also be greatly missed in the endeavors to avert new dangers for civilization and to turn the great advance to the lasting benefit of all humanity.

Bohr sent his articles "Energy from the Atom" (*Times*, August 11, 1945) and "A Challenge to Civilization" (*Science*, October 12, 1945) with his letter to Kapitsa and asked him to show them to mutual friends: "I shall be most interested to learn what you think yourself about this all-important matter which places so great a responsibility on our whole generation."

Bohr's articles were about the "mortal threat to civilization," about the necessity of a new approach to international relations, and about the important role to be played by contacts among scientists. But not contacts such as those arranged by Beria.

Terletsky's visit to Copenhagen has now been studied in detail.[6] Beria planned this operation principally in order to demonstrate his efficiency. Although Bohr, after secret work at Los Alamos, was fully aware of whom he was dealing with, he acted in concert with Western security services and did not communicate any atomic secrets to Beria's emissary. He tried using this unique—for those conditions—opportunity to connect with his Russian friends Kapitsa and Landau, to support them (primarily Landau, who had been imprisoned just a few years before), and to use this connection to prevent the nuclear "mortal threat to civilization."

Bohr gave Terletsky a letter of response to Kapitsa, in which he proposed organizing an international conference of scientists in Copenhagen to discuss the situation and the opportunities for working together in science: "If you and several of your colleagues could come, I am sure that a whole group of leading physicists from other countries would join us."

Terletsky returned to Moscow on November 20. And five days later Kapitsa sent off a letter to Stalin in which he requested to be relieved of participation in atomic work, and criticized in detail the organization of this work and its chief organizer, Beria.

Kapitsa's request was granted, and then some—he was dismissed from all his positions, including directorship of the Institute for Physical Problems that he had created. For seven years—until Beria's ignominious end—Kapitsa could pursue physics only at his dacha near Moscow, in the home laboratory he built. The reasons that prompted Kapitsa to take a mortally desperate step—to complain to Stalin about his chief comrade-in-arms and executioner—were an alloy that included both the genuine absence of his own scientific problem within the atomic project's framework and an original scientist's antipathy toward replicating other people's paths and copying ready-made models. But, most important, it was the feeling that he was being manipulated and his definite unwillingness to remain in serflike dependency on an ignorant and crude master. It was not his way of living.

In 1921 he had left the USSR to visit Rutherford's laboratory in England, with the goal of working there for several months. He was found very suitable at Cambridge, and in 13 years there he went from being an unknown Russian physicist to a member of the Royal Society and the director of a first-class physics laboratory. Believing in the bright socialist future of his country, he kept his Soviet citizenship and did a great deal to help Soviet science.

However, in 1934, when he returned to the USSR on vacation as usual, the Soviet government forbade him to return to England—by right of force. Deeply offended, Kapitsa nonetheless did not break down and didn't even abandon his socialist ideals. He compared himself "with a woman who wants to give herself in love, but whose lover insists on raping her."[7] He used the expression "our idiots" for the Soviet leaders, and both "our" and "idiots" here are equally important:

> I am sincerely well-disposed toward our idiots, and they do remarkable things, and this will go down in history . . . But what is one to do, if they understand nothing in science? . . . They (the idiots), of course, could wise up tomorrow, but maybe only in 5–10 years. There is no doubt that they will wise up, since life will force them to do it. But the whole questions is— when?[8]

"Our idiots" built Kapitsa an institute, having bought out the equipment from his Cambridge laboratory for him; he got the work of the institute going and discovered superfluidity there, which brought him—many years later— the Nobel Prize.

Kapitsa wrote more than hundred letters to the Kremlin leaders during Stalinist times. Through his participation in building socialism, he secured such an influential position in the socialist system that he was able to help get several physicists, including the anti-Stalinist Landau, out of the abyss of 1937.

But Kapitsa could not find a common language with Beria, and had no desire to. However, this didn't mean that he morally condemned Kurchatov, who was successfully doing his work. Anna Kapitsa, wife and friend, talked about it this way:

> Kurchatov was an unusual man in the respect that he knew how to talk to these people [Soviet leaders]. Pyotr Leonidovich didn't; he preferred to write to the top leaders, to those whom he always called "senior comrades." But Kurchatov excelled by having a perfectly brilliant way of talking to them. He found an utterly particular tone . . . And then he possessed a great deal of charm, Kurchatov did, so he could work with them. But then you see, he perished very quickly![9]

Kurchatov, the "Great Diplomat"

Kapitsa's name is missing from the KGB document dated July 8, 1945, about potential candidates for the position of president of the Academy. Kurchatov is 18th (of 22) on the list, and it is said of him: "At the present time, the leading scientist in the USSR in the field of atomic physics. Possesses great organizational skills, is energetic. By character a secretive man, careful, shrewd and a great diplomat."[10]

Kurchatov was totally informed about atomic matters, and it was impossible for him to close his eyes, for instance, to the fact that Beria's right hand—his Gulag hand—helped his left hand—his atomic hand. Visiting the installations of the atomic archipelago, Kurchatov couldn't help but see the long gray columns of prisoner-builders escorted by guards with German shepherds. Did he peer into these columns expecting to see one of his acquaintances who "disappeared" in 1937, or did he comfort himself with the hope that the columns consisted only of common criminals? He knew that Beria's right hand sifted through the prisoner-slaves in order to winnow out scientific and technological specialists and use them in special institutes, or *sharashki*, "in the first circle" of the Gulag.

How did his moral sense reconcile itself to this? And if it did, then how did he leave such universally fond memories? In the Soviet state, everything that promoted the most rapid victory of communism—universal happiness on Earth—was considered moral. Different modes of life could be supported by such a philosophy. Safest of all was to entrust the leadership with deciding what promoted communism and what didn't, and then rationalize the decision sent down from above—through dialectics, demagoguery, or autosuggestion, depending on the circumstances.

Kurchatov placed the interests of the cause above personal feelings, but often took it upon himself to decide where those interests lay. In the interests of the cause, the "managed" Kurchatov, as we will see, also used the levers of management in reverse: to defend physics from "Lysenko-ization" and militant ignorance, defend specific physicists from the party-police apparatus, substantiate the inevitability of peaceful coexistence in the nuclear age, promote the revival of genetics, and, finally, support Sakharov when he started to broaden the range of his thoughts from nuclear-military physics to the politics of the nuclear age.

Here are two seemingly contradictory examples. Both are from the time after the successful test of the first Soviet atomic bomb, when Kurchatov became a figure of state magnitude.

At the end of 1950, Academician Ioffe, soon after his 70th birthday, was dismissed from the directorship at PhysTech under demonstratively humiliating conditions. His favorite student Kurchatov, who was also obliged to Ioffe for his advancement to the directorship of the atomic project, did not prevent or ameliorate this fall.

At the same time, a prominent Moscow physician, a Jew, was arrested and then sentenced under the "counterrevolutionary" statute to 10 years in the camps. His son Boris Yerozolimsky (a classmate of Sakharov's) worked at the Kurchatov Institute at that time. It was a time when state anti-Semitism was on the rise (reaching its peak in the "Doctors' Plot" in 1953). And Kurchatov, along with other scientific directors, had to agree to "purge" his institute of some Jewish associates.[11] But Kurchatov resisted the efforts of security service to dismiss Yerozolimsky for more than two years—until Stalin's death.[12]

What was behind the "cowardly ingratitude" in one case and the "courageous defense" in the other? It was true that 70-year-old Ioffe, Kurchatov's teacher, could no longer cope with the scientific directorship of PhysTech. An evidence of this was the awarding of the 1949 Stalin Prize to Ioffe's associate for a work in nuclear physics, which Ioffe was showing off, and which was soon refuted.[13]

As for Yerozolimsky, Kurchatov saw with his own eyes that this son of "an enemy of a people" was a first-class physicist who was selflessly devoted to science. There are always few such workers, and Kurchatov knew how to value them.

The interests of the cause: What was the cause that Kurchatov served? Of course, the magnificent goal of science for "the benefit of humanity" was always visible on the horizon. But in the 1940s it was first and foremost a military cause—the country's defense, and according to Sakharov, Kurchatov used to say then: "We are *soldiers*."

The American atomic explosions helped instill this simple soldier's psychology in the minds of Soviet physicists who were engaged in the atomic project. Stalinist propaganda had plenty of material with which to forge a completely new attitude toward a recent ally in the war on fascism. The Americans created a superpowerful weapon in secret from their Soviet ally who endured the main brunt of the war on fascism and used it against the civilian population of two Japanese cities. This was the picture that rose from the pages of the newspaper *Pravda*—and it contained a fair amount of truth. Because the Soviet physicists had no access to other sources of information, the American atomic monopoly was easily perceived as a direct military threat to their native land—and to the entire bright future of the planet.

A soldier is not supposed to think about a commander's qualities beyond his military duties. Direct evidence of how Kurchatov perceived his commanders Stalin and Beria has not come down to us. Kurchatov met with Stalin only twice (in 1946 and 1947); he kept the piece of paper with the record of his impressions from the meetings in his personal safe until the end of his days, and an enormous portrait of the leader remained in its place in Kurchatov's office even after the Twentieth Congress at which the cult of personality was exposed by Khrushchev.[14]

Kurchatov had contact with Beria frequently enough to form an idea of who he was. Personal acquaintance with Beria disappointed even Terletsky, a physicist on KGB assignments, in the fall of 1945. Terletsky was struck by how Beria's aide—"a two-hundred-kilogram egg-shaped tubby," Kobulov—talked about Moscow and Stockholm girls while he was waiting to be summoned by Beria. "This was his most important assistant and, as I found out later, an executioner and sadist, who personally took part in torturing prisoners." And Beria himself—"aging, with a slightly narrowing skull at the top, with severe facial features, without a shadow of a smile or any warmth, made an impression different from what I expected, having seen his portraits before then (as a young energetic *intelligent* in a pince-nez)."[15]

How did Kurchatov manage to maintain the feeling of being a soldier of communism under such a commander? In those military conditions there was no time to think; one had to act. Besides, Kurchatov was not a thinker by nature. When in 1959 Sakharov responded with admiration about Kurchatov to a colleague, the latter warned: "Do not overestimate his closeness to you. Kurchatov is first and foremost 'a doer,' moreover a doer of the Stalinist era; that was precisely when he felt himself a fish in water."[16] Sakharov saw a grain of truth in those words, but only a grain. After all, he himself also felt involved in the same cause—securing peace for the country after a terrible war. Having given much to this work, he "unwittingly," in his own words, created "an illusory world to justify myself."

The ability to create illusory worlds to justify one's behavior is human nature. And the participants in the Soviet atomic project were no exception. The rarest exception was Landau for whom there simply was nothing from which to create illusions; he worked for Beria's "left" hand out of fear, based on his personal knowledge of the punishing right hand gained during a year in prison.

For other physicists, Beria was chiefly a customary portrait on the wall in the pantheon of other Soviet leaders. Between the physicists and the portrait was Kurchatov—cheerful, charming, full of infectious enthusiasm for research, somehow connecting the world of neutrons and mesons with the loftiest interests of the state and humankind. Kurchatov knew the Special Marshal representing these interests too well to limit himself to illusions. But he could completely accept Beria as a doer who served the same cause he did, and who answered, as Kurchatov did—with his head—for the success of this cause. Sakharov saw him in the same way: "For me, Beria was part of the state machine and, in this capacity, a participant in the work of 'utmost importance' in which we were involved."

The success of the Soviet atomic project also speaks of Beria's businesslike qualities. Mortal fear of the secret police and the enormous resources of gulag slave labor were not the Marshal's only contribution to the creation of nuclear weapons. He had to make important—thus, bold—decisions, going far beyond his education as a construction technician.[17] He needed to know how to understand—if not nuclear physics, then the people who understood it.

Landau's election to academician in 1946, bypassing the rank of Corresponding Member, clearly speaks of Beria's efficiency. It was Beria's order that released Landau from prison in 1939, despite his anti-Stalinist leaflet. Beria believed Kapitsa that Landau "is a major specialist in the field of theoretical physics and could be useful to Soviet science in the future."[18] Niels Bohr added his word to this in the fall of 1945. After returning from Copenhagen, Terletsky reported to Beria and had to communicate Bohr's opinion that Landau was the most talented young theorist of all those who worked for him. Avoiding giving direct answers to "special" atomic questions, Bohr said that "qualified physicists, like Kapitsa and Landau, were capable of solving the problem, if they already know that the American bomb had been detonated."[19] Beria also believed the famous Danish physicist. And even though Kapitsa refused to work for the "atomic problem," Beria allowed the anti-Soviet criminal Landau to be elected academician.

Besides court intrigues and inventive politicking, to Beria's credit there were also bold initiatives. For example, right after the end of the war in 1945 Beria presented a project to Stalin of limiting the punitive force of extralegal "justice" (as if foreseeing that the comrades were to deal with him using precisely this special justice in eight years). Stalin declined the People's Commissar's project.[20]

Klaus Fuchs and Others

Beria was in command not only of Soviet nuclear physics but also of a certain part of Western physics. One of his priceless freelance associates was Klaus Fuchs. It was precisely because of Fuchs that the most important information from the very depths of the American atomic project wound up on Kurchatov's desk.

How can we explain the phenomenal success of Soviet intelligence in the 1940s, when it managed to get enormous scientific and technological information gratis? The biographies of Soviet agents of that time make clear that they were in fact not recruited by professional intelligence agents but by the ideals and myths of socialism. And it was primarily German fascism that assisted this.

In 1933, soon after Hitler rose to power, the twenty-one-year-old student-physicist Klaus Fuchs fled Germany to England under threat of death. He grew up in the family of a Protestant pastor with socialist sympathies. Renouncing religion and joining the Communist Party, Klaus retained the family tradition of moral idealism and social service. At the beginning of World War II, the antifascist Fuchs was nonetheless interned in England as a German, and it was only thanks to his scientific repute that he was freed, after spending half a year in a camp. He began working on the British atomic project, and then with the British group that joined up with the American atomic project. He worked in Los Alamos and was initiated into the main atomic secrets. Fuchs contacted Soviet intelligence of his own free will and passed on scientific and technological results—in whose achievement he himself was a participant—on more than one occasion.[21]

In the 1940s a grand total of nearly 200 Americans supplied the USSR with priceless espionage; not all of their names have been disclosed to this day. The thirty-three-year-old British-German Klaus Fuchs didn't know that, alongside him in Los Alamos, were two twenty-two-year-old Americans who independently felt it their duty to inform the Soviet Union about American work on the atomic bomb. In 1944, when one of them, along with his nonscientist friend, offered their help to the homeland of socialism, Soviet intelligence gave them the code names Mlad [young] and Star [old].[22] The only expression in the living Russian language in which the archaic word mlad appears is star i mlad [young and old], meaning "everyone without exception." It's possible that this paired pseudonym reflected the professional Soviet agents' amazement at the abundance of volunteers.

After fascism was destroyed, the Cold War began too quickly. The atomic monopoly of the United States replaced the fascist threat. The Iron Curtain was opaque in both directions: people fell under the spell of the idea of socialism in the USSR out of ignorance about the Western world and in the West out of ignorance of the reality of Soviet life.

The bewitched physicists on both sides of the Iron Curtain developed Soviet nuclear weapons in record time under the command of Marshal Beria and Generalissimo Stalin.

8

RUSSIAN PHYSICS AT THE HEIGHT
OF COSMOPOLITANISM

The late 1940s for the Russian intelligentsia was not as bloody a time as the late 1930s but was probably even more suffocating. It was easier for those who knew about uranium and neutrons. The expanding Cold War was the reason. The primary Soviet symbol of the Cold War was the American atomic bomb. Now it is quite clear that the confrontation between the Stalinist empire and the Western democracies was inevitable after the disappearance of their common enemy, German fascism. The structure of the two social systems differed too much for them to maintain stable relations in contact with each other—like cold and hot liquids. The isolating material was the Iron Curtain, described by Churchill in the small American town of Fulton, Missouri, in March 1946. Perhaps it might be more accurate to call this curtain "reinforced concrete," considering that it would later develop into the Berlin Wall of 1961.

The atomic sword and shield, embedded in this curtain from its very creation, doesn't seem to have been so inevitable. What if Churchill had heeded Niels Bohr's advice (during their meeting on May 16, 1944) and informed Stalin about the atomic project without any technical details?[1] What if Roosevelt had lived a few more months and refrained from the atomic bombing of Japan, and thus from the rapid application of "atomic diplomacy" in relations with the USSR?

Knowing what Stalin was, one simply couldn't imagine advising something like this to Western leaders. But looking at the course of events in hindsight, one must admit that Soviet intelligence learned about Anglo-American atomic affairs without any permission and long before Bohr's visit with Churchill and—what's more—including many technical details. And American atomic diplomacy turned out to be ineffective, unlike Soviet atomic propaganda. The moral unanimity of the Soviet atomic scientists and their volunteer Western helpers attests to this.

How would the postwar political game have played out if the United States had not shown its atomic ace in Potsdam, in Hiroshima and Nagasaki? How would this have affected the Iron Curtain—would it have been so difficult to lift? And what would Stalin have used then to fortify his dictatorship?

These questions are easier to ask than to answer. And in 1948, when Sakharov began thinking about the hydrogen bomb, even asking the questions was difficult. The country was isolated from the Western world. State ideology gave the isolation an underpinning: American imperialism took the place of German fascism in propaganda, and a bomb marked with the letter "A" replaced the swastika in political caricatures.

Cosmopolitanism in Life and Science

The Soviet ideological machine used Russian nationalism, the euphemism for which was "patriotism," as the foundation for a new worldview, while its antonym became "cosmopolitanism." The steering wheel had to be turned very sharply in order to discern the sworn enemies within the ranks of recent military allies. But in fact it was the second turn of the wheel—in the opposite direction—after four years of war, a return to a path already emerging by the late 1930s.

Lenin left a legacy of the expectation of world revolution and all-encompassing internationalism. Lenin and his accomplices saw tsarist Russia as a "weak link in the chain of imperialism" and saw Soviet Russia as a match that would light a purifying revolutionary fire in the near future worldwide and establish socialism on a world scale, primarily in the industrially developed nations, which were most ready for socialism. This internationalist view was inculcated surprisingly deeply in public life, if one recalls the ethnic tensions in tsarist Russia. Sakharov's generation grew up in the most internationalist or, you might say, the most cosmopolitan period of Russian history.

In the 1930s the *Great Soviet Encyclopedia* explained the word "cosmopolitanism" wholly sympathetically: "the idea of a homeland bordering on the entire world," which originated in antiquity with the ideologues of "the impoverished, oppressed masses." There was also a reminder that "the homeland for the working classes of all nations is the country in which the dictatorship of the proletariat has been established. While patriotic toward its socialist homeland, the working class also strives to turn the entire world into its homeland."

The popular film *Circus*, released in theaters in May 1936, depicts this mood more graphically. The film shows a white American circus performer who is forced to leave her country because she gives birth to a black child. A disgusting impresario with a German fascist accent blackmails and exploits her. Only in the USSR does she find friendship, love, and a new homeland. At the end of the film, a lullaby is sung to the black child in five languages, including a couplet in Yiddish by Solomon Mikhoels, the famous actor and creator of the State Jewish Theater.[2] And for the finale, the main characters march off to a May

Day rally and sing (beneath a portrait of Stalin) the song "Broad Is My Native Land," which became extremely popular and served as Radio Moscow's call sign for a long time.

But Stalin, who had information on the world situation from sources other than Soviet newspapers and films, could not console himself with the vision of world revolution, already 20 years overdue. In December 1936, Stalin's constitution fixed a new picture of the world in place: the Soviet Union was no longer the spark of the world revolution—socialism could be built, as Stalin put it, "in one separate country," and that was why our country was the strongest one, not the weakest; it was the avant-garde of humanity.

Such a reorientation provided the foundation for a revival of national pride that easily grew into arrogance toward peoples who had not yet succeeded in carrying out socialist revolutions in their own nations. Russian patriotism had already become a subject for propaganda back in the prewar days, and the word "Russian" began to replace "Soviet." During the war years, government propaganda also used the theme of the fraternity of the Soviet ethnic groups and friendship with the non-Soviet but antifascist peoples of the United States and Great Britain. However, this was covered up by the notion of the Russian nation as the "older brother" to whose lot has fallen both the main burden on the battlefield and—accordingly—the main responsibility for the fate of humanity. At the celebration of victory over Germany, Stalin made a toast to "the great Russian people" and then introduced into his speech, in his strong Georgian accent, the expression "we Russians."

The name "United Nations" emerged in 1942 as the designation for the countries allied against the Axis of fascism. The new political term was instituted by Charter of the United Nations at the San Francisco Conference in April 1945. The uneasy notion of united capitalist and socialist—or democratic and authoritarian—countries revealed its problems in just a few months. The real postwar disunity of the United Nations made the life of party ideologues simpler—they were back on track. The Soviet people were told that cosmopolitanism "is propagandized by reactionary ideologues of Anglo-American imperialism, whose goal was to establish its world hegemony," that cosmopolitanism "is the reverse side of aggressive bourgeois nationalism and hostile antithesis of proletarian internationalism." This is how postwar dictionaries explained the new word in postwar Russian life.[3]

From early 1949 you didn't have to look this word up in dictionaries. On January 28, *Pravda*, the country's main newspaper, explained it in an editorial "About an Anti-Patriotic Group of Theater Critics."[4] Other newspapers followed *Pravda*'s suit. Other professions followed the theater critics. The list of the names of exposed "rootless cosmopolites" was eloquent enough for folk wisdom to record it in a couplet: "Not to be taken for an anti-Semite, / Call a kike a cosmopolite."

Although the return of the old Russian affliction was completely unexpected for many, it is entirely explainable.

The Jewish Question
in Soviet Physics

In the last decades of tsarism, in addition to the word "intelligentsia," Russia gave the Western world the gift of the word "pogrom." Jewish pogroms, the trial of Mendel Beilis accused of a ritual murder of a Christian child, the "Protocols of the Elders of Zion," which was fabricated by the tsarist police—in the twentieth century all these elements were added to the long-standing "legal" restrictions on the right of residency (the Jewish Pale of settlement), education, and employment. In the tsarist "prison of peoples," Jews wound up in the toughest isolation cell.

This aroused ardent compassion from the Russian intelligentsia, and it was active compassion. Justice triumphed at the trial of Beilis, who was acquitted by a jury in 1913 after he spent a few years in prison under investigation. Four years later the autocracy fell, and the provisional government repealed all ethnic restrictions.

During Soviet Russia's first two decades, anti-Semitism virtually disappeared from government life. State power and society were successfully dealing with the vestiges remaining in everyday life. There was no ethnic discrimination in the admission process to institutions of higher learning.

Let's entrust ourselves to the observant naturalist Vernadsky, who examined social tectonics with no less perspicacity than geological ones. He succeeded in combining science with public life; no wonder he was elected in 1906 to both the Academy of Sciences and the Government Council (the highest advisory organ of the Russian empire). One of the founders of the Constitutional Democratic Party, he responded to the Jewish pogroms and fundamentalist "patriotism" in his newspaper writings, and he understood why the participation of the Jews in the revolutionary movement in Russia was disproportionately great. Percentage quotas, which were an obstacle to Jews entering universities, increased the percentage of Jews in the revolutionary movement.

The revolution's repeal of ethnic restrictions opened the road to spheres of life that not been previously accessible to Jews—big cities, the government administration system, science, and culture. The more rapidly a social sphere developed, the more quickly the relatively more literate and energetic inhabitants of the Jewish Pale would head into it.

In 1927 Vernadsky wrote to a friend: "Moscow is in places Berdichev [a predominantly Jewish town in provincial Ukraine]; the power of the Jews is horrifying, while anti-Semitism is growing unstoppably (in communist circles, too)."[5] And in March of 1938 he noted in his diary:

> Destruction by ignoramuses and people on the make is taking place. The people in the Academy publishing house all these years are below average. A rich gallery of types from Shchedrin-Gogol-Ostrovsky. Where do they get

them? Jews who have attained power and authority are the new type in this gallery. For all my philo-Semitism, I can't ignore this.

It was Vernadsky's "philo-Semitism" in a diary entry the next month, when, in remarking on "Mandelshtam's interesting and brilliant lecture at the Academy," he summed up: "A noble Jewish type of ancient Jewish culture." But those who knew Leonid Mandelshtam saw "depth and subtlety of thought, breadth of scientific and general erudition . . . irresistible charm, a man of genuine culture in the European manner . . ."[6] Mandelshtam was a man of the European culture, rather than Jewish—that is how the circumstances of his life turned out.

Philo-Semitism was not that rare a response to historical Russian anti-Semitism among the Russian intelligentsia. A good example is Kapitsa, a friend of Mikhoels, who used to attend plays with his wife at the Jewish Theater (where "a word-for-word" translation was whispered into their ears) and spoke out at a rally of "representatives of the Jewish people" in August 1941. The speech by Kapitsa, Acting Member of the Academy of Sciences of the USSR, member of the English Royal Society, and Stalin Prize laureate, became part of a collection published by the main political publishing house with the improbable (at any other time) title of *Brother Jews of the Whole World*![7]

Andrei Sakharov's aversion to anti-Semitism was entirely practical, while his philo-Semitism was rather theoretical. It manifested not in indulging any particular Jew, but in attentive curiosity and generalizations, at times highly exaggerated. For example, Sakharov wrote about his childhood friend, whose "national Jewish intellectual gentleness—I don't know what to call it: perhaps spirituality, which is often evident even in the poorest families—was appealing. By this I don't mean that there is less spirituality in other peoples; sometimes, perhaps, there is even more, and yet there is something special and deep in Jewish spirituality." The "national, seemingly, sad ancient tact" of a comrade from his student years appealed to him.

It's hard to imagine that Sakharov didn't run into slobs and scoundrels of Jewish descent. The theoretical philo-Semitism of a Russian *intelligent* is easier to explain as revulsion toward practical anti-Semitism. Sakharov's practical compassion for the Crimean Tatars and Russian Germans, when he came into contact with their Soviet troubles, was of a similar origin. The only difference was that the tradition of anti-Semitism was stronger—the laws of tsarist Russia restricted only the rights of Jews on the basis of ethnicity.

Sakharov received from Tamm his "foolproof way of determining if a person belonged the Russian intelligentsia—a Russian *intelligent* was never an anti-Semite; if he had a touch of this illness, he was something else, terrible and dangerous." But the very need of such a test arose only in the postwar years, when the "struggle with cosmopolitanism" exploded. How did state anti-Semitism return to Russia after thirty years of the Soviet regime?

The general policy of Soviet "proletarian internationalism" stumbled at the German-Soviet Friendship Treaty of 1939, in the first weeks of World War II. But officially Soviet anti-Semitism began its state existence in the second year of the Great Patriotic War—as World War II was called in the USSR after the German army, by order of Stalin's recent friend Hitler, invaded Russia in 1941. On August 17, 1942, the Central Committee of Communist Party issued the report "About the Selection and Promotion of Cadres in Art." This document exposed the "appalling distortion of ethnic politics," as a result of which "too great a number of non-Russians (predominantly Jews)" were found in many of Russian art institutions. The Bolshoi Theater was the first such "institution" in the document.[8]

It would seem that the second year of the war with Nazism was an ill-suited time for such a campaign. But that is merely at first glance. The directive about purging Jews came to the Central Committee, one could say, from Hitler. In the propaganda that the Nazis conducted to demoralize the Soviet Army, there was one simple motif: "Russian soldier, do you know who rules Russia? Kikes!" followed by a list of names. It's difficult to say whether these leaflets had an effect on the soldiers, but when they made their way to the Central Committee, they definitely worked.[9] A few years before, the Stalinist Great Terror had done a fair job of "purging" the highest party cadres of a generation of internationalist old guard. A new generation of party martinets—less burdened by the traits of the intelligentsia—had already been in the service of Stalinist patriotism for several years. Now they had to knock the anti-Semitic weapon from enemy hands, and then they took it into their own hands, dismissing Jews from socially prominent positions.[10]

It began with music but went off in all directions and exploded with particular force in "the struggle against cosmopolitanism." Administrators and enthusiasts began to count up the percentage of (Jewish) "impurities" in the cadres. The quota system was being revived, only its very existence was secret, in contrast to tsarist times.

The new tendency aroused the indignation of the socialist idealists and the satisfaction of Soviet materialists, who had been handed a new tool to advance their careers. These two categories, however, did not exhaust people's actual diversity.

Alexei Krylov, a prerevolutionary academician and tsarist general, mathematician and shipbuilder, translator of Newton from Latin and expert in the boatswain's dialect of Russian, had practical knowledge of the various methods by which Russia was governed. He had grounds to consider the tsarist bureaucracy less than ideal, but if necessary, he could also cite it as an example for the Soviet bureaucrats. He understood human nature and did not amuse himself with dreams of instantly creating the ideal society with the aid of science.[11] He simply did his work in science and accepted the government as an unavoidable reality of this world, like the climate and the weather, like thunder and lightning, like floods and earthquakes, with which one must live.[12]

This could be called common sense or healthy cynicism, but in any case, he didn't try to "keep in step with his time." He can't be easily placed in time from his speeches and articles but can easily be defined as an individual, unwilling to depend on current times.

In December 1944 Academician Krylov gave this kind of speech—independent of the time—in the House of Scientists, at a meeting commemorating Mandelshtam. Krylov began his speech with "Leonid Mandelshtam came from a wealthy Jewish family," and then, after a short description of his scientific career, continued:

> Leonid Mandelshtam was a Jew. There are many Jews who follow the iron rule of the Old Testament of Moses and the Prophets literally: "An eye for an eye, a tooth for a tooth," forged from millennia of persecution by government authorities, slavery, inquisition, dukes and feudal lords.
>
> Two thousand years ago the voice of a great idealist proclaimed the New Testament: "Love your enemies as yourself; if they strike you on the left cheek, turn your right one to them." Everyone reads these words, no one follows them; Leonid Mandelshtam didn't follow them either, but he did approach this ideal in many ways. Naturally he didn't love his enemies, but because of the purity and loftiness of his character, he virtually had no enemies.
>
> He was notable for his directness, honesty, a complete absence of obsequiousness and slyness, and he earned the special respect of the better part of the Moscow University; but in the last two years a cohesive group of physicists caused him much distress in the scientific arena.
>
> Leonid Mandelshtam was in the front ranks as a scientist, an academician and a professor. May his soul rest in peace, for he was a righteous man![13]

Such a free use of the word "Jew," when people had begun half-whispering it as if it were nearly indecent, created an indelible impression on those present.[14] For Krylov, the "Jewish question" was not at all an object of some particular attention, but he saw the realities of life. And it appears that he used the word "Jew" to spite the state's bad weather, which was brewing into a storm.

In January 1948, Stalinist lightning killed the famous Jewish actor and public figure Mikhoels; a group of Jewish cultural activists, members of the Jewish Anti-Fascist Committee, was next. And finally, under the peals of anti-Semitic propagandist thunder, "the Doctors' Plot" was launched and—had Stalin lived a few more months—would have led to a bloody deluge.

University Physics versus
Academy Physics

But what was this "cohesive" group of physicists which caused Mandelshtam "much distress"? When Krylov said this, he was probably looking straight at these people in the auditorium. He gave his speech at a joint meeting of the Academy of Sciences and Moscow University, and this group had coalesced right there at the physics department of Moscow University and had distressed Mandelshtam's associates and students throughout the entire preceding Stalinist decade.

This confrontation began in 1937 after Gessen's arrest when Aleksandr Predvoditelev (1891–1973) was appointed to Gessen's position at Moscow University—director of the Physics Institute and dean of the physics department. The new director was not favorably disposed toward the Mandelshtamites. He had established himself at the university in the years of the post-Lebedev decline. Perhaps that is why his ambitions surpassed his scientific potential. The disparity was especially noticeable in the presence of the Mandelshtam constellation of first-class physicists. However, Predvoditelev himself did not wish to admit this, and strayed far from technical physics beyond his area of expertise into theoretical physics with astonishing conceit.[15] Tamm could not ignore his ignorant publications and responded to them in 1936 in a critical article.[16]

When he took over Gessen's position at Moscow University, Predvoditelev declared:

After enemy of the people Gessen was exposed, the Institute's Party organization conducted extensive work exposing and liquidating the consequences of his wrecking. In the course of this work, several professors (those upon whom Gessen depended) underestimated enemy of the people Gessen's damaging activities. These professors were forced to leave their administrative positions.[17]

Nonetheless, the Mandelshtam group continued to teach at the university until the war started, although less intensively. The evacuation of the Academy and university to different cities helped Predvoditelev and his old gray guard to get rid of them. After the return to Moscow from the evacuation in 1943, Landsberg and Tamm, formerly the heads of the departments of optics and theoretical physics, were unable to return, and Khaikin and Leontovich were forced to leave Moscow University. All four were the authors of the best university textbooks—and all were Mandelshtam's closest associates.

In the summer of 1944, fourteen physicist-academicians, including Kapitsa, wrote a letter to the government regarding the abnormal situation at the physics department at Moscow University.[18] After additional letters from Kapitsa, and as a result of the work of a specially created commission, Predvoditelev was dismissed in 1946 from his position as dean.

On the whole, however, the bastion of university physics stood its ground and remained unassailable until Stalin's death. Those defending this bastion were pretty different people. The only thing they shared in common was a dissatisfied claim to a high evaluation of their scientific achievements and a commensurate position in scientific life. It has to do with a person's professional and moral qualities intersecting. A "mathematical" method of expressing this connection as a fraction was formulated long ago as EC/SE, where EC is evaluation by colleagues and SE is self-evaluation. For those in the university group, this fraction was pretty small—self-evaluation was substantially greater than evaluation by colleagues. Predvoditelev himself later tried to explain his actions by "directive from Party organizations," but in fact it was a common effort.

There were two physicists in this university group, Anatoly Vlasov and Yakov Terletsky, who had studied at Mandelshtam's school and left. Highly skilled professionals, they both turned out to be at the mercy of their inordinately great ambitions that induced them to join the university guard and impede Tamm and Landsberg's return to Moscow University. Mentally unbalanced Vlasov was mostly the object of manipulation; Terletsky, as secretary of the department's Party Bureau, was an entirely active individual. His break with his mentors and the party loyalty he displayed made him a suitable candidate when Beria needed a physicist for special assignments in the fall of 1945. A physicist capable of serving in the KGB has to be capable of a great deal.

The building of Stalinism "in one separate country" damaged ethics in the science community. In a dictatorship, those who tried to achieve scientific recognition at any price appealed to state authority. And one had to speak to power in its own language, picking up all its latest words: idealism, Trotskyism, kowtowing to the West, cosmopolitanism. In the process, of course, the more energy an individual spent on activities outside science, the less energy he had for science, and the greater the gap between scientific ambitions and real achievements. But on the other hand, such extrascientific activists had more opportunities to fill the ears of the party administrators of science with their words. The Moscow University physicists surpassed the Academy ones in this art and successfully defended their "autonomy" until the end of the Stalinist era, despite all the scientific might of the Academy.

Numerous vestiges of university intrigues against the Academy physicists are preserved in the Soviet archives. One of the intrigues emerged in the newspaper under the banner of the struggle against "kowtowing to the West."

The campaign against harmful Western influence began in the summer of 1946, and literature became the first objective of patriotic reeducation. However, the natural sciences—by their nature the most international ones—could not help but wind up in the area under fire. It was only a matter of what would serve as the first occasion. It became Anton Zhebrak's short review article "Soviet Biology," published in the American magazine *Science* in 1945.[19]

Read through patriotic eyes, this article made its author a character in an entirely different, and devastating, article: "Against Kowtowing" published in 1947 in the wholly official *Literaturnaya Gazeta*. The patriot-author couldn't comprehend how Soviet biologist Zhebrak could have managed to write: "We are developing a unified world biological science along with American scientists"! "With whom," the patriot fulminates, "is that Zhebrak developing a unified biology? With those fascistic genetic 'scientists'??!!"

The article in *Literaturnaya Gazeta* was signed by the rector of the Timiryazev Agricultural Academy, and this is understandable, because Zhebrak was a professor at that academy. What is not understandable, however, is how a second character, the physicist Vitaly Ginzburg, wound up in this article. He worked in a different academy—the Academy of Sciences— but the patriotic article blasts three of his publications, "which discredit our Soviet science":

> Professor Ginzburg's "About the Atomic Nucleus," a pamphlet published in 1947 in a mass print run, completely omits the names of Ivanenko and Gapon. Ivanenko and Sokolov developed the meson theory of atomic forces in 1940–1941. Yet Dr. Ginzburg's recent survey article "Theory of the Mesotron and Nuclear Forces" is also shamelessly silent about this achievement of Soviet physics . . . And Dr. Ginzburg's most recent article is complete and ridiculous boot-licking of American science . . . Doctor Ginzburg, speaking about the radiation which interests us, is silent about Ivanenko's authorship . . . You can't go much farther than this shameful silence about the discoveries of Soviet science, this rubbing out of Soviet authors.[20]

The physics terms here, partially replaced by ellipses, were used surprisingly competently by the rector of the Agricultural Academy. He obviously studied the physics literature carefully. But the systematic mention of Dmitri Ivanenko as the victim of harmful kowtowing raises a red flag. He was the most distinguished theorist in "university physics," whereas Ginzburg—a student and closest associate of Tamm's—belonged, naturally, to "academic" physics. It was clear who was behind the article in *Literaturnaya Gazeta*.[21]

The ironic physicists then introduced a method of measuring kowtowing in scientific publications. The angle of incline toward the West was measured by the ratio of the number of foreign references in the bibliography to native ones. Thus, an impeccable verticality could be achieved only by zero number of foreign references.

However, the patriotically vertical physicists were deadly serious and, following the example of biology, devised an intrigue on a much grander scale, aimed at the "Lysenko-ization" of Soviet physics. This became no laughing matter for the Academy physicists.

The Aborted All-Union Meeting

The name of Trofim Lysenko (1898–1976) acquired worldwide fame after a session of the All-Union Academy of Agricultural Sciences when, by the will of the country's dictator, in August 1948 he was declared the dictator of Soviet biology and subsequently "introduced order" in the science subject to his rule.

Until the mid-1930s, biology in the USSR was evolving no less rapidly than physics—until Lysenko appeared on the scene. Stalin played the key role in his political promotion. They were two of a kind: unburdened by education and gifted with powerful political intuition—skill in manipulating people to achieve their personal goals that justified any means. Their incomplete education made it easier to believe in the miracles of science than to see the limits of its possibilities. Both needed miracles: Lysenko in order to refute all of contemporary genetics, and Stalin to build "socialism in one separate country." Lysenko used as a foundation the postulate that if you try hard enough you can "raise" one living organism from another, one species from another—rye, for instance, from wheat. Stalin needed a similar postulate to cultivate a "new—socialist—man" and to populate the country with this new species. The limits of "so-called scientific genetics," which are against this, should not impede the process. Even if social genetics did not exist then, Stalin might have thought, it would be good if it could rely on Lysenko's science.

Stalin's personal sympathy toward Lysenko was also evident in high decorations (Stalin Prizes 1941, 1943, 1949), in government positions, and in—most astonishingly—the fact that Stalin personally edited Lysenko's crowning 1948 speech.[22]

There was no Lysenko to be found in physics. The university group could propose several candidates who were prepared to replicate Lysenko's feat—to grab power in physics for themselves and their understanding of this science by assault, using resources indiscriminately. Sakharov put it this way: "Lysenko's laurels robbed many people of their sleep. Terletsky was, apparently, one of them." According to Terletsky himself, Vlasov was being proposed as a Lysenko clone.[23]

However, they both lacked important qualities to be appointed a Lysenko-in-physics, primarily scientific ignorance—both were too competent as physicists to propose some sort of miracle. Physics, luckily in the Soviet case, was more remote from life (and from party life, in particular) than biology, where, as in art, anyone can have his own opinion, especially if he is a member of the Politburo.

The entire university group as a whole tried to play Lysenko's role in physics. By the time of Lysenko's triumph in August 1948, they had already purged the university of the academic cosmopolites and were ready to step out into the vast spaces of the whole country. Four months later, in the Ministry of Higher Education, "the need to organize a broad public discussion of the main methodological issues in the field of physics" had come to a head.[24]

"The Organizational Committee of the All-Union Meeting of Physicists," established on December 17, 1948, was headed by the Deputy Minister of Higher Education. The preparations for the "free" discussion took three months: more than 40 meetings, a hundred participants, and thousands of pages of shorthand transcripts. The "bathhouse" (as the Organizational Committee was called by Academy physicists) was at the disposal of the Moscow University physicists. They were on the attack, and their accusations against the Academy physicists became the basis for the draft of a resolution: "It is imperative to mercilessly extirpate all shades of cosmopolitanism—Anglo-American imperialism's ideological weapon of diversion." Only the Academy physicists were accused personally in the resolution: Landau and Ioffe "are servile before the West"; Kapitsa "preached open cosmopolitanism"; Frenkel and Markov "accept Western physics theories uncritically and propagandize them in our country."[25]

The words of this draft were chosen under the direction of the head of the Central Committee's Department of Propaganda and Agitation. The patriots from Moscow University, not shy in their expressions in the "bathhouse," were not shy at all in their letters to the Central Committee. The most militant, Professor N. Akulov (1900–1976), simply accused the Mandelshtam group of espionage and sabotage. He began his letter with the fact that the founders of this group, Mandelshtam and Papalexi, had even before the revolution "crossed the Russian border in different places and penetrated into Petrograd," where "they organized radio contact with Germany." According to Akulov's information, Mandelshtam worked in Soviet times with Trotskyites, Zionists, and "physicists of German background": "his closest assistant became B. Gessen, who was in touch with Trotsky's son, as well as with the German General Staff"; "Mandelshtam's group had the closest links with espionage and sabotage groups in Leningrad and Kharkov." And further on, Akulov lists, on ten pages, physicists who fell under Mandelshtam's influence, including Vavilov, the president of the Academy of Sciences, and Kapitsa, who "was groomed [in England] for 12 years, where they established his authority, and then he was transferred back into the USSR."[26]

The Academy physicists were forced to respond to the accusations. "Professor Ivanenko extremely insistently accuses all those who forgot or didn't think it necessary to mention him or his associates, to the point that he considers this or that scientist's attitude toward Ivanenko's papers to be the touchstone of Soviet patriotism."[27]

Aleksandr Andronov explained to the anti-cosmopolites why he considered his deceased teacher a Russian physicist:

> Anton Rubinstein is a Russian musician, Levitan is a Russian painter [they both were of Jewish descent], and Mandelshtam is a Russian physicist. If a Jew tells me that Mandelshtam is a Jewish physicist, I will reply to this Jew that he is a Jewish nationalist. If a Russian tells me that Mandelshtam is a Jewish physicist, then I will tell this Russian that he is a Russian nationalist and chauvinist.[28]

The last pages of the Organizational Committee's transcripts are dated March 16, 1949. An invitational ticket to the "All-Union Meeting of Physicists" from March 21–26 is part of the same archival folder.[29] But the meeting itself did not take place—despite the passions of the confrontation, the enormous expenditures of time, and party effort for its preparation. No documents were found in the archives that intelligibly explain why the meeting was canceled. However, it's clear that this could have only been done with the knowledge of Stalin. An oral legend has come down that physics was saved from a pogrom by the atomic bomb.

Vitaly Ginzburg, a "kowtower and cosmopolite" who was earmarked to become one of the meeting's victims, retained this version in his memory: "Kurchatov told Beria: 'All our work on the atomic bomb is based on quantum mechanics and the theory of relativity—if you start abusing them, then close up our shop first.' The latter apparently reported this to the Great Leader and Teacher, and the meeting was canceled."[30]

This explanation doesn't seem quite adequate. After all, the atomic project had begun a few years before the meeting, and the meeting's major targets were not quantum mechanics and the theory of relativity at all, but specific people among the Academy physicists. And all the instigators from Moscow University were aiming primarily at the Mandelshtamites. If, in the fall of 1948, an unproduced atomic bomb was not enough to prevent the authorities from "bringing order in physics" and from loading the university guns, why would it have been enough to make them drop the guns a few months later?

A few months later something new emerged in addition to the A-bomb. By spring of 1949, Kurchatov could already speak about an H-bomb. By that time, a very promising project of the hydrogen bomb had been conceived by Mandelshtamites in Tamm's FIAN group. The kowtower Ginzburg had proposed the second main ideas for the project—his secret report is dated March 3, 1949. And the first key idea was advanced back in the fall of 1948 by another of Tamm's students, Andrei Sakharov, his report dated January 20, 1949. Two chicks from Mandelshtam's nest had made too significant a contribution to state power to have this nest destroyed.

The results of the work of Tamm's group were entirely approved by the scientific directors of the atomic project. Kurchatov found that he was right in trying to get hold of Sakharov back in 1947. However, the fact that Kurchatov failed then helped him save physics in the spring of 1949. The salvation of Soviet physics from the pogroms of "Lysenkoization" can be called the first use of thermonuclear energy for peaceful purposes.

As for the role of personality in this story, Sakharov, unbeknownst to himself and others, played no less a part. The young theorist, who couldn't have perturbed any of the "patriotic" physicists, might not have even noticed the threat hanging over physics in the early months of 1949. He was too busy with thermonuclear physics and was too young and "sinless" to be invited to the Organizing Committee meetings.

On March 17, 1949, a few days before the planned meeting, Yuli Khariton, scientific chief of nuclear weapons development, asked Beria to permit Tamm access to the intelligence data. Beria's office decided that "the data should not be turned over to I. Tamm, in order not to attract unnecessary persons to these documents." They permitted only certain experimental data to be shown to him, without reference to their source.[31]

But how did "unnecessary person" Tamm and his students end up in the nuclear project and become the founders of the thermonuclear one?

9

THE HYDROGEN BOMB AT FIAN

A- and H-, or Nuclear
and Thermonuclear

How does the hydrogen bomb differ from the atomic bomb? Thermonuclear energy from simply nuclear energy? The difference didn't present much difficulty for Soviet cartoonists: they began to label the bomb—which the fat American imperialist, ever-present fat cigar in his mouth, clutched under his armpit —with the letter "H" instead of an "A." Even the words "atomic" and "hydrogen" written out in full do not say much, and they do not explain why the expression "superbomb" came to be used only for the hydrogen bomb. The atomic bomb that was dropped on Hiroshima was 20,000 times more powerful than the largest ordinary bomb exploded during World War II. Isn't that sufficient for the label "super"?!

The difference between simply "nuclear" and "thermonuclear" played too great a role in Sakharov's life and the life of mankind to be reduced to etymological explanations or examples of cartoons. A full explanation can only be given in the language of physics, using the alphabet of mathematics. But if the nucleus of an atom had absolutely nothing in common with the world of everyday experience, people couldn't have penetrated so far beyond the limits of this experience.

A few chapters ago the first portion—just a bit—of nuclear physics explained what it was that so interested Tamm about the atomic nucleus proper. Now we'll need a little more. If you have ever had the occasion to drop a mercury thermometer, and you did it when you were still young and carefree enough, then you probably didn't throw away the shiny drops of mercury but observed them for a while. You must have noticed that the smallest drops, when touched, merged readily, but that the biggest ones did the opposite—separated into smaller ones just as readily when disturbed in the slightest.

This observation is enough to give a quick idea of how the H-bomb differs from the A-bomb—because atomic nuclei resemble drops. There are 92 forms of nuclei, or chemical elements, in nature as arranged in Mendeleev's table. Hydrogen has the smallest nucleus, and uranium the largest. Small nuclear drops readily merge when they are put into contact with each other, but the big ones readily separate, and these two quite different nuclear reactions are *fusion* and *fission*. The word "readily" means that, after the fusion of small drops or the fission of large drops, energy is released. Just how much becomes a matter of equations, the most important of which is Einstein's famous $E = mc^2$.

Here is how the equation works. In fusion, the total mass of two *readily merging* nuclear drops is larger than the mass of the nuclear drop that forms as a result of their fusion. If the initial mass is larger than the final mass by the amount m, then in this nuclear reaction, energy E equal to mc^2 is released, where m is the "missing" mass. Similarly, in fission, the mass of one *readily separating* nuclear drop is greater than the total mass of the resulting separate drops, and again the energy E released is equal to mc^2.

Sometimes people speak about the transformation of mass into energy. This is about as correct as saying that when a person shops, the money that disappears from his wallet has turned into the package of rice in his shopping cart. The difference is that the coefficient between the money and the number of grains—the price of a single grain—can differ from shop to shop. But in physics the energetic value of a unit of mass is an enormous and constant amount. This value is always equal to c^2, where c is the speed of light, and c is so great that light can fly around Earth in a fraction of a second. Lebedev had trouble in his experiments measuring the pressure of light precisely because you need to divide by this large quantity, whereas in nuclear processes you need to multiply by it. The law $E = mc^2$ operates in all physical processes; however, outside nuclear physics—even during the explosion of TNT—the decrease of the mass that "pays" for the energy of the explosion is no bigger than a billionth part.

In nuclear fission and fusion, this part is millions and billions of times larger. This means that nuclear explosives can be more powerful by just as many times. All you need to do is figure out how to have a huge number of nuclear droplets separate or fuse simultaneously. The first case is the A-bomb, or the fission bomb, where the nuclear explosive is a substance with large nuclei such as uranium. The second one is the H-bomb, or the fusion bomb, filled by substances with small nuclei, for example, isotopes of hydrogen—deuterium or tritium.

But how do you make all the separate nuclear droplets separate or fuse, not one at a time, but all at once—collectively? For the fission reaction, nature hinted at a way a few months after the discovery of fission itself in 1939 (and a few months before World War II began). It turned out that, during the fission of uranium nucleus droplets, along with fragments, some neutrons also get released; each of these neutrons is capable of making another nucleus fission, and so on—a chain reaction. One needed only gather a sufficient amount of uranium in one place, and an atomic explosion, the likes of which the world had not yet seen, was guaranteed. But amassing a sufficient amount of this rare substance, more expensive than gold, was difficult. Uranium deposits had to be found and the extracted ore refined through complex processes. That is why billions of dollars and several years of effort of many thousands of people were required before the world saw such an explosion in 1945.

As for the nuclear reaction of fusion, it has been going on since time immemorial in full view, producing sunlight and making other stars shine in the sky. By whimsical historical coincidence, in that very same year, 1939, physicists were able to explain exactly how the energy of the sun originates during the continuous fusion of nuclei in the sun's depths. The nuclear fuel for the solar energy generator is the most abundant element in nature, hydrogen, of which there are two atoms in every molecule of water.

But reproducing this natural process on Earth turned out to be much more difficult than creating the unnatural process of "collective" fission. The reason for the difficulty is that nuclear droplets—in contrast to ordinary ones—are electrically charged and thus repulse one another. This helps large nuclei separate but interferes with the fusion of small ones. After touching, small nuclei fuse energetically, but enormous power is necessary to make them touch.

By speeding up separate nuclear droplets in an accelerator, physicists were able to make them touch the target nuclei and observe that fusion was actually taking place in the process. However, it was only one nucleus at a time and not with an appreciable amount of matter.

Gravitation helps stars, including the sun, deal with this job—the matter in the center of a star is compressed by its own weight. With temperatures in the millions of degrees, the numerous particles of matter inside stars have speeds comparable to those attained by a few particles in an accelerator.

Nuclear reactions occurring in such high thermal conditions were named "thermonuclear." Naming isn't hard, but how do you reproduce stellar conditions on Earth? It's easier to answer the question of why physicists began to use the prefix "super-" for the thermonuclear bomb long before it appeared. The point is, you can only collect up to a specific limit of several kilograms of uranium in one place. This limit is called the critical mass, and when it is reached, the explosive chain reaction of fission begins by itself. There's no critical mass for the fusion reaction, and this means that the power of the thermonuclear explosion can, in principle, be any amount—larger by any amount than the monstrous explosion that reduced Hiroshima to ashes. Hence talk of a super-bomb. That is why the A-bomb was still just a bomb, while the H-bomb was the superbomb—the bomb unlimited.

Special Energy at FIAN

The thermonuclear superproblem was the main subject in Tamm's article "Intra-atomic Energy," published in the newspaper *Pravda* in the spring of 1946.[1]

But for him, it wasn't at all the explosions that made the problem significant: "It is fairly certain that the use of intra-atomic energy will transform the economic and technological foundation of human existence in the near future."

Tamm informed the readers of *Pravda* that the energy of the sun arises from a nuclear reaction in which "as much energy as burning 15 tons of gasoline is released when a single gram of hydrogen is transformed into helium." From 1919, when Rutherford conducted the first artificial nuclear reaction, "more than a thousand different nuclear reactions have been produced in laboratory conditions," but "until recently, their practical use was not possible." Using uranium's intra-atomic energy turned out to be possible only in the atomic bombs, "dropped by Americans on Japan."

> But the potential of atomic energy in other elements is inexhaustible, and if we don't know to use it yet, then we must not forget that the splitting of uranium was discovered only seven years ago. We are just beginning scientific advancement into new fields of unknown phenomena, and an unusually rapid development of physics will undoubtedly open new unexpected opportunities for mankind.
>
> The path to this lies first of all in developing pure research along the entire frontier of physics . . . [because] we cannot predict beforehand what the next stage in man's mastery of the forces of nature will be. The great new force of nature can transform the economic and technological foundation of human existence. It must be directed not at destruction, but at the universal good.

In the spring of 1946 the "universal good" hadn't yet become a risky cosmopolitan goal, and the tone of the article breathes optimism. And the content does show that the author was not involved in the Soviet atomic project. The project had an absolutely practical goal at that time—to make an atomic bomb, rather than to investigate "the entire frontier of physics" just because "we cannot predict beforehand what the next stage in man's mastery of the forces of nature will be."

Completely predictable scientific and technological elaborations were necessary in order to make an A-bomb. Man mastered the bow and arrow long before studying the laws of mechanics and made quite a decent metal axe without studying the microscopic properties of metal. An incomparably large amount of scientific work had to be accumulated before the A-bomb could be made. What was required was primarily the work of inventive engineering and applied physics. Tamm was not engaged in this work, and he probably didn't aspire to it. This was not his kind of physics.

The rapid drive to transform pure science into technology, as well as its social potential, aroused general enthusiasm. In the fall of 1945 Tamm gave nine talks on the subject "Atomic Energy."[2]

The object of FIAN's main theorist's enthusiasm for lecturing was a more practical subject for FIAN's director Vavilov. He aspired to include his institute in the project, to open up better opportunities for government financing to develop "the entire front" of FIAN physics. But doing this was not simple, even for the president of the Academy of Sciences.

Although the Kurchatov's Institute was called the No. 2 Laboratory of the Academy of Sciences, his atomic empire was independent of the Academy. From the atomic project's very inception in 1943, Kurchatov relied on Ioffe's Leningrad school "graduates." Aside from personal ties that developed at the Leningrad Physics and Technical Institute (PhysTech), there was no particular reason why Kurchatov would attract the FIAN theorists. After all, they were oriented toward fundamental physics, whereas the project essentially dealt with applied physics in which the PhysTech school had no equals in the country.

Initially, starting in 1944, Vavilov managed to involve only a few people from FIAN in auxiliary uranium research. Among them was Ilya Frank, who coauthored a 1937 paper with Tamm that won the Nobel Prize in Physics in 1958. Two secret rooms appeared at FIAN, from which other researchers were barred.[3]

Secret services, of course, did their work as they did it throughout the entire country, vigilantly checking and rechecking the personnel. In 1946 Tamm filled out a newly introduced, enormous personnel questionnaire. He reported that his brother Leonid, arrested in Moscow in the fall of 1936, "died in prison in 1942," and that his father, arrested in 1944, was acquitted in the absence of *corpus delicti*, the essence of a crime.[4]

Academy President Vavilov was not part of the management of the atomic project but regularly participated in meetings of its Scientific-Technical Council. In April 1946 he presented a memo, "On the Organization of Research in Various Scientific Fields Related to the Use of Atomic Nuclear Energy," the first concrete proposal on the peaceful use of atomic energy.[5]

Even more direct evidence of his efforts is a FIAN document dated September 24, 1947, and addressed to GosPlan, the State Planning Committee. It is a general list of scientific objectives with an annotated note.[6] On the document there is the customary notation for secret record-keeping showing that the rough draft was destroyed, signed off by Frank, who was, therefore, involved. The document itself is anonymous, meaning highly official; evidently, it was accompanied by a letter from Vavilov.

The document is innocently titled "Basic Problems of Scientific Research," but what these problems are is obvious from its opening lines: "Research into fission of heavy elements and the possibility of using them for special energy, research into the reactions of light elements and possibility of using their synthesis for special energy." Simply put, or reading through Beria's eyes, it's about atomic and hydrogen bombs. The rest of the document concerns "the entire frontier of physics," ranging from theory of nuclear forces to astrophysics and from computer theory to biophysics.

The list of "basic problems" was accompanied by an explanatory note in whose content Tamm's views can be seen. Regarding the second form of special energy, it says laconically: "The synthesis of light elements is undoubtedly one

Экз.№ 1

ОСНОВНЫЕ ПРОБЛЕМЫ НАУЧНО-ИССЛЕДОВАТЕЛЬСКОЙ РАБОТЫ.

а) Научно-исследовательская работа по развитию и усовершен-
ствованию известных в настоящее время методов получения
спец. энергии.

б) Исследования расщепления других тяжелых элементов и
возможности их использования для получения спец. энергии.

в) Исследование реакций легких элементов и возможности использования их синтеза для получения спец. энергии.

ОБ'ЯСНИТЕЛЬНАЯ ЗАПИСКА
К СПИСКУ ОСНОВНЫХ НАУЧНЫХ ПРОБЛЕМ.

При разработке долголетнего плана научно-исследовательских
работ, связанных с проблемой атомной энергии, необходимо в
первую очередь руководствоваться анализом последнего этапа
развития ядерной физики.

Развитие ядерной физики характеризуется прежде всего:

1) чрезвычайно быстрым проникновением науки в совершенно
новую и в высшей степени своеобразную область физических явле-
ний;

2) рекордно малыми сроками между научным открытием и
реализацией его технических применений;

Figure 9.1 Excerpts from the FIAN "special" letter about "special energy" to GosPlan, 1947.

of the sources of stellar energy. However the methods to practically accomplish such a synthesis in laboratory conditions are unknown at present."

FIAN proposed not ruling out yet another possible source of special energy— "the possibility of using the internal energy of elementary particles themselves, out of which the nucleus is made, in processes analogous to the annihilation of the positron by the electron." Only a very pure and optimistic theorist could write this. The annihilation of matter and antimatter, if it could be stuffed into a bomb shell, would justify the name "maxibomb." A bomb's effectiveness is determined by the part of the mass that is transformed into energy according to the formula $E = mc^2$. For a uranium fission bomb, this is tenths of a percent; for a hydrogen fusion bomb; this is one percent; and for the annihilation bomb, it is practically 100%—hence the maxi. There was only one hitch—a complete lack of knowledge of how to build it. But what is clear is that the maxibomb would have been much closer to the theory of elementary particles about which the FIAN theorists dreamt. So the maxibomb, even if it were not feasible, would have helped to study remarkably interesting things—for example, why the electron is 1,836 times lighter than the proton.

Vavilov's efforts were not in vain. On June 10, 1948, a government resolution committed FIAN

> to organize research on formulating a theory of the combustion of element "120" according to assignments from the Laboratory No. 2 of the Academy of Sciences USSR (Comrades Khariton and Zeldovich), to create within a period of two weeks a special theoretical group directed by Corresponding Member of the Academy of Sciences USSR Tamm and Belenky (Deputy Director of the group) and with the participation of Academician Fock.[7]

The number 120 was the code name for deuterium, and the "combustion of deuterium" is the thermonuclear fusion of light nuclei.

The Academy president was not aware that Klaus Fuchs had helped him reach his goal. On March 13, 1948, Fuchs met with a Soviet agent in England and passed information related to the superbomb over to him. This was not the first such transfer. The first had taken place in the fall of 1945, when Fuchs was still working at Los Alamos.[8] He returned to England in June 1946 and made a detailed report about the American work in the spring of 1948.

On April 20, 1948, the Russian translation of the materials from Fuchs wound up on Beria's desk, and Beria instructed Kurchatov to analyze them. The conclusions presented on May 5 became the basis for the government decisions of June 10. The new espionage, it seemed, spoke about substantive American advancement, but was silent about the fact that this advancement occurred before the summer of 1946, when Fuchs left the United States. In passing on the scientific and technological information, the Secret Service did not reveal its sources—when and from whom the information came.

Fuchs considered the prospects of the superbomb more real than they were in the United States in 1948. He had left before American work had come virtually to a standstill. Atomic monopoly did not induce American government concern about the dubious superbomb project.

But the Soviet leadership, alarmed by the report about American advancement, regarded it seriously and obliged Kurchatov, along with FIAN, to verify the possibility of a hydrogen bomb. Even unverified, that possibility had been assigned the name "RDS-6." At that time, more than a year remained before the atomic bomb RDS-1 was tested experimentally. Two popular explanations of the acronym RDS have come down to us—*Reaktivnyi Dvigatel' Stalina* [Stalin's Jet Engine] and *Rossiya Delayet Sama* [Russia Does It Herself], both incorrect: there was no jet engine in the atomic bomb, and Russia built it with substantial, although covert, help from the United States.

So FIAN was to work on "assignments from comrades Khariton and Zeldovich." Yuly Khariton was the scientific director of KB-11—the Design Bureau for the Development of Nuclear Weapons located far from Moscow. Yakov Zeldovich was the main theorist of the Soviet atomic bomb and had already worked at KB-11.[9] Back in 1945 he received the first set of American "secrets" about the hydrogen superbomb, and in 1946 his group at the Moscow Institute of Chemical Physics began studying the special energy of light nuclei. But Zeldovich's main work remained the A-bomb—as it remained for the entire Soviet atomic project.

And now in June 1948 here was Tamm's theoretical group being created in Vavilov's FIAN to assist Zeldovich. But perhaps this group was not created due to Vavilov's efforts, but only as a result of the intelligence from Fuchs? Or was it simply because there were no other theorists to help Zeldovich?

That wasn't the case. Back in September 1945, Yakov Frenkel, the main theorist at PhysTech, wrote a letter to Kurchatov:

> It would be interesting to use high temperatures—in the billions, which develop during atomic bomb explosions, for conducting synthetic reactions (for example, the formation of helium from hydrogen), which are the source of the energy of stars and which could raise the energy liberated during the explosion of basic matter (uranium, bismuth, lead) even higher.[10]

Although Frenkel did not know that the heavy nuclei of bismuth and lead could not fission as uranium does, he in fact proposed the principle of the thermonuclear bomb in a schematic form—to create "stellar" conditions for fusion of light nuclei by means of an atomic explosion. In any case, it would seem that with his thermonuclear initiative, Frenkel, a distinguished physicist and author of an important 1939 paper on nuclear fission, would have been a viable candidate for the thermonuclear project. But when the government decided to organize an additional theoretical group, he was left out.[11]

And yet it was Frenkel's books on quantum mechanics and the theory of relativity that Sakharov read with such passion as a student. In addition to his books, Frenkel was known for a broad grasp of physics and thinking quickly on his feet on very diverse matters, from nuclear physics to why tram wires spark. Such qualities are more important for nuclear weapons work than Tamm's concentrated attention on the fundamentals of matter. It seems as if Hans Bethe, the American atomic project's chief theorist, also thought this: in a 1946 article about the prospects for the creation of the A-bomb in other countries, he named Kapitsa, Landau, and Frenkel (and not Tamm) as the potential "fathers" of the Soviet A-bomb, and indicated five years as the timeline.[12]

Close friends since the early 1920s, Frenkel and Tamm were contemporaries and of comparable scientific status. Why was Tamm chosen to assist Zeldovich and not Frenkel? The leadership hardly knew that one of Tamm's graduate students, Sakharov, tore himself away with difficulty from invention in cartridge technology for the sake of pure physics and that he was so ideally suited for the new task in which engineering inventiveness had to be combined with pure science.

It appears that the main reason was, in fact, Academy President and FIAN Director Vavilov's persistence. Kurchatov had to respond to the intelligence information received from Beria, and at the same time he could accommodate the Academy president. Moreover, the hydrogen bomb situation did not look promising after two years of the Zeldovich group's efforts, and the job of creating a reliable atomic bomb required major focus. Besides, the role assigned to Tamm's group was exclusively secondary—working on Zeldovich's assignments.[13]

"Extremely Witty": The First and Second Ideas

Sakharov, as already mentioned, twice resisted the temptation to leave FIAN for the atomic project. In 1948, the project itself came to FIAN, and this is how it looked to Sakharov:

> With a mysterious look, Igor Tamm asked me and another of his students, Semyon Belenky to remain after the seminar. He shut the door tightly and made a stunning announcement to us. By a resolution of the Council of Ministers and the Central Committee of the Party a research group is being created in FIAN. He was appointed the group's director, and both of us—its members. The group's work was to verify and refine the calculations made by the Zeldovich group at the Institute of Chemical Physics.
>
> A few days later, after he recovered from the shock, Belenky said in a melancholy way: "And so, our assignment is to kiss Zeldovich's ass."

The independence of FIAN's assignment can be gauged by the independent and not overly ambitious researcher Belenky's response. Soon after, three more of Tamm's students were included in the group—Vitaly Ginzburg and graduate students Yuri Romanov and Efim Fradkin.

One of the reasons that Sakharov was included in the Tamm group is recorded in the text of the above-mentioned government resolution: "to provide housing as a high-priority measure" to the seven participants of the work, including the last one on the list, "Sakharov, A. (a room)."[14]

What did this mean for him? Infinite happiness measuring 14 square meters:

> We didn't have a dinner table (no space for it), and ate on stools or the windowsill. Ten families lived off the long hallway with a small kitchen and a single toilet on the staircase landing for two communal apartments; of course, there was no separate bath. But we were infinitely happy. We finally had our own place—not capricious landlords who could kick us out at any time. And so one of the best and happiest periods of our family life with Klava began.

The summer of 1948 in Sakharov's memory was family happiness in the dacha on the banks of the Moscow-Volga Canal, "the sparkling water, sun, the fresh greenery, the sailboats gliding over the reservoir" and intense work in the theoretical department of FIAN: "The world in which we were immersed was strange and fantastic, in stunning contrast to everyday and family life beyond our work rooms, and to ordinary scientific work." The most apparent contrast was related to secrecy:

> We were assigned a room which no one else had the right to enter. The key was kept with security. We had to make all our notes in special notebooks with numbered pages, which had to be stored in suitcases, sealed with our personal seals, and turned into security at the end of the day. All this formality was probably a bit flattering at first, but then it became routine.

This unusually secretive work was very interesting. Theorists, armed with paper, pencils, and invisible mathematical instruments wound up, you might say, in the depths of the stars and were supposed to predict how matter would behave in temperatures in the tens of millions of degrees, unattainable in the laboratory. And so the only practical verification of their calculations would be a thermonuclear explosion—or the lack of one.

In those stellar conditions the most fundamental laws of nature moved to center stage, while the particulars that complicate the physics of everyday reality receded behind the scenes or off the stage entirely. This both simplified the life of theorists and at the same time presented them with fascinating new problems. "Superb physics," was what Enrico Fermi, a superb physicist on the other side of the Iron Curtain, said about nuclear explosion.[15] Sakharov expressed himself even more powerfully: "paradise for a theorist."

What were they actually doing in this paradise? The FIAN theorists received from Zeldovich a sketch of a (possible?) hydrogen bomb. It was a tube filled with thermonuclear fuel—deuterium. The igniting devise, an atomic bomb, was located at one end of the tube. The supposition was that the atomic explosion would ignite the thermonuclear reaction, and the latter would spread farther down the tube. The longer the tube, the more powerful the explosion.

Zeldovich's group had already been tinkering with this plan for two years without success. They could have gone on trying out different variations: changing the dimensions of the tube and the composition of the material in it, and recalculating the processes again.

At first, Sakharov immersed himself in these calculations, too. But within a few months he felt the inventor within himself and devised an entirely different design. It was visually different—no longer a tube, but a sphere. The spherical layers formed something like a nut in which the nutmeat is an atomic bomb, surrounded by a sophisticated shell of several alternating layers. Sakharov's construction was named "Sloyka," just because "layer" in Russian is *sloy*. At the same time *sloyka* means a puff pastry roll, like a croissant. Unfortunately, this etymology doesn't help much to explain the recipe for FIAN's thermonuclear roll.

Sakharov conceived his recipe by relying on the basic laws of physics. In order to help light nuclei to fuse, they must be brought close together—that is, compressed. It's clear without studying physics in depth that the more particles of matter there are in a vessel, the more pressure is in it. If you are blowing up a rubber balloon, "more particles of matter" means "more air molecules." The structure of the molecule itself isn't important; you don't need to know that an oxygen molecule consists of two identical atoms, each of which contains a nucleus and eight electrons. All these subparticles are "packed and sealed" inside the molecule, and the balloon wall doesn't suspect the very existence of those inside subparticles. The pressure depends only on the number of free traveling molecules.

The thermonuclear Sloyka's action starts with its A-bomb nutmeat going off. A picture of the horrifying mushroom cloud instantly appears before our eyes. But such a cloud forms seconds after the explosion, while Sakharov was thinking about what happened microseconds after the explosion began, after the A-lighter clicked. A microsecond is a millionth of a second, or a millionth of a blink. So Sakharov came up with a device that works within just a few microblinks.

Zeldovich's Tube, the cigar filled with deuterium, was supposed to be lighted by an atomic lighter attached to the tip of the cigar. In Sakharov's spherical Sloyka, the lighter is located in the center. The A-bomb lighter is surrounded by a layer of substance, capable of fusion (like hydrogen). The next layer is made from a substance with heavy nuclei and, thus, with a large number of electrons in each atom. For example, from lead, each atom of which contains eighty-two electrons.

The explosion of the atomic core releases enormous energy in the form of a burst of radiation—neutrons, photons, and other particles. This splash of radiation on its way out within microseconds transforms the lead layer not simply into steam, but into plasma—the state of matter in which all interatomic links are torn apart. The powerful radiation tears the atom's electrons from the nucleus, and instead of one particle—an atom of lead—there are eighty-three: its nucleus and eighty-two electrons. But if the number of particles grows instantly by a factor of eighty-three, then the pressure grows by that much as well.

The enormous pressure in the heavy-nuclei layer compresses the adjoining hydrogen layer, heated to stellar temperatures by the same radiation, and the stellar (thermonuclear) reaction ignites in the hydrogen—the nuclear fusion reaction. This is the explosion of the hydrogen bomb.

The method of compression devised by Sakharov was named "Sakharization" by his colleagues. The Sakharized Sloyka, "sugared croissant" from the Russian word for sugar, *sakhar*, already looked very appetizing from its very first theoretical sketches.[16] And the recipe became even more promising after Vitaly Ginzburg came up just a few weeks later with a novel filling—a new substance for the "hydrogen" layer.

The need for hydrogen in a hydrogen bomb is apparent only at the linguistic level. But at the level of physics this element is not used in a hydrogen bomb at all. Hydrogen is the lightest element but not the most given to fusion. The conditions in which fusion is possible differ quite a bit for different nuclei and is most achievable not with hydrogen itself but its isotopes—deuterium and tritium, D and T.

Deuterium is mixed in natural hydrogen (although in small amounts), and it was already being separated into a pure form back in prewar times. That is why the June 1948 government resolution mentioned the "combustion of deuterium." Tritium virtually doesn't exist in nature, and obtaining it is very difficult or, to be more precise, very expensive. Moreover, tritium is radioactive and decays with time. The properties of deuterium, let alone tritium, had not been sufficiently studied to do precise calculations. But it was known that deuterium and tritium are gases. How, then, do you make a layer to surround the central atomic sphere in the Sloyka out of a gas? It's difficult.

Ginzburg proposed using a much more suitable substance—solid and not radioactive—for the "hydrogen" layer: a chemical combination of deuterium and lithium, lithium deuteride, LiD in chemical symbols. A nonchemical diminutive suffix ["-ochka"] was soon added to these symbols, and the new thermonuclear substance became known by the affectionate women's name *LiDochka* [darling Lydia].

Solid LiDochka is easier to work with. But Ginzburg proposed LiDochka for a different reason, not realizing himself just how good this thermonuclear explosive was. Initially he focused on the fact that lithium, radiated by neutrons from the igniting atomic explosion, adds a certain amount of energy and in doing so warms up the thermonuclear layer more, rendering it more capable of fusing nuclei. That is, he spoke about the reaction

$$Li^6 + n \rightarrow T + He^4 + 4.8 \text{ MeV},$$

or lithium + neutron → tritium + helium + energy, and mainly focused on the energy. Several months later he would realize that the tritium component was much more important.

Ginzburg formulated his proposal in his secret report dated November 20, 1948, and in this report he mentioned Sakharov's Sloyka for the first time. Sakharov himself didn't describe his Sloyka idea and its related calculations until a January 20, 1949, report. In addition to mathematics, experimental physics was needed for the calculations—the interaction of the appropriate nuclei had to be measured. At that time they had a fair idea of how deuterium nuclei interact with one another, that is, what D + D equals. But tritium was too new. The experimenters received instructions from the theorists, but the measurements themselves required time. As Sakharov wrote in his report, "The reactions D + D and D + T have not been experimentally studied and all evaluations about how they interact are guesswork." In expectation of the measurements, Sakharov assumed that D + T interact approximately the way D + D do.

Ginzburg also used this as a basis in his March 3, 1949, report, "The Use of Li^{6}D in the Sloyka." But now, he turned his attention to the fact that LiDochka—when irradiated during ignition by an atomic explosion—creates tritium, which can immediately be used as thermonuclear fuel. That is, there's no need to produce tritium, which is expensive, radioactive, and inconvenient to use. One can supply a much more convenient semifinished product into the bomb, from which the initial atomic explosion itself will prepare everything needed for a thermonuclear explosion.

If only the FIAN theorists had known that D in fact interacts with T a hundred times more readily than it does with D, that this interaction had already been measured by their American colleagues, and that thanks to Klaus Fuchs, the results of these measurements had been in Beria's safe for nearly a year.

The leadership informed them about this espionage only after they were sure that the FIAN theorists had made a theoretical breakthrough; they informed Tamm on April 27 about the data from the American experiments about the interaction of D + T without citing the source. However, these precautions were no longer necessary—this information had been declassified in the United States and published in the main physics journal of that time, *Physical Review*, two weeks earlier—on April 15, 1949. Familiarizing himself with this publication, Ginzburg realized that the thermonuclear explosive that he had proposed was a hundred times better than he had thought.

On May 8, Khariton supported the work on the Sloyka in his review, noting that "the proposal's main idea is extremely witty and physically lucid."[17]

Sakharov could be pleased with himself. Although Aleksandr Kompaneyets, Zeldovich's deputy, had initially doubted his project, Zeldovich "immediately appreciated the seriousness" of the new proposal. It was decided that Tamm's group would work on the Sloyka exclusively, while Zeldovich's group would continue working on the Tube and at the same time assist the FIAN group. So the situation that had initially so depressed Belenky was radically changed.

It is now easier to appreciate Sakharov's physical intuition. Much later, he surmised that the main idea of the Tube project was based on intelligence information. No one knew in 1948 that it would take another two years for the Americans to admit that their prototype "Classic Super" was a dead end. It would take another five years for the Soviet version to close down.

Now, had the detailed information furnished by Fuchs in the spring of 1948 been reported to the FIAN group, it would have probably hampered their realization that this plan was a dead end. The content of Fuchs's reports confirms the independence of Sloyka, since they have nothing in common.

Thus, the FIAN thermonuclear bomb project emerged unplanned and, you might say, outside the atomic project—in a certain sense, inadvertently. But then, discoveries in pure science are always, in a sense, made by accident.

There's no point in discussing the main reason for success—talented people. The human factor is inexplicable. You can, however, talk about the conditions

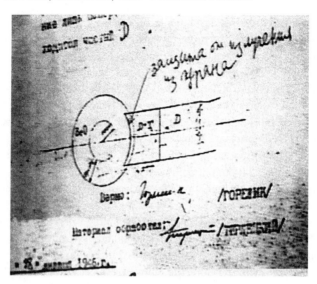

Figure 9.2 A description of the American Classic Super based on intelligence information prepared in Sudoplatov's department in January, 1946, with Yakov Zeldovich's handwritten notes.

in which talented people operate. Tamm's group continued living the usual scientific life: seminars, scientific news. And they dealt with the applied bomb problems in the same free spirit as the purely theoretical ones.

This manifested itself even at the language level—compare the playful FIAN terms Sloyka and LiDochka with RDS, the official name of the nuclear weapons projects (with both its popular explanations—"Stalin's Jet Engine" and "Russia Does It Herself"), or with the Tube, which dryly emphasizes geometry. American physicists had enough of a sense of humor to introduce the terms "quark," "strangeness," and "charm" into pure physics, but in bomb physics, created behind the high walls of the Manhattan Project, they used the more solemn terms "Classical Super" and "Alarm Clock" (to wake up the world).

Tamm, the group leader, in many ways determined the freewheeling and friendly character of FIAN's thermonuclear research. His incurable enthusiasm helped overcome the feeling of hopelessness that characterized the thermonuclear problem in 1948.[18]

All that remains to be said is that the words Sloyka and LiDochka themselves (not to mention their meaning in terms of physics) remained secret while Sakharov lived—they were declassified only in 1990, after his death, for the commemorative issue of the journal Priroda [Nature].[19] For this reason, Sakharov in his Memoirs limited himself to only the following:

Two months [after Tamm's group was formed in June of 1948] I made a serious change in my work: namely, I proposed an alternative design for a thermonuclear charge which was entirely different—in its physical processes during explosion and even in its basic source of energy release—from the one being worked on by Zeldovich's group. I call this proposal the "First Idea" below.

Soon after, Vitaly Ginzburg proposed a "Second Idea," which supplemented my proposal substantially.

How did Sakharov feel while working on thermonuclear invention? He reminisced about this three decades later: "The thermonuclear reaction—this mysterious source of the energy of the stars and the sun, the source of life on Earth and the possible cause of its destruction—was already within my grasp, taking place right on my desk!"

Was he really thinking about the destructiveness of thermonuclear energy for his native planet back in 1948? It was only after thermonuclear explosions were conducted at U.S. and USSR test areas in the late 1950s that this thought became commonplace. In the very first years of the atomic age only the most far-seeing theoretical physicists realized that this was not simply a new powerful bomb but a weapon to end the world.

This is how Einstein stated the problem in 1948: "Is it really unavoidable that, because of our passions and our inherited customs, we should be condemned to annihilate each other so thoroughly that nothing would be left over which would deserve to be conserved?"[20] This question was expressed in an article by Einstein, in response to an open letter of four Soviet scientists. And although Einstein's article was not published in the Soviet press, the ghost of mankind's destruction was already roaming Earth. Einstein also answered his own question: "The goal of avoiding universal mutual annihilation must have priority over all other goals."

At that time, however, neither side of the world confrontation was ready for this answer. Sakharov talked about his attitude "toward the moral, human side of that work," in which he participated, and his "total absorption" in this work in the first years:

Most important for me and also, I think, for Igor Tamm and the others in the group was the inner certainty that this work was indispensable. I couldn't ignore how horrible and inhuman our work was. But the war that had just ended was also inhuman. I wasn't a soldier in that war, but I felt like one in this scientific and technological war. With the passage of time, we learned about or we invented ourselves such concepts as strategic parity, mutual thermonuclear deterrence, etc. I still think even now that these global ideas truly intellectually justify (perhaps not in a fully satisfying way) the creation of thermonuclear weapons and our personal part in it. At that time we experienced all this more on an emotional level. The monstrous destructive power, the enormous effort required for development, the resources taken away from an impoverished, hungry country ravaged by war, the human casualties at harmful manufacturing plants and in forced hard labor camps—all this intensified our sense of tragedy, forcing us to think and work so that all these sacrifices (implicitly considered inevitable) would not be futile. This truly was a war psychology.

The psychology of war was created by Soviet propaganda, and it had plenty of material to work with. In 1949 the philosopher and mathematician Bertrand Russell wrote:

> Unless the Soviet Government changes its mind, which does not seem at all likely, I am afraid we must conclude that no approach to unification will be possible until after the next world war . . . If—what I devoutly hope is not the case—only war can prevent the universal victory of communism, then I, for my part, would accept war in spite of all the destruction that it must involve.[21]

Six years remained before the Einstein-Russell manifesto of 1955, which initiated the Pugwash Movement of scientists for peace and nuclear disarmament. This period spans the genesis of the hydrogen bomb and the death of Stalin.

Sergei Vavilov's Burden

By the will of history, the creation of the FIAN hydrogen bomb took place against the background of Lysenko's destruction of biology and an analogous threat looming over Soviet physics. While Sakharov was writing a report about the Sloyka, Ginzburg, his closest associate in the thermonuclear work and a branded "kowtower," was preparing a presentation for the All-Union Meeting of Physicists, that was intended to "Lysenko-ize" physics.

On January 15, 1949, Ginzburg sent the seventeen-page text of his speech with a note to Vavilov: "It seems appropriate that I send you the text of my proposed presentation for the debate on your speech. It was suggested that I make this presentation, and its proposed draft was discussed in our group at FIAN."[22]

Three-quarters of Ginzburg's presentation is about "the philosophy of contemporary physics," and the last quarter is "about the struggle for honor, dignity and the primacy of Soviet science." Ginzburg admitted his own guilt for "writing papers and not giving any thought to whether I had emphasized or indicated the primacy of Soviet work." And his presentation ends with a quote from Stalin: "in the near future not just catch up to, but to surpass the scientific achievements attained beyond the borders of our country." Ginzburg wrote this while "catching up and surpassing" the West in thermonuclear work, in particular with respect to the use of LiDochka.

Things were more difficult for Vavilov, the main speaker. He, the president of the Academy of Sciences, had to rewrite his speech twice, picking out ideological formulas that would satisfy the overseers from the Central Committee. Fortunately, Vavilov the president could thank himself, Vavilov the director of FIAN, for getting him out of this situation: the bomb achievements of his theorists helped him avoid disgrace, the meeting was canceled, and all three drafts of his speech were sent to the archives.

However, this was the only disgrace Vavilov avoided. The meeting's chief organizer—a zealous scientific bureaucrat, Deputy Minister of Higher Education Aleksandr Topchiev—was immediately made the Chief Scientific Secretary of the Academy and, three months later, was "elected" an academician, skipping the rank of Corresponding Member. On May 24, 1949, Vavilov chaired a meeting of the Scientific Council at FIAN, "On Errors of Cosmopolitanism, Committed by the FIAN Researchers." Ginzburg was one of the four "exposed" cosmopolitans. Vavilov respectfully named the sinners by their first names and patronymics, the sinners pronounced the appropriate ritual words, and for them it ended there. But not for Vavilov.

In 1949–1950 the journal *Doklady Akademii Nauk* (Proceedings of the USSR Academy of Sciences) published four articles by A. P. Znoyko.[23] In the journal, which is intended for the urgent publication of the most significant new, succinctly described scientific results, we read:

> Eighty-one years ago, the great Russian chemist D. Mendeleyev formulated the basic natural law of nature . . . It is known that Mendeleyev's brilliant predictions came true . . . It is known that one of the authors of this article (A. P. Znoyko) succeeded in finding the relationship between the changing qualities of nuclei, their specific charge and their structure . . . The depth of the Mendeleyev method is clearly evident, the method which stimulated the development of atomic chemistry and nuclear physics, the method which makes it possible for present-day science to penetrate deeper into nature's secrets. Element 96 is followed by element 97 [what a surprise!], which two years before its detection [by American physicists] was predicted on the basis of the periodic system's discovery of atomic nuclei using Mendeleyev's method. We propose naming this element "mendeleviy" and establishing the symbol Md.

Articles in the *Doklady* were published only when presented by an Academy member. Now, who had recommended Znoyko's pseudoscientific prattle? Chief editor and president of the Academy of Sciences Sergei Vavilov himself.

We do not know what forces the now-forgotten Znoyko mobilized for this. But these forces were undoubtedly very great, if in September 1949, a 42-year-old engineer, a specialist in corrosion—that is, rusting—became the head of a secret nuclear laboratory at Moscow University created by a special Government resolution.[24] He also headed the uniquely named "empirical section" of the laboratory.

How could Sergei Vavilov carry such a heavy burden of shame of being the president of the Stalinist Academy of Sciences?

In November 1948 he received an open letter from Sir Henry Dale, president
of the London Royal Society. The British biologist and Nobel laureate asked to
be removed from the list of foreign members of the Soviet Academy. Lysenko's
pogrom of Soviet biology a few months earlier put a dark end to the mysterious story of the disappearance of a foreign member of the British Academy of
Sciences—Nikolai Vavilov. This is what his brother Sergei read in Dale's letter
from London:

> Since the time of the threats that forced Galileo into his historical renun-
> ciation, many attempts have been made to repress or distort scientific truth
> in the interests of this or that belief devoid of science, but not one of these
> efforts has had any lasting success. The last person to endure this failure
> was Hitler. Considering, Mr. President, that you and your colleagues
> are now working under analogous coercion, I can only express to you my
> sympathies.[25]

Galileo, in renouncing scientific truth, was saving his own life. For Sergei
Vavilov, it was a matter of saving the life of his country's science. He became
president of the Academy, assuming the new burden of responsibility in the
summer of 1945, several weeks after the defeat of Nazi Germany and before the
start of the Cold War. There was no warning that to this burden would so soon
be added another—that of shame. But the responsibility remained for his institute and for the Academy of Sciences.

The brothers Nikolai and Sergei Vavilov were similar in their sense of
responsibility but differed primarily in their moral endurance. The ability to
choose between the two life principles—"It's better to die standing, than live
on your knees" and "Better to defend your life's work on your knees than to
die standing"—depends on one's innate resources.

Sergei Vavilov had endurance. But the burden of shame he was forced to
bear grew larger with time. His term was the shortest of the presidents of the
Academy of Sciences—five years; he was the most broadly educated yet delivered the most shameful speeches. He did not act out of personal considerations.
Those who knew him attest to his exhausting efforts to defend people he took
care of. He transcended his "ego" not only to satisfy the powers-that-be. For
example, Ginzburg recalls how Vavilov treated a FIAN associate who showed
great promise but was

> rather ill-bred, . . . was irritating because of his nervousness (it was per-
> ceived as impertinence) and, finally, . . . sometimes talked obvious nonsense.
> It is known that intelligence and abilities are not the same thing. And so, I
> remember the expression on Sergei Vavilov's face on several occasions: he
> undoubtedly saw everything, was dissatisfied, but did not react either in
> words or deeds and, most importantly, he helped this man, defending him
> when necessary.[26]

The fact that "this individual" was Ginzburg himself adds particular significance to this testimony.[27] Vavilov helped people who were devoted to science. Science for him was an unconditional value, the main instrument of human progress.

Nonetheless, even Vavilov's endurance had limits. Those around him got the impression that he, by nature physically strong, by refusing medical help for his heart problem in the last months of his life, "consciously went to meet his end."[28] In any case, such a departure did not look like a political step, which would have subjected the cause he served to more danger. But serving no longer made any sense.

The inclusion of FIAN into the atomic project no longer looked like a success. After all, Vavilov had aspired to strengthen the broad frontier of academic research. But "the highest government interests" required that FIAN's main theorist Tamm and his very promising student Sakharov leave FIAN and leave fundamental science in 1950, departing "in an unknown direction," to devote their creative energy to military work.

We have Anna Kapitsa's testimony that Vavilov was worn out near the age of 60. As she put it, during the years when Pyotr Kapitsa was out of favor, "Vavilov surreptitiously did much good for him," although Kapitsa never hid that he did not rate Vavilov very highly as a physicist. The Kapitsas were all the more surprised when they received an invitation to dinner at the Vavilov home in early 1951:

> We never visited the Vavilovs at their home, but we left Nikolina Gora, spent the evening with Sergei Vavilov and his wife, and were completely astonished—we couldn't understand why they had invited us and why he was so outrageously candid about things people didn't talk to one another about then. We understood perfectly well, and so did he, that his house was bugged. And despite this, Vavilov was maximally frank. Pyotr and I never completely understood it all. At that time we just sensed that things were very difficult and serious for Vavilov. And we were not surprised that he died soon after.[29]

Sergei Vavilov died from a heart attack on January 25, 1951, on the day marking the eighth anniversary of his brother Nikolai's death in the Saratov prison.

PHOTOGRAPHS

1
Pyotr Lebedev (1866–1912).

2
The Physical Institute, founded for Pyotr Lebedev in 1912
and built in 1916. The Physical Institute of the Academy
of Sciences (FIAN) moved into this building in 1934.
Andrei Sakharov's path in science began here in 1945.

3
Ivan Sakharov and Maria Domukhovskaya, Andrei Sakharov's grandfather and grandmother, in 1882.

4
Aleksei Sofiano, Andrei
Sakharov's maternal
grandfather, in 1905.

5
Ekaterina Sofiano, future mother
of Andrei Sakharov.

6
Dmitri Sakharov, future father
of Andrei Sakharov.

7
Andrei Sakharov with his younger brother Georgi (Yura),
1930.

8
Leonid Mandelshtam (1879–1944).

"Igor Tamm's study had the same furniture, which I then saw for decades; a desk, strewn with dozens of numbered pages covered with calculations unintelligible to me, dominated everything, and above the desk—a large photograph of Leonid Mandelshtam, who had died in 1944, and whom Igor Tamm considered his teacher in science and life."—SAKHAROV

9
A session of the student physics club, Moscow University, 1940. These young people, soaring in the clouds of theoretical physics, had been living in Stalin's Great Terror only two years before.

10
Andrei Sakharov, 1943.

11
Andrei Sakharov's father
and his students at the
Moscow Pedagogical
Institute, 1949.

12 *facing page top*
Corresponding Member of the Academy of
Sciences of the USSR Igor Tamm in the Caucasus
Mountains, 1947. A year later he would head the
secret group at FIAN that would invent the first
thermonuclear bomb. Ten years after that he would
receive the Nobel Prize (for work done in 1937).

13 *facing page bottom*
Igor Tamm at a seminar.

*"His true passion, which tormented him all his life
and gave his life a higher meaning, was fundamental
physics. He said a few years before his death, already
gravely ill, that his dream was to live long enough to see
the New (capitalized) theory of elementary particles
that answered 'the damned questions' and to be in a
state to understand it."*—SAKHAROV

14
Andrei Sakharov with his daughter, summer 1948.

'That summer is memorable for the sparkling water, sun, the fresh greenery, the sailboats gliding over the reservoir. Although it was summer, we all worked very intensely. The world in which we were immersed was strange and fantastic, in vivid contrast to everyday city and family life beyond our work rooms, and to ordinary scientific work.'—SAKHAROV

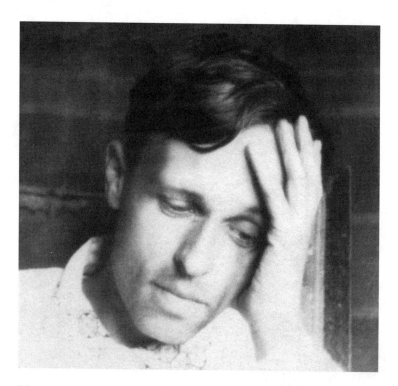

15
Andrei Sakharov in 1948.

16
Vitaly Ginzburg in 1947, when
the newspaper article "Against
Kowtowing!" berated him
for his publications, "which
discredit our Soviet science."
A year later he came up with
the "second idea" for the
Soviet H-bomb, and in 1950
he did his work on the theory
of superconductivity, which
brought him the Nobel Prize in
2003.

17
The monastery town of Sarov, nearly 500 kilometers from Moscow, which was transformed into the Installation—a secret city where nuclear weapons designers lived and worked. Andrei Sakharov spent 18 years here, from 1950 to 1968. The town was removed from all maps and surrounded by rows of barbed wire. The last of its code names was Arzamas-16.

18
Andrei Sakharov
in the early 1950s.

19
The Installation's theorists worked in this building.

20

From left to right: Yakov Zeldovich, Andrei Sakharov,
and David Frank-Kamenetsky at the Installation.

21

Evgeny Lifshits, Lev Landau, and Igor Tamm meet in the mountains.
Lifshits and Landau arrived by car, Tamm on foot.

22

The newlyweds Klavdia Vikhireva and Andrei Sakharov, 1943.

"We were married on 10 July. Klava's father blessed us with an icon, made the sign of the cross over us, and said a few words of advice. Then, holding hands, we ran across the meadow to the side where the registry office was situated. We lived together 26 years until Klava died on 8 March 1969. We had three children—our elder daughter Tanya (born 7 February 1945), our daughter Luba (28 July 1949), and our son Dmitri (14 August 1957). The children brought us great joy (but of course, like all children, not only joy). We had periods of happiness in our life, sometimes years at a time, and I am very grateful to Klava for them."—SAKHAROV

23
A typical picture of Stalin's Russia, where brainwashing began in kindergarten
(the author is one of the children). The year is 1953, when Andrei Sakharov,
preoccupied with the first Soviet H-bomb, mourned Stalin's death. He soon
grew ashamed of his reaction, and that shame was a step toward his staunch
opposition to the Soviet system created by Stalin.

24
Andrei and Klavdia's
home at the Installation.

25
Andrei Sakharov and
Klavdia Vikhireva
near their home at the
Installation.

26
A photograph of the Sakharov family in the mid-1960s (the only photograph of the entire family). From left to right: Luba, Tanya, Dima, father, and mother. Andrei Sakharov used a timer, which is why he is in the background and the proportions are distorted (the camera was too close).

27 *facing page top*
Two-time Hero of Socialist Labor Academician Andrei Sakharov and three-time Hero of Socialist Labor Academician Igor Kurchatov.

"My meeting with Kurchatov was in September 1958 in his house on the grounds of the institute. Part of our conversation took place on the bench near his house under spreading leafy trees. Kurchatov listening to me closely and basically agreed with my points. He said, 'Khrushchev is in the Crimea now, vacationing by the sea. I'll fly out to see him and I'll present your views to him.' Our conversation lasted about an hour. At the end, Kurchatov's secretary came over with a camera and took pictures of the two of us from various angles; in some pictures you can see the Kurchatov dog, which was constantly underfoot."—SAKHAROV

28 *facing page bottom*
"I saw this photograph for the first time soon after we met [in 1970]—a supremely self-confident young man. And that's what I told him: 'I don't like you in this photograph.' He said nothing, but at our next meeting, he returned to it. 'You didn't like me the way I am in the photograph with Kurchatov. That's all in the past. All those valences have long been filled.' And I said, 'Are you sucking up, is that it?' And he replied in the same manner, 'A bit.'"—ELENA BONNER, JANUARY 31, 1997

29
A traditional evening game of chess between
Andrei Sakharov and his first wife, Klava.

30
Andrei Sakharov.

31
Aleksandr Solzhenitsyn.

"Aleksandr Solzhenitsyn and Andrei Sakharov had differences of opinion in philosophy, history, and politics. But they respected each other profoundly. And naturally, Solzhenitsyn publicly defended Sakharov several times. And naturally, Sakharov immediately supported his dissident contemporary as soon as the rumor of Solzhenitsyn's arrest flew through the town on the evening of 12 February 1974."
—LYDIA CHUKOVSKAYA

32
Tea party in the Sakharov
kitchen: Andrei Sakharov,
Ruth Bonner, and Lydia
Chukovskaya, 1976.

33
Lydia Chukovskaya in
her room, which held the
photographs of Andrei
Sakharov and Aleksandr
Solzhenitsyn.

34
Andrei Sakharov and Elena Bonner before their first press
conference, August 21, 1973.

35

*"I am no volunteer priest of the idea, but simply a man with an unusual fate. I am
against all kinds of self-immolation (for myself and for others, including the people
closest to me)."* —FROM ANDREI SAKHAROV'S DIARY, MAY 27, 1978.

36
Andrei Sakharov learned he had
been awarded the Nobel Peace
Prize on October 9, 1975.

37
Elena Bonner accepting the Nobel Peace Prize on behalf of Sakharov,
December 10, 1975.

*"Speaking at one of the dinners [during the Nobel ceremony in Sweden], Lusya repeated
her line about Russian women [using the word 'baba,' which literally means peasant
woman, but is used colloquially for women] on whose backs Russia plows and threshes
and grinds flour, and added that now they also accept prizes. Maria Olsufyeva, who was
interpreting for Lusya, had difficulty translating the phrase. And really, how do you say
'baba' in English?"*—SAKHAROV

38
Andrei Sakharov in 1979, a few months before his exile from Moscow to Gorky.

21 мая 1981 года
АНДРЕЮ ДМИТРИЕВИЧУ САХАРОВУ
исполняется 60 лет.
Сейчас он живет в г. Горьком
проспект Гагарина д.214 кв.3

39
Photo distributed in Moscow on Andrei
Sakharov's sixtieth birthday when he was in
exile in Gorky. (The caption below the photo:
"On May 21, 1981, Andrei Sakharov will be
60. He is now living in Gorky, 214 Gagarin
Prospect, Apt. 3.")

40
Andrei Sakharov and Elena Bonner, Gorky, September 1984,
three months after a hunger strike.

41
Sakharov at the train station upon his return to Moscow from exile,
December 23, 1986.

42
Rally at the Presidium of the Academy of Sciences, February 2, 1989.
The nearest sign reads, "WHO IF NOT SAKHAROV?"

43
Sakharov speaking at the Congress of People's Deputies, May 1989.

Конституция Союза Советских
Республик Европы и Азии

1. Союз Советских Республик Европы и
Азии (сокращенно: Европейско-Азиатский Союз Совет-
ский Союз) — добровольное объединение суверенных
республик Европы и Азии

~~2. [Важнейшая цель Деятельности и стремления
цель народа Союза Советских Республик
Европы и Азии и ее органов власти
направлены на создание условий счастья]~~

2 Цель народа Союза Советских Республик
Европы и Азии и их органов власти —
— ~~счастье~~ счастливой, ~~достойной~~ полна

44
Andrei Sakharov's manuscript, draft Constitution
for the Union of Soviet Republics of Europe and Asia.

отказываются от многомерной
модели. Ну ладно, подождём. Будущее
покажет, кто прав, покажет - всем нам
и многое другое. К счастью, будущее
непредсказуемо (а также — в силу
квантовых эффектов) — и неопределённо.
10/V-82; С наилучшими пожеланиями, А.С.

45

"Fortunately, the future is unpredictable and also—because of quantum effects—uncertain."

This was the consolation Sakharov offered to physicist and human rights activist Boris Altshuler in a letter written during his exile in Gorky (May 10, 1982).

Для меня бог — не управ-
ляющий миром, не творец мира или
его законов, а гарант смысла бытия
— смысла вопреки видимому бессмыслию

46

"For me God is not the ruler of the world, not the creator of the world or its laws, but the guarantor of the meaning of existence—meaning despite all the apparent pointlessness."
—FROM ANDREI SAKHAROV'S DIARY, APRIL 27, 1978.

47
Sakharov in his last year—1989
(photo by Yousuf Karsh, © Yousuf Karsh/Retna Ltd.).

III

In the Nuclear Archipelago

10

THE INSTALLATION

Just a few weeks after the report on Sloyka (January 20, 1949), Sakharov found out that he was to leave not only pure science, but FIAN and Moscow as well, because it was "essential for the successful development of the project." Boris Vannikov, Beria's deputy and head of the First Main Directorate, as the Soviet atomic empire was called, informed Sakharov and Tamm about it.

Tamm began heatedly explaining that limiting the work of the young, talented theorist Sakharov to mere applied research was not right, not in the state's interests. To which Vannikov said only that Comrade Beria "begs you to accept our offer." How could one refuse?

Sakharov was required to leave for the Design Bureau No. 11 (KB-11), which was established away from Moscow. Its location remained a state secret until the end of Sakharov's life—he never revealed it to anyone or uttered the word Arzamas-16, the most famous of its pseudonyms, or the name "Sarov," the historical name returned to this small town near Nizhny Novgorod in 1996.

The wits of Arzamas-16 came up with the nickname "Los Arzamas" for their town to match the American location Los Alamos, and in 1994 these towns became sister cities. In his *Memoirs* Sakharov calls this place the Installation, which is what its residents called it for nearly half a century. It was one of the first islands in the nuclear archipelago, which dotted the nation's map with islands of blank spaces.[1] The new archipelago was joined with the Gulag Archipelago and was built mainly by it. The unlimited slave labor of the Gulag, which easily implemented secrecy, and the single authority administering both archipelagoes made this union inevitable in the nation of Stalin's socialism.

A place held sacred by the Orthodox Church was set aside for the creation of the atomic bomb. The Sarov Monastery, in the central region of Russia for over two hundred years, was the home of St. Serafim of Sarov (1778–1833), who was canonized in 1903. The Soviet regime closed the monastery and first housed a "children's labor commune" for orphans in its buildings, then a corrective labor camp. The prisoners were corrected at a small plant they themselves had built. The plant first produced artillery shells, but switched to the famous "Katyusha" rocket launchers during the war.

Sarov disappeared from the nation's maps in 1946. Along with a fair-sized tract of adjoining land, it was encircled by barbed wire, and a branch of the Kurchatov Laboratory No. 2 named KB-11 was established there. Khariton became its chief designer, and the first project was the "Special Jet Engine" [*Reaktivnyi Dvigatel' Spetsial'nyi*] or RDS-1, the code name for the atomic bomb, perhaps in honor of the Katyusha rocket launcher [*reaktivnyi minomyot*].

Zeldovich, the main theorist of the atomic bomb, worked at the Installation from 1948 on, while continuing to lead his group in Moscow. In February 1949, a theoretical group was organized at the Installation to work on the hydrogen bomb.

"Trial Communism"

Sakharov wasn't aware of any of this in the summer of 1949, when he was urgently summoned on his day off. Sakharov left for the Installation that same day in the personal railroad car of Vannikov, the head of the First Main Directorate, with only one other passenger. The 28-year-old Sakharov got a glimpse of his new social status and his first impression of the strange world to which his "extremely witty" Sloyka idea had brought him.

He got the required pass to the assigned railroad car at the "Vegetable Warehouse," a basement building used by the Office of Passes as camouflage. They rode on the train all night, then drove in a car past semiruined huts, and finally through "two rows of barbed wire on tall posts, with a strip of tilled soil between them ('our own barbed-wire home,' as we later called it when returning to the Installation zone from our travels)."

Vannikov brought Sakharov to a meeting on nuclear weapons development progress. Kurchatov was already there. The main issue was the upcoming test of the atomic "device" (in the language of the Installation) in a few months. But a plan about hydrogen devices was also approved and the decision made to conduct work on both Sloyka and the Tube.[2]

Sakharov learned "important and surprising things about atomic charges (one wasn't supposed to even talk about such things beyond the confines of the Installation)." The "superb physics" of the A-bomb was inventively embodied in engineering. This found strong resonance in the soul of the inventor-physicist who had a long road ahead of him from the physics of the Sloyka to engineering the H-bomb. Later on, Sakharov "took pleasure in personally repeating how these atomic and thermonuclear charges worked to newly arrived researchers, and watching their amazed looks."

But then, during his first weeklong trip, it was Zeldovich who familiarized Sakharov with the Installation's work. Zeldovich also acquainted him with the theoretical department researchers, among whom Sakharov recognized his classmate Yevgeny Zababakhin. They had parted after the Air Force Academy medical commission when the war began. The medical commission had rejected Sakharov, while Zababakhin graduated from the Academy and later wrote a dissertation that attracted Zeldovich's attention when he received it for review. Different roads led to the Installation.

Sakharov got his initial information about the history, geography, and "physiology" of the island-town from the Installation theorists. He then spent 18 years of his life in this town. Here's how he tells it:

> The peasants of the nearby impoverished villages saw a solid barrier of barbed wire which encircled an enormous area. They found a highly original explanation for what was going on—"Trial Communism" was being built there. This trial—the Installation—was a kind of symbiosis between an ultra-modern scientific research institute with production facilities for the experiments and test areas—and a large camp. In 1949 I still heard stories about the time when it had been just an ordinary prison camp with a mixed-prisoner population (including long-term convicts), probably not much different from the "typical" camp described in Solzhenitsyn's *One Day in the Life of Ivan Denisovich*. The plants, testing grounds, roads, and housing for future researchers were built by the prisoners.

Longtime residents of the Installation talked about an attempt by a group of prisoners to organize an escape. The rebels were mercilessly killed; after this, inmates serving long sentences and therefore having nothing to lose were removed from the Installation. The authorities had a different sort of problem with prisoners who were nearing the end of their sentences: they knew too great a secret—the Installation's location. The problem was solved simply: freed prisoners were exiled to "eternal settlement" (the legal term at that time) in places so remote that there was no one to whom they could tell secrets.

For three years, until Stalin's death, this was the picture before Sakharov's eyes: "Every morning long gray columns of people in quilted jackets, guarded by German shepherds, passed by our curtained windows." The Installation residents helped some of the prisoners with food and clothing. Although these prisoners weren't "lifers," some desperate ones would try to escape anyway and hide somewhere within the Installation grounds for a time. There would be a state of high alert during searches for them, and the residents felt greatly distressed behind their drawn curtains.[3]

What did the "free" residents of the Installation think about the forced labor of the convicts? In a nation where coercion was so universal, the idea of corrective labor did not raise objections. The free residents couldn't do many things, either. A regime of secrecy was added to the usual socialist unfreedoms: "No one could go on vacation, visit a relative (even one who was very ill or dying), go to a funeral or travel on official business without permission from the security department. 'Townspeople' were granted such permission only in exceptional cases, virtually never. Young specialists could not leave during their first year of work."

Even Sakharov—for all his strategic importance—was not permitted to travel to Moscow for seven months after moving to the Installation in March 1950, with no communication by telephone or mail. He could only find out about his family from those who were allowed to travel. He celebrated his first birthday at the Installation without his loved ones.

Later, when his family got settled in their new place, the entire family would visit Moscow for a month or two—as his work required it. The children were registered at a Moscow school, but they learned not to tell their classmates or teachers about where they traveled. An answer was fabricated: "We go away to our dacha, where our Dad works." The name "Sarov," the code names Arzamas-70, Arzamas-16, and even "the Installation" were forbidden. The parents explained to the children that they even had to avoid answering the question about how long it took to get there.[4]

Sakharov verified from his own experience that the Installation was very vigilantly guarded when, during his first months there, he headed off with several associates for a walk in the woods surrounding the town. While conversing, they didn't notice that they had walked close to the guarded boundary. They were arrested and forced to sit with their legs stretched out in the back of a truck. They were told: "We will shoot without warning if you attempt to escape or pull up your legs." When they got out of the truck, they had to face a wall while their identities were checked.

The outcome of another incident, which took place on the other side of the same boundary in the 1960s, was worse. The KGB chief at FIAN was once informed that a researcher at his (Moscow) institute was arrested during an attempt to cross the barrier into the Installation, so far away from Moscow. It became clear from the investigation that this researcher, the son of a priest and himself a believer, had decided to go pay his respects to the relics of Serafim of Sarov, in what had been a traditional place of pilgrimage in pre-revolutionary times. He had found a book that described the location of this holy place and he set out for it. A check showed that he really had requested the book in question from the main State Library shortly before his pilgrimage—his name was listed in the requesting record book. He was not punished, but the psychic trauma he suffered as a result was so severe that he never recovered.[5]

But the scientific inhabitants of the zone of "trial communism" were not depressed by the prison state of their lives:

I think that the Installation's atmosphere, even the proximity of the camp and the security's "excesses" helped us concentrate on our work. We saw ourselves at the center of a vast endeavor for which enormous resources were allocated, and we saw this being achieved at very great cost to the nation. This made many people, it seems to me, feel that these sacrifices and difficulties should not be in vain. At the same time we could not doubt the importance, the absolute vital necessity of our work. And there was nothing to distract us: it was all beyond our world, outside our double row of barbed-wire fence, somewhere very far away. Undoubtedly, the very high salary level, the government awards, and the other privileges of high status were also a considerable support.

But in addition to this, they had interesting work for their creativity.

Making Sloyka

Although the June 1949 meeting at the Installation approved the work plan of the H-bomb, Tamm's group continued working at FIAN. President Harry Truman's directive, publicly announced on January 31, 1950, "to continue its work on all forms of atomic weapons, including the so-called hydrogen or Super Bomb," changed the situation. This directive was in response to the first test in August 1949 of the Soviet A-bomb that brought the American atomic monopoly to a shocking end. The U.S. Central Intelligence Agency had assured the president just a few weeks before the test that 1953 was the most probable date for the first Soviet test of an A-bomb.[6]

President Truman didn't know that the USSR had tested a copy of the American A-bomb, nor did he know that test of a more effective domestic Soviet A-bomb design was postponed to play it safe—Beria was more fearful of a first test failure than Kurchatov.

Truman also didn't know that, just a few weeks after his directive, American physicists would conclude that the "Classical Super" design (the "Tube" in Russia) wouldn't work, and that it would take another year to create a fundamentally new H-bomb design.

However, this wasn't known in the Soviet Union, either, and that is why Truman's words were taken deadly seriously. A few days after the U.S. president's announcement, the Special Committee reviewed the H-bomb issue, and on February 26, 1950, a government resolution required "the organization of computational, theoretical, experimental and construction work on the creation of the devices RDS-6s and RDS-6t." Sloyka and Tube are indicated by the letters "s" and "t." A model of the Sloyka had to be made by May 1, 1952, and tested on a site in June. The first real Sloyka had to be done in 1954, but before that the production of lithium deuteride—LiDochka—had to be set up.[7]

Khariton was appointed scientific director of the research and development of both projects, with Tamm and Zeldovich as his deputies. A theoretical group for work on Sloyka under Tamm's directorship was created at the Installation. Sakharov and Romanov arrived there in March, and Tamm arrived in April. Three other members of Tamm's group—Ginzburg, Belenky, and Fradkin—remained at FIAN for various reasons.

In 1946, Ginzburg chose a wife who was unsuitable for the Installation. Nina Yermakova, who had been arrested in 1944 and convicted under the "counterrevolutionary" article 58, was pardoned in 1945 but with restricted rights of residence. Ginzburg sent requests to permit his wife to live in Moscow to the authorities. Although he was the inventor of LiDochka, the security agencies forbade him to travel to the Installation with such a wife.[8]

Belenky remained at FIAN due to poor health (he died at the age of 40 in 1956). Fradkin's situation was the most difficult. He asked Tamm not to take him to the Installation and not to question his reasons. Tamm carried out both these requests. At FIAN they knew that Fradkin's entire family had been killed by the fascists, which is what he wrote in his personnel questionnaires. But he didn't write that his father had disappeared in the Gulag a few years before the war began. And he was afraid that if he submitted the paperwork for a security clearance higher than the level he had at FIAN, they would discover that he was hiding a "criminal" fact.[9]

Fradkin's apprehensions were entirely justified. The security services, for instance, studied Sakharov's wife's biography for half a year before granting her permission to visit her husband at the Installation with her children. "Later Klava's father said that in the summer of 1950, the Ulyanovsk KGB had intensively studied his connections with his relatives." Half a year also elapsed between Sakharov's meeting with Vannikov and his first trip to the Installation, so perhaps it was the standard time period for investigating family background.

The members of Tamm's group who remained at FIAN were called the "support group" and worked on problems that came up at the Installation. The theory of stability, developed by Belenky and Fradkin, proved to be particularly significant, supplying a method of searching for the best versions of nuclear "devices."[10]

At the same time, the FIAN people still had the opportunity to pursue pure theoretical physics. Ginzburg did his best-known work on the theory of superconductivity (which brought him the Nobel Prize of 2003).[11] Fradkin wrote papers about which Sakharov said: "Of the entire group Fradkin was the only one who became the type of a highly professional 'cutting-edge' theoretical physicist that we all dreamt about."

Sakharov had no opportunity to pursue pure science. The burden of tasks related to Sloyka was too great. The properties of the thermonuclear device, as they appeared at FIAN, were only a "very preliminary, very incorrect understanding of many things." It's not easy to discern the properties of a mature, hitherto-unseen plant in a seed. However, nothing whatsoever can be grown without the seed.

What the team of theorists headed by Tamm and Sakharov had to do was to find the conditions that would turn the FIAN seed Sloyka into a thermonuclear mushroom of monstrous power. This meant an enormous volume of computational work. The assignments were given to the experimenters, then their results were used in new calculations. And then came more experimental measurements, and new calculations. The physics in which Tamm and Sakharov's team was engaged was not theoretical physics—it was technological physics and computational mathematics.

Work on the Tube project continued in parallel under Zeldovich and David Frank-Kamenetsky. There were about twenty theorists in the two groups.[12] The prominent theorists Isaac Pomeranchuk and Nikolai Bogolyubov arrived at the Installation to help them. Several groups in Moscow and Leningrad, two of which were headed by the future Nobel laureates Lev Landau and Leonid Kantorovich, helped them in their calculations.

There's a simple explanation why complicated mathematical computations were necessary—experimentation by trial and error was virtually ruled out. Atomic explosive material was too expensive and too scarce to waste on "errors." This very need for complex thermonuclear calculations stimulated the creation of early computers. However, it wasn't until 1954 that the first computer calculations were begun in the USSR; until then, work was done by hand on desk calculators. Many dozens of women operators, following instructions that meant nothing to them, added and multiplied incomprehensible numbers without knowing that the result of their work could answer the question of how powerful a thermonuclear explosion would be and whether it would take place at all. The physicists and mathematicians had to come up with the computing plan that would give them a fail-safe outcome.

"Want makes wit" is close to a popular Russian proverb, and the Soviets managed to do the calculations on their first thermonuclear bombs by hand. Americans postponed the real calculations until the creation of an adequate computer in 1952.

But the people involved in the special mathematics assigned by Sakharov had their own special Soviet problems.

Valid Grounds for Dismissal

Describing the first months at the Installation, Sakharov mentioned the mathematics group affiliated with the theoretical group, which

> was headed by Mates Agrest, a disabled war veteran, a business-like if rather
> eccentric, person. He had a huge family, taking up an entire cottage, where
> I visited several times. His father was a tall, striking old man, who reminded

me of Rembrandt's Jews; he was very devout, as was Mates . . . Agrest had
to leave the Installation, ostensibly some relatives of his had been discovered in Israel; at the time it seemed like very valid grounds for dismissal to
all of us (including me); the only thing I could do for him was let his family
use my empty apartment while he looked for another job.

What took just a sentence for Sakharov to describe was for Agrest one of
the most dramatic turns in a life that was hardly boring to begin with.[13]

Six years older than Sakharov, Agrest was born in Belorussia into the family
of an elementary school teacher of Judaism. His father taught him at home until
he was eleven, when he was sent to a Jewish religious school. The school was
closed in 1930, and at the age of 15 the expert in the Torah and Talmud set
out for Leningrad to study secular sciences. He was especially attracted to astronomy, but for all his intellectual development, he had little education beyond the holy books. He covered the five-grade equivalency in a few weeks and
enrolled in a factory training school. A lathe turner, fourth class, he independently prepared himself in the higher grades and began taking his exams for
Leningrad University. He passed his exams in mathematics, but got a "D" in
Russian, which differed too much from Hebrew and Aramaic.

But given his capabilities in the exact sciences, he was accepted at the university—*as a Jew*, for whom Russian was not a native language. That was 1933,
about 10 years before state anti-Semitism was launched. After graduating
from the university, he began graduate work in celestial mechanics, studying Saturn's rings, but the war broke out. He was called up to the barrage balloon service. An emergency situation was blamed on him; military tribunal;
death sentence commuted to serving in a penal battalion, where he was seriously wounded; hospital; invalid status of the second category.

The retired captain returned to Moscow, defended his dissertation in 1946,
and began solving mathematical problems for the Zeldovich group at the Institute of Chemical Physics. He wound up at the Installation as part of this group
in 1948 and was involved in calculations for the bomb devices until January
13, 1951, when he was suddenly given 24 hours to leave the Installation.

It was a desperate situation. There were eight people in Agrest's family: the
youngest was several months old; the oldest, his father-in-law, was over 70.
There was nowhere to go: no home, no work, and beyond the Installation's
barbed wire—savage anti-Semitism under the euphemism the "struggle with
cosmopolitanism."

His colleagues came to his aid—Tamm, Bogolyubov, and Frank-Kamenetsky
went to the leadership and succeeded in having the sentence mitigated, replacing the 24 hours with a week to get organized. But there was still no place to
go. And in this hopeless gloom, "Andrei Sakharov appeared, like an angel from
the heavens . . ."[14] Sakharov came to the Agrest home on numerous occa-
sions—just to chat or to help uproot tree stumps on their plot. Agrest doesn't

recall what they talked about then, but he remembers Sakharov's manner of engaging in conversation.

Agrest was short, and it was usually difficult for him to talk to people of Sakharov's height, but the latter always managed to make things comfortable. In Agrest's words, Sakharov spoke little and unhurriedly and took so long choosing his words that you even wanted to prompt him. Sometimes one would hint at a word, but he wouldn't accept it. But then, the sentences he expressed could go straight to press. As one spent time with him it became clear that the mechanism of his thinking worked according to some very unique laws. Unexpected turns of thought were combined with a very calm, unpretentious manner of exposition.

After he learned what happened, Sakharov came to Agrest's house and quoted a Russian proverb about not appreciating things until they're gone. He suggested to the Agrest family that they live in his Moscow apartment for a while, until the situation cleared up, and gave them a piece of paper with his parents' address and telephone number.

Agrest was left with the impression that Sakharov had made his offer cautiously and was, perhaps, influenced by Tamm, who had publicly announced at work that he was planning to go over to help the Agrests. Sakharov's caution would be understandable: he was not an academician or a Hero of Socialist Labor yet, nor did he have the superbomb to his credit—he was still only a Ph.D., whose ideas were just beginning to be implemented, without a guarantee of success.

But when you look with 1951 eyes at Sakharov's 1951 note, carefully preserved by Agrest, it's difficult to agree with his impression. This is, after all, material evidence that Sakharov trusted a man whom the regime no longer trusted. And it was not at all clear how the Agrest affair would end. "The unmasking of cosmopolites" was in full swing.

What was the real reason for his expulsion? Atomic secrets were not the only secrets inhabiting the top secret Installation. The reason for Agrest's banishment was also secret. The security department did not bother with explanations. There were more than a few Jews in the nuclear archipelago. And it wasn't until Agrest read Sakharov's *Memoirs* that he found out that "ostensibly, some of his relatives were discovered in Israel"—Agrest hadn't even been aware of them.

Sakharov didn't know about the much more probable and practical reason for Agrest's dismissal. At the end of 1950, a son was born to Agrest. The ups and downs in Agrest's life did not shake his religious beliefs, much less the thousand-year-old norms of religious behavior. One of them requires that a boy must be circumcised on the eighth day after his birth. This ritual was performed by Agrest's father-in-law, who lived with them (Agrest's own father, mother, brother, and sister died at the hands of the Nazis in the fall of 1941).

Soviet health care was not only free but was sometimes mandatory. The local pediatrician couldn't help noticing the small change in the baby's anatomy during a routine required examination of the boy. The doctor was, as Agrest recalls, very nice, but a circumcision in a center of scientific and technological progress during the height of the struggle with cosmopolitanism was a news event that was too amusing not to be shared with others. From mouth to mouth . . . and the news got to those with the biggest ears.

To them it wasn't just a glaring remnant from the past, it was a dangerous level of asocial, or more aptly—antisocialist—behavior in a militantly atheistic society. Simply put, a challenge to the existing order. It's entirely possible that the guardians of order at the Installation would have reacted just as strictly to an Orthodox Christian baptism as they did to the circumcision.

Among those who stood up for Agrest, only Bogolyubov—a preeminent mathematical physicist—could feel sympathy for his religious sense. The son of an Orthodox priest, he assimilated religious convictions from his family for life.[15] Agrest found out about this by accident. He once decided to drop in on Bogolyubov on some matter. He approached the house and found the door half open. The sounds issuing from behind the door astonished him—it was a radio broadcast in Hebrew. Not knowing what to do, he knocked on the door anyway. Bogolyubov came out, and when he noticed Agrest's amazement, he simply and cheerfully explained that he was listening to foreign radio, adding the word *mitsraim*, which means Egypt in Hebrew. Bogolyubov had learned some of Jesus Christ's native language from his father the theologian, who wrote about the history and philosophy of religion, including that of Judaism, Islam, and even Marxism.

This event created mutual trust between the two scientists, two believers of diverse religions. Based on this trust, a seminar soon began (it was a secret not shared with nonparticipants) at which the Christian Bogolyubov and the Jew Agrest, who were not at all inclined toward merging their religions, discussed religious and philosophic issues at the top secret Installation of science and technology.

Sakharov was not invited to this seminar. Like most of the physicists of his generation, he was an atheist. In his later years, this son of a physicist, grandson of a lawyer, and great-grandson of a priest was to come to a new—and free thinking—view of religion. However, it appears this subject did not engage him in the 1950s. He, like Tamm, was simply a humanist for whom it was enough to show his compassion for a fellow man, even if he considered that the requirements of secrecy justified dismissing this person from the secret Installation.

"I Really Don't Like All This"

The name Lev Landau (1908–1968) symbolized the power of Soviet theoretical physics for decades. He and Tamm were the only Soviet theorists to be awarded a Nobel Prize in Physics. He put his pedagogical gift into practice by creating a powerful school, and the ten-volume *Course of Theoretical Physics*, known to physicists around the world, brought him even more glory.

Sakharov shared this high regard. Discussing Tamm's desire that Landau be his official dissertation opponent in his *Memoirs*, he remarked that the latter "fortunately, refused; I would have felt very awkward because I realized the dissertation's inadequacies." Sakharov also talked about his failure in pure physics in the summer of 1947, and how Pomeranchuk (his dissertation opponent) did "a hatchet job" on the same problem, while Landau dealt with it "in an elegant and productive way." This gave Sakharov the basis to humbly "formulate a system of inequalities: L > P > S" (L for Landau, P for Pomeranchuk, S for Sakharov).

Nonetheless, in the early 1950s, Landau worked on Sakharov's assignments. True enough, that work was in computational mathematics, not theoretical physics. Odd "material evidence" of this appears in Landau's *Collected Works*: placed between the 1958 article about fermions and the 1959 article about quantum field theory is the lecture "Numerical Methods of an Integration of Partial Equations by a Method of Grids." It was published in 1958 but, as it indicates, describes the methods developed in 1951–1952.[16]

When you look at the article's unexciting formulas, it's difficult to imagine what's behind them. What's behind them, among other things, is the first thermonuclear bomb in the world and the suicide of the head of the security department.

In those years Sakharov

for some reason came to the Institute of Physical Problems, where Landau headed up the Theoretical Department and a separate group doing research and calculations for "the Problem." After we finished discussing our work, Landau and I walked out into the Institute garden. This was the only time we talked without witnesses, heart-to-heart. He said: "I really don't like all this." (The context was nuclear weapons in general and his participation in this work in particular.)

"Why?" I asked somewhat naively.

"Too much fuss."

Landau usually smiled a lot and easily, baring his large teeth, but this time he was sad, even mournful.

Sakharov didn't know why Landau was sad. It's possible that at that time he also didn't know that Landau had spent a year in prison, from which Kapitsa, the founder of the Institute for Physical Problems, had rescued him and that, on the eve of May Day, 1939, Beria had bailed out the remarkable theorist for the remarkable experimenter. Things such as this were not discussed in those years.

It's entirely feasible that Sakharov didn't know at all—simply due to secrecy—that Beria had accepted the anti-Soviet criminal Landau into the atomic project soon after he became its head (and got the additional "recommendation" from Bohr via the KGB physicist Terletsky). Landau's group did the calculations for the 1949 A-bomb, for which he received an Order of Lenin and a Stalin Prize of the Second Degree.

Landau's contribution to the hydrogen bomb was even greater, judging by the fact that he was awarded the title of Hero of Socialist Labor and a Stalin Prize of the First Degree. Landau's group managed to complete the Sloyka calculations "by hand"; it was the problem akin to the one the Americans postponed until computers appeared. This required devising an entirely new method of calculation.[17]

The processes of a thermonuclear explosion are much more complicated than an atomic one, if only because it includes the atomic one as its first step. Numerical calculations using old methods would have taken years, but the problem had to be solved in months, which ensured a new method needed to be found. However, while developing it at the Institute for Physical Problems, theorists found a serious mathematical problem—the stability of the calculations. Without solving it, they couldn't be sure that the calculations, no matter how precise, would actually have any relationship to physical reality. The new method solved this problem. But the mathematics group directed by Andrei Tikhonov, which had been created in parallel as a failsafe, denied the problem's very existence.

Dissent and discussion are common in science, but in this case the science was top secret and super-urgent. Beria could not wait for the problem to be resolved in a free exchange of ideas, so a meeting was convened under the chairmanship of Mstislav Keldysh, the future president of the Academy of Sciences. It lasted for several days and the discussions ended in an unusual way: based on Keldysh's opinion, the top leadership gave the order regarding which interpretation was to be considered scientific truth—the top leadership was Nikolai Pavlov, the KGB general in charge of nuclear weapons development. And Tikhonov's group switched to the new method of calculation.

The assignment for the Sloyka calculations sent to the Landau group was "a piece of graph paper, handwritten on both sides in green-blue ink, and it contained all the geometry and data of the first hydrogen bomb."[18]

This was possibly the most secret document in the Soviet project—and it could not be entrusted to any typist. After a mathematical assignment was prepared on the basis of this document at the Institute of Physical Problems, it was sent on to the Institute of Applied Mathematics where Tikhonov's group worked. And the page disappeared there. Perhaps it was mistaken for a rough draft—it was a single handwritten page—and it was destroyed along with other drafts. But this action was not recorded, which is what led to the tragedy Sakharov describes:

> The head of the Security Department from the Ministry—a man whose mere physical appearance, his stare from under drooping eyelids, elicited physical dread in me—came to investigate the extraordinary incident. Former

head of Leningrad State Security during the so-called "Leningrad Affair," when about 700 top leaders were executed there, he spent nearly an hour on Saturday with the head of Institute Security. The Institute official spent the next day, Sunday, with his family; they say he was cheerful and very affectionate with his children. He came to work on Monday 15 minutes early and shot himself before his co-workers arrived.

This incident was extraordinary. But secret record-keeping was the quite ordinary background of the special science. The strict rules of conduct and document storage, which were to some extent justified and logical, also included the administration's senseless and burdensome inventiveness. Thus, for example, the "most secret" physics terms had to be necessarily encoded in reports: the neutron had to be called "zero point," and plasma was "dregs."[19] The mandatory requirement to read texts with key words in code was added to the scientific difficulties. Breaking established rules was fraught with the most unpleasant consequences. The proximity of the Gulag to the nuclear archipelago was a reminder of that.

People get accustomed to many things. They get accustomed to thinking freely about science in unfree conditions.

The Thermonuclear Doughnut

Sakharov had trouble sleeping on a June night in 1949 during his first trip to the Installation. There was no more need to think about the physics of the thermonuclear explosion—the principal method had been discovered. But a thermonuclear explosion is artificial to the ultimate degree: such phenomena do not exist in nature. At the same time, a natural thermonuclear reaction is constantly occurring right before people's eyes—the sun, "the source of life on Earth."

That night in 1949, Sakharov thought about "a new problem, which came to [him] this night—a controlled thermonuclear reaction." The problem was how to attain the fusion of light nuclei, not by means of an atomic explosion but in a peaceful way, and not in the center of the Sun but on Earth. What forces could bring light nuclei closer together, not singly but en masse so that the energy contained in them would move in a controlled flow? What container could stand the temperature of many millions of degrees required for a thermonuclear reaction? No solution came to Sakharov that night.

The 28-year-old Candidate of Sciences from the center of academic physics wasn't the only one in Russia who was thinking about this problem. Ten thousand kilometers from Moscow, 23-year-old Sergeant Oleg Lavrentiev, a radio telegrapher in an artillery division on Sakhalin Island, was also thinking about it. He had access to a very good library, an enormous interest in science, and the ability to educate himself. His unit commander noticed his unusual interest and abilities and instructed him to give a lecture on the atomic problem to his fellow servicemen. He overfulfilled the assignment, devising a way to create a thermonuclear bomb and "a design for using nuclear reactions between light elements for industrial purposes."[20]

This wasn't his first invention. He had already sent one nuclear proposal to the Academy of Sciences and another one on surface-to-air missiles to the Ministry of Defense. The response was discouraging, which explains why he kept his new ideas to himself until 1950, when he read in a newspaper about the American decision to create a hydrogen bomb.

Sergeant Lavrentiev knew how to make a hydrogen bomb, and wrote about it to Stalin.

> It was a short note, literally a few sentences, about the fact that I knew the secret of the hydrogen bomb. I did not receive an answer to my letter. I think that it drowned in the flood of greetings on the occasion of Stalin's 70th birthday on December 21, 1949. Waiting for several futile months, I wrote a letter with the same content to the Central Committee of the Communist Party, that I knew the secret of the hydrogen bomb. The response to this letter was instantaneous. At headquarters, a guarded room was set aside and I got the chance to write my first paper on thermonuclear synthesis. One copy of the paper was typed and sent on July 22, 1950, by secret post to the Central Committee. The rough draft was destroyed. It was sad to see the recently written pages into which I had put two weeks of intense work warping and burning up in the stove.[21]

The discharged sergeant arrived in Moscow on August 8 to enroll at the university, unaware that, by that time, Beria had already sent his proposal on to Sakharov for review at the Installation.

The proposal's main point was that for a container to hold billion-degree plasma, it would have to be made of electrical forces. In his response of August 18, 1950, Sakharov wrote that "the author raises a very important and not hopeless problem," but also pointed out the difficulty that in his opinion made Lavrentiev's proposed solution unfeasible. He concluded his review this way: "Changes that could correct this difficulty are not ruled out. I consider necessary a detailed discussion of Lavrentiev's proposal. Whatever the results of the discussion, the author's creative initiative must be noted now."[22] And, as Sakharov said in his *Memoirs*: "My first, still-vague thoughts about magnetic thermal insulation occurred to me while I read the letter and wrote my reply." The idea for the magnetic thermonuclear reactor grew out of these thoughts.

In accordance with the laws of electrodynamics, it is impossible to confine electrical charges to an electric container, no matter how cleverly you design the device, because the charges necessarily come into contact with the walls of the container. Much better material for a container is a magnetic field: it makes the charge travel in a circular motion, and the stronger the field the smaller the circle. A magnetic container can prevent charged particles from coming into contact with the walls by keeping them in circular motion. For example, a container can be made in the shape of a doughnut filled with a magnetic field and the charged particles set moving along the central part of the filling.

In the language of geometry a "doughnut" is a "toroid," and Sakharov called the new project a TTR (toroidal thermonuclear reactor). Then the more general name MTR (magnetic thermonuclear reactor) came into use, but the toroidality returned in the term Tokamak, which is the Russian acronym for "Toroidal Chamber with Magnetic Coils."

The creative "electrical-to-magnetic" impulse that Lavrentiev's proposal generated in Sakharov is reminiscent of another impulse during Sakharov's work at the munitions plant. While inventing a device to magnetically test bullets, he asked himself: what if the magnetic forces were replaced by electrical ones? The answer to that question became his first problem in theoretical physics. In 1950 he asked himself a reverse question: What if the electrical forces were replaced by magnetic ones?

These questions reflected the deep connections between electricity and magnetism, as well as how profoundly Sakharov understood them. And there was no more suitable person, perhaps, than Tamm with whom to discuss the new idea. Tamm was the author of *The Theory of Electricity*, the best Russian textbook on electromagnetism, reissued many times. An example of a magnetic doughnut is examined in this textbook, too. "At the beginning of August 1950, Igor Tamm returned from Moscow. He expressed enormous interest in my ideas—the entire future development of the idea of magnetic thermal insulation was produced by us together."

The rapid development of MTR initially generated enormous hope, enhanced by the innate optimism of the authors: "In 1950 we hoped to have the MTR working in ten, maximum, fifteen years. I'm speaking about Igor Tamm and myself and other hotheads from the LIPAN group." The name of the Kurchatov Institute at that time was LIPAN—the Laboratory of Measuring

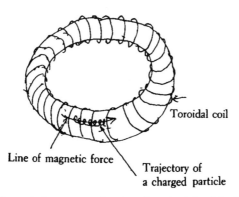

Figure 10.1 Sakharov's drawing explaining the principle of the Tokamak.

Instruments of the Academy of Sciences, and it is where the thermonuclear work was focused.

Sakharov soon discovered that the potential source of unlimited peaceful energy was also a potential source of cheap plutonium. A thermonuclear reactor, once functioning, would become a powerful source of neutrons, with the help of which it would be possible to produce plutonium much more quickly than in ordinary uranium reactors. The security level on the MTR work was immediately raised. As a result, Ginzburg was no longer permitted to see his own previous reports—his level of clearance wasn't high enough.[23]

A sentence that Sakharov uttered at LIPAN with black humor a few minutes before the beginning of a routine meeting on the MTR recalls that time: "Two months of the Big Model, and it's curtains for world imperialism!"[24] What was meant was that in two months the "Big Model" (of Tokamak) would have produced so much cheap nuclear fuel, and thus, nuclear weapons, that it would put an end to world capitalism. Only Sakharov himself could determine the amount of truth in this joke, but undoubtedly there was some. Fortunately for capitalism, serious problems kept coming up one after another on the path to making the MTR a reality; the initiators overcame only the first few. There were enough problems for a hundred other physicists. Half a century later the design of the International Thermonuclear Experimental Reactor (ITER) was begun, with the participation of European Community nations, Russia, the United States, and Japan. There is hope that this project, based on the idea of Tokamak, will finally demonstrate the feasibility of a peaceful thermonuclear reaction.

Regarding Oleg Lavrentiev's scientific life, it appears that Sakharov's reply, with its recommendation of "all possible support and assistance," hindered rather than helped: The very top leadership's attention and the focus on a problem of exclusively government importance stifled the free scientific development of "a very enterprising and creative person."

11

THE "HEROIC" WORK
AT THE INSTALLATION

In his *Memoirs*, which Sakharov wrote in the 1980s during his exile in Gorky, he called the period of work at the Installation until the mid-1950s "heroic," in quotation marks. He would hardly have needed the quotation marks in the 1950s when he received two of his three Hero of Socialist Labor stars. This labor so engrossed him that it hampered his seeing what was happening in his country beyond his desk. Sakharov looked at the world through the prism of his work. Prisms break light up into its spectral components; staring too closely at one of the spectral lines makes it easy to neglect the others.

One of Sakharov's associates from that period remembers "a really direct physical sense of how intensely Sakharov worked, sitting at his desk, grasping his head in his hands, his gaze focused on a drawing."[1] Heroic concentration on his own work may, in part, explain why Sakharov himself did not completely understand his own grief over Stalin's death:

> I already knew a great deal about the horrible crimes—the arrests of innocent people, the torture, starvation and violence. I couldn't help but think of the guilty with indignation and disgust. Of course, there was a lot I didn't know and I didn't put it all together in one picture. Somewhere at the back of my mind was the idea induced by propaganda that brutalities are inevitable during major historic upheavals. As the saying goes "When you cut wood, chips fly" . . . On the whole, I see that I was more impressionable than I would like to be.
>
> But what was primary to me was my feeling of commitment to the same goal I assumed was Stalin's—building up the nation's strength to ensure peace after a devastating war. Precisely because I had already given so much to this cause and accomplished so much, I was unwittingly—probably like any one else would in the situation—creating an illusory world to justify myself.

He had given so much and achieved so much. The 32-year-old physicist became an academician in October of 1953, skipping the level of Corresponding Member. The youngest academician physicist in the history of the Soviet Academy of Sciences before him was 38-year-old Lev Landau, elected in 1946. However, the academicians who elected Sakharov knew far less about him than they did about Landau, primarily what was contained in the evaluations of his colleagues Kurchatov, Khariton, and Zeldovich:

> Andrei Sakharov is both an exceptionally gifted theoretical physicist and a remarkable inventor. As a result of this combination of an inventor's initiative and purposefulness with a depth of scientific analysis, A. Sakharov has achieved the most outstanding results within the short period of six years, placing him at the forefront of the foremost field of physics in the Soviet Union and the entire world.
>
> After beginning work in physics in 1948, A. Sakharov proposed entirely new ways to solve one of the most important problems. This proposal was distinguished by its boldness and profundity; its significance was immediately acknowledged by specialists. There followed years of intense work implementing the proposal, which was crowned with brilliant success in 1953.
>
> Of great importance to the state, Sakharov's proposal was implemented by a large team of scientific workers, engineers, and builders. The Institutes of the Academy of Sciences also played an esteemed role in its implementation; many academicians and Corresponding Members were enlisted to develop the proposal. Nonetheless, even within this group, Sakharov remained the true scientific leader of the work, grasping its scope as a whole and successfully directing the development of its individual projects throughout the course of the work.
>
> In 1950 and 1952 A. Sakharov launched the two new directions in physics which he had proposed; they are presently being developed by a large group of scientists and engineers.
>
> In recent years and into the near future, A. Sakharov's ideas continue to determine the course of the foremost field of Soviet physics.
>
> A. Sakharov's election to a Full Member of the Academy of Sciences is but a true acknowledgement of Sakharov's great achievements in Soviet science and for our Homeland. Sakharov's youth, his enormous initiative and talent give us the confidence to anticipate great achievements from him in the future.[2]

As will be made clear below, this evaluation omits specific information, but the academician-electors had already learned that "one type of hydrogen bomb" had been successfully tested from an August government report in the newspapers.[3] And so the academicians could make educated guesses about a thing or two, and they did elect a group of scientists involved in "the foremost field of Soviet physics" to their ranks.

The unknown information in this testimonial has now been spelled out. "The bold and profound" proposal of 1948 was, of course, Sloyka. The idea of the magnetic thermonuclear reactor came into being in 1950. Experiments on producing superstrong magnetic fields by using explosions (based on an idea proposed by Sakharov in 1951) were begun in 1952. And the "brilliant success" was Sloyka's test on August 12, 1953.

Sloyka aka RDS-6s aka "Joe-4"

The successful RDS-6s was named "Joe-4" in the United States for the already deceased Stalin (Uncle Joe) and its place in the sequence of Soviet nuclear tests.

On the last part of the road to the success, Zeldovich's team, whose Tube (RDS-6t) showed no promise of success, joined the Tamm-Sakharov team. On June 15, 1953, Tamm and Zeldovich signed the final report on Sloyka along with Sakharov. In a routine autobiography for the personnel department dated September 13, Sakharov wrote: "For the past few years I have been working on special research and development projects with the assistance and guidance of I. Tamm and Ya. Zeldovich. I defended my doctoral dissertation on a special research topic in June 1953."[4]

In the June report the theorists estimated the yield of the upcoming blast at 300 plus or minus 100 kilotons, that is, from 200 to 400 kilotons; during the August test Sloyka yielded about 400. That meant the theorists had done their job well. It would have been excellent from a purely scientific standpoint had the test yielded exactly 300.

Most significant from the government's standpoint was that the Soviet Union had a weapon that was twenty-five times more powerful than the one America dropped on Hiroshima. And the government fully appreciated Sakharov's contribution: a congratulatory telephone call and kiss from Georgi Malenkov, then head of the Soviet government, were conveyed to him right after the explosion. Malenkov had his own reasons to feel happy. A week before the test, while speaking before a session of the Supreme Soviet, he had declared to the entire world that the USSR had its own hydrogen bomb—and he turned out to be right.

Summing up the test results, Kurchatov gave special thanks to Sakharov for his, as he put it, "patriotic deed of heroism." A few months later, Sakharov, like the other "fathers" of the Soviet superbomb, received an enormous prize of five hundred thousand rubles (approximately 40 annual salaries of a physician) for this feat and his first Hero of Socialist Labor star. Two-story single-family dachas were built for the Heroes near Moscow.

His first reward was election to the Academy of Sciences. Sakharov's scientific status had been acknowledged even before the successful test, and in the spring of 1953, according to Kurchatov's instructions, Sakharov, then still with just a Candidate of Sciences degree, was nominated for the upcoming elections of Corresponding Members of the Academy. However, after the triumphant test, Kurchatov revised his plan and nominated Sakharov for Full Member of the

Academy. He became an academician a month before he was confirmed a Doctor of Sciences.

In that same election Tamm (after 20 years as a Corresponding Member) and Khariton (the scientific director of the Installation) became Full Members, as did several other scientists engaged in H-bomb physics. Ginzburg became a Corresponding Member, and he understood only much later that LiDochka, his "Second Idea," was the deciding factor, not his achievements in pure science.[5]

But Zeldovich (a Corresponding Member since 1946) was not promoted to academician status. Regarding this, Sakharov wrote: "It was completely unfair, distressed me very much and placed me in a false position." From all appearances, this unfairness simply reflected the personal sympathies and antipathies backstage in the academic elections, which Sakharov had the chance to observe for the first time. Zeldovich was not elected until 1958.

In the eyes of the government, Sakharov became the number one H-bomb physicist in the fall of 1953. While still at the testing site, Tamm requested from the government permission to return to pure science. It was granted and he was even permitted to take several new associates (including one from the Installation, Vladimir Ritus) with him to the FIAN theoretical department.

Sakharov became head of Tamm's theoretical department at the Installation. Why didn't he follow his teacher into pure science at that time? Perhaps he felt that it was "a pity to abandon my work as an inventor just when I was starting to get results," just as he had felt ten years earlier at the munitions plant. There was no doubt that his thermonuclear innovation was starting to yield results, but he also couldn't admit that the Sloyka was his crowning glory. Soon after the test when Zeldovich asked him what he was planning to work on, prompting the expected answer, MTR [controlled thermonuclear reactor], Sakharov answered, "No, I should finish working on the device." To understand Sakharov's attraction to this lethal but highly scientific invention, we would need to look closer at its design. But to comprehend Sakharov, it's even more vital to understand the nature of passion for invention.

The passion for scientific invention is like passion for a woman, for poetry, for playing cards, for music. And that the object of scientific passion seems more rational and socially useful makes little difference. Passion is impossible to explain, but in and of itself, it explains a lot. For example, it can explain why even the sudden fall of the top government official in charge of the nuclear project did not disturb the rhythm of preparations for the first thermonuclear test. At the end of June, Beria was arrested, declared an enemy of the party and the Soviet people, and the First Main Directorate was immediately renamed the Ministry of Medium Machine-Building. The Soviets abbreviated the camouflage name to MedMash, *SredMash* in Russian. A biographer of Sakharov can afford to take a step away from historical Soviet reality by coining the name "Mad-Mash" with its suggestion of Mad Machinery. This name would correspond to Sakharov's later attitude, which combined some irony and self-pity with acceptance of the inevitability or even necessity of making the Soviet Big Bomb.

Even dealing with the machinery of the H-bomb, it is easy to give yourself up to passion of invention if you are certain that it is for the benefit of your homeland and for all that is good in the world. Sakharov still had a few years of such cloudless certainty. If Sakharov had any reason to doubt his importance, it must have vanished in November 1953, when he was asked by Vyacheslav Malyshev, the Minister of MedMash, to delineate the direction of future thermonuclear weapons work.

Without realizing the importance his recommendations would be accorded, Sakharov set forth his view on improving Sloyka and indicated the parameters of the next "device," based on the idea that seemed most promising to him at that time. The nation's leadership so trusted the thirty-two-year-old academician that his rough draft was immediately used as the basis for the government's resolutions. One of these resolutions, "On the Development of a New Type of Powerful Hydrogen Bomb," dated November 20, 1953, committed the Ministry of MedMash to creating the new thermonuclear bomb in 1954–1955, the design that Sakharov so casually announced and that had already been named RDS-6sD.[6] Another government resolution ordered the creation of an intercontinental ballistic missile capable of carrying this bomb. The size and weight of the bomb indicated by Sakharov became the base parameter used to determine the entire scale of the missile. This was the missile that launched both the first artificial Earth satellite and the first person into space.

Thirty years later, in reflecting on the reason for the level of trust placed in him, Sakharov cited several factors: "My self-confidence, at such a peak after the test, Malyshev's faith in my talent, impressed upon him by Kurchatov, Keldysh and many others, and reinforced by the successful test and my demeanor at that time—outwardly modest, but in fact quite the opposite."

Economics of the H-Bomb

But compare these achievements with the American ones, and frankly, what grounds were there for self-confidence? Could Sakharov have believed at that time that "we were ahead of the entire planet" in his professional field?

The answer greatly depends on the year referred to in the question. He couldn't have believed this in 1953. But he might have thought that Soviet Union had caught up to the United States in this project. Back on December 2, 1952, Beria sent off a note to Kurchatov in which he mentioned the first American H-bomb test (conducted on November 1, 1952, under the code name "Mike"):

> The decision regarding the creation of RDS-6s [Sloyka] is of the first priority. Judging from certain information that has reached us, there have been tests related to this type of device. Upon your departure with A. Zavenyagin to KB-11, inform Yu. Khariton, K. Shchelkin, N. Dukhov, I. Tamm, A. Sakharov, Ya. Zeldovich, E. Zababakhin and N. Bogolyubov, that we must exert all

efforts to insure the successful research and development related to RDS-6s. Also tell L. Landau and A. Tikhonov.[7]

Later it became known that Mike was an 80-ton laboratory construction, not a bomb like Sloyka that was "deliverable," or rather "air-droppable" (the first American H-bomb was air dropped in 1956). So the Soviet bomb makers could say that the first H-bomb was Soviet.

But the American bomb makers responded by refusing to consider Sloyka an H-bomb at all, calling it merely a "boosted A-bomb." After all, their Mike was many times more powerful than Sloyka.

The point wasn't so much the differing terminologies of the two nations, but their different thermonuclear histories. Soviet thermonuclear history cannot be measured by an American yardstick. Nothing similar to Sloyka had been built in the United States. Even though the idea had been proposed, the Americans did not develop this project because it did not promise a large enough leap in yield compared to an A-bomb. The most hopeful appraisals would yield an output "only" 50 times greater than Hiroshima.

There were two opposing tendencies among American bomb physicists. Most of them considered that there were more than enough A-bombs and were against any type of H-bomb, whereas the minority, spearheaded by Edward Teller, set their sights on a real superbomb a thousand times more powerful than Hiroshima. This is what they expected from the "Classical Super" or, in Russia, the Tube.

But the Soviet goal was simply to catch up to America, and that is why Sloyka's output of twenty-five Hiroshima bombs made the game fully worth the candle. The question of "what is worth what" is not very relevant in science. But it is unavoidable in the social history of science.

No matter how insulting this may sound for pure-science adherents, scientific-technological progress can be seen as a process of reducing the prices of material goods. Electricity emerged from the science lab into public daily life only after its production became cheap enough. This happened thanks to research and development in science and technology.

In American debates about the hydrogen bomb, along with discussions about the morality of this type of weapon, there were considerations about a gram of thermonuclear explosive costing pennies, while uranium explosives cost hundreds of dollars. "Damage area per dollar" was discussed.[8] In the nation of Soviet socialism, the strategic goal of parity in military power with America made time more valuable than money. However, when it came to mass producing a new type of weapon rather than building a single experimental "device," resource limitations became a strategic parameter. To put it simply, an H-bomb is a lot cheaper than an A-bomb for the damage it causes. And the temptation of a thermonuclear weapon was not so much increasing the yield of a single explosion as making this yield less expensive.

Making an A-bomb requires not only mining the rare element uranium, but extracting the very rare uranium isotope U-235, about 1 in 140, from the natural uranium, mainly U-238. This task is more difficult than the one in the fairy tale in which the stepmother mixed millet with the poppy seeds and asked her stepdaughter to separate them. The different isotopes of uranium are far less distinguishable from one another than millet from poppy seeds. The separation of isotopes is an extremely costly process. The artificial element plutonium can be used instead of U-235; although cheaper, it is still quite expensive.

In Sakharov's eyes of the 1950s, his invention saved a great many resources for his country, ravaged by war.

The American physicists who did not consider Sakharov's Sloyka a thermonuclear bomb were using its small yield as the basis of comparison with their own first thermonuclear device. But a small apple is still an apple and, given proper care, can produce an apple tree that bears large fruit.

In American terminology, a boosted atomic bomb means that a small amount of thermonuclear explosive is placed inside the atomic bomb, which will add a small amount of its thermonuclear energy to the process of fission. The amount cannot be significantly increased without disturbing this process. But in Sloyka, the thermonuclear explosive is placed outside the atomic bomb. Because there is much more space on the surface of the sphere than in its center (for the same thickness), Sloyka can be made highly thermonuclear.

Besides producing energy, the thermonuclear fusion also releases neutrons, which are far more energetic than those produced in fission. And Sakharov, while devising a new method of compression for Sloyka (Sakharization), concurrently found a use for these fast neutrons. From the outset, he suggested making the layer of matter with the heavy nuclei required for Sakharization from natural uranium (rather than from lead, as was said, for the sake of simplicity, in the chapter "The Hydrogen Bomb at FIAN"). Natural uranium, which consists almost entirely of U-238, is unsuitable for an A-bomb but works in a thermonuclear one. There is just as much energy in U-238 as there is in a U-235, but only a fast thermonuclear neutron can extract it; the neutrons of an atomic explosion are too slow.

During Sloyka's explosion, the A-bomb trigger released only one-tenth of the energy. The rest was thermonuclear in origin: the energy from fusion (about two-tenths) and the energy obtained from the natural uranium by thermonuclear neutrons (about seven-tenths). This is why the veterans of the Soviet project did not doubt that Sloyka was a thermonuclear device. Whether or not it can be called an H-bomb is a matter of terminology.

But there's another question: could Sloyka be built with a smaller amount of expensive uranium? (Uranium is expensive even in its natural state.) For example, they could make the heavy-nuclear layer from cheap lead and increase the yield by using more (cheaper) thermonuclear explosive—LiDochka.

The theoretical—you might say geometric—possibilities of doing this are limited. The room for LiDochka is limited. The only way to increase it is by increasing the thickness of the LiDochka layer, but only up to a certain point. Indeed, Sloyka, like an A-bomb, starts its performance by detonating the ordinary explosive surrounding the whole Sloyka. This explosion compresses the A-bomb core, which results in its chain reaction. And only then does the thermonuclear reaction in the layer of LiDochka ignite. But the thicker the shell, the more difficult it is to compress the A-core.

All of this is due to the finite geometry of the sphere. The cylinder is a different matter: its length is theoretically infinite, meaning as much thermonuclear fuel can be placed in it as desired. This was exactly the cylinder—the Tube— that the Zeldovich team had already spent seven years pondering without any success. It wouldn't burn, no matter how little or how much fuel they used. The thermonuclear flame, ignited by an A-lighter at the tip of the "cigar," did not behave and refused to spread further along the "cigar," as if the thermonuclear tobacco were wet.

How Physics Can
Outsmart Geometry

In November 1953 when the government asked Sakharov to outline the next step in nuclear weaponry, he was certain that the idea of the spherical Sloyka had not yet been exhausted. But by January 1954 the theorists realized that the methods Sakharov was expecting to improve Sloyka were not working. And, at the same time, it became clear the Tube project had no future and work on this track was discontinued.

The double impasse was a kick-start for their imagination. Zeldovich and Sakharov's January 14, 1954, report "On the Use of a Device to Compress the Super-Device RDS-6s" can be considered the beginning of the new phase. In their consideration of how to compress the Sloyka, the key idea was that if they replaced the ordinary explosive on the outside with an atomic explosion, then much more LiDochka could be placed inside Sloyka. The key term in this report was "atomic compression."

A few months later, "a fundamentally new idea" grew out of this reflection on "atomic compression," which Sakharov, constrained by the restrictions of secrecy, called the Third Idea in his *Memoirs* (after the First and Second Ideas of 1948—Sloyka and LiDochka). He noted that "'the third idea' had been discussed earlier in a certain form, more as a wish," but he didn't indicate that he himself had mentioned the possible "use of an additional plutonium charge for a preliminary compression of 'Sloyka'" in his very first report on Sloyka in January 1949.[9] This possibility remained in the theorists' sights and was even included in their work plans, but two specific projects—the Tube and Sloyka—left little energy for other dubious possibilities. The double impasse in early 1954 freed the hands—and brains—of the theorists.

But why "dubious" and why did it take a five long years from the wish stage to fulfillment? Atomic compression even in its most general form staggers the imagination. For it means that the all-annihilating atomic explosion in the first millionth of a second—before turning into the horrifying mushroom—must perform a certain task, symmetrically compressing another intricately designed device located, let's say, a meter away.

Vladimir Ritus, a young physicist, worked on thermonuclear calculations with his colleagues under Sakharov's command. He left the Installation in May 1955, and his recollections capture the atmosphere of that time. When he found out about the new idea, he was astonished by its recklessness: "What? Won't it blow everything away?!"[10] However, the wishful idea itself is completely understandable. The efficiency of the fusion depends on the degree to which the fusionable material is compressed. The denser the material, the more it resembles intrastellar matter and the more intensely the stellar reaction will take place in it. Naturally, pressure from an atomic explosion is far more powerful than that from the ordinary explosive surrounding the Sloyka. The problem was how to make this atomic compression symmetrical, how to make the thermonuclear material compress simultaneously from all directions.

By that time, one of the directors of the nuclear project, Avraamiy Zavenyagin, a KGB general, proposed a simple solution himself: Surround Sloyka with several A-bombs from all sides and blow them up simultaneously. His design was called the "candelabra," in which the external atomic candles had to ignite the central thermonuclear one. The bold engineering is a credit to the KGB general.

However, physicists couldn't take his candelabra idea seriously. The A-bomb had already posed a difficult technical problem: an atomic explosion begins with a compression of the atomic charge by an ordinary explosive. They borrowed an English word for this "inward bursting"—implosion. But introducing the term was much easier than exploding all the parts of the explosive simultaneously to obtain the symmetrical compression of the atomic charge. The signals for the explosion had to be delivered to these different parts with a discrepancy of less than a thousandth of a second.

An atomic explosion occurs thousands of times faster than an ordinary one, but the simultaneous explosion of several atomic charges isn't just a thousand times more difficult—it's impossible. The unsynchronized delivery of these signals would, in fact, result in the explosion of just the first atomic candle—the rest would scatter unexploded. Instead of a symmetrical candelabra, you'd be left with a single crooked candlestick.

Sakharov understood that a single source for the atomic compression was necessary when he made his "wish" in 1949. But how could a single atomic explosion compress Sloyka from all sides if it was located on only one side of it?

Geometry would resist it. The physicists had to outsmart geometry. And it was a very difficult problem.

There is a simple sketch on the first page of the above-mentioned January 14, 1954, report about "device compressing Super-device" drawn by Zeldovich. Between the "S" of Sloyka and the additional atomic charge "A" there is a partition "D" to shield Sloyka from the atomic explosion at least for a millionth of a second, to make it possible for the explosion "fragments" to compress Sloyka from the sides, after they curved around the edges of the partition.

The experimenters were asked to test the possibility of such a course of events, but nothing came of it—the sphere simulating Sloyka flattened into a pancake.[11] The Russian proverb says, "The first pancake turns out lumpy." Here the first lump turned into a pancake. So the idea remained as a wish.

A careful look at the sketch, drawn by Zeldovich, can reveal his coauthor who wrote the second half of the report and who expressed this wish back in 1949. The letter "D" for the deflector might also refer to Viktor Davidenko, who proposed this version of atomic compression. But maybe Zeldovich was tempted

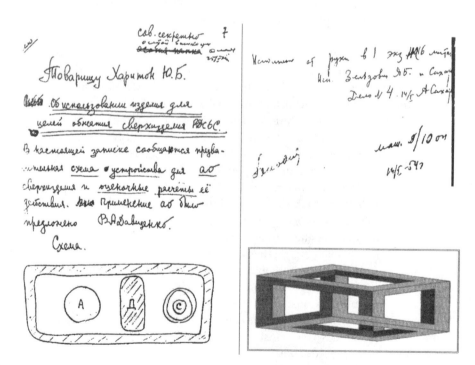

Figure 11.1 *A-D-S* design. On the back of the last page of the report, dated January 14, 1954, written in Sakharov's hand: "One handwritten copy on 16 pages. Executed by Zeldovich Ya. B. and Sakharov A. D." The non-secret mind-boggler is added to the right to present an up-to-date summary of the 1954 proposal.

by the emerging combination of letters—"A-D-S"—which matched Sakharov's initials (first name, patronymic, surname). This would be entirely in Zeldovich's playful style; he once slipped an acrostic maliciously baiting a colleague into a serious review article.

Sakharov and Zeldovich's coauthorship of this report continued in a close collaboration that nurtured the Third Idea over the next months. The event that might have expedited the conception of this idea occurred on March 1, 1954, many thousands of miles from the Installation. The United States conducted a thermonuclear explosion on the Bikini atoll on that day. The day went down in world history primarily because of a miscalculation the American theorists made—the yield of the explosion actually exceeded the expected yield by two and a half times. As a result, a considerably larger area than expected was exposed to radioactive contamination. The entire world learned about the Japanese fishing boat *Fukuryu Maru* [Lucky Dragon], which was located in the radiation zone. The fishermen received high doses of radiation, and one of them subsequently died.

One could conclude just from the size of the radioactive fallout zone alone that the yield of the American explosion was much larger than Sloyka was capable of producing. In fact, it was 40 times greater than that of Sloyka—or 1,000 times larger than that of the Hiroshima bomb. It's true that this was only one and a half times higher than the yield of the Mike test in November 1952, but there was no reliable information about Mike yet. In March 1954, the genie was out of the bottle. The bottle turned out to be too small, or more precisely, American theorists did not appreciate the size of the genie.

The American physicists who created the A-bomb said that its main secret was that it was possible. This secret was revealed in Hiroshima. The 1954 American test on Bikini Atoll revealed the analogous secret of the thermonuclear bomb with the official name of "Bravo."

Analyzing Sloyka's capabilities inside out, and knowing that a method of producing an explosion of much larger magnitude was possible, made it easier for Sakharov to discover the path to "a fundamentally new idea."

The Third Idea

According to the testimony of Sakharov's closest associate, the Third Idea emerged in the spring of 1954. "It began when Sakharov brought the theorists together and set forth his idea about the high coefficient of reflection of impulse radiation from the walls made of heavy material."[12] Let's try to translate this from the language of theoretical physics.

The January *ADS* diagram assumed that the fragments of the atomic explosion *A*, after curving around the partition-deflector *D*, would compress Sloyka *S*. Alas, the laws of physics proved to be against this assumption. And it appears that Sakharov then thought about the fact that an atomic explosion is first of all a flash of light brighter than a thousand suns. He asked himself whether it was possible to use this very flash for "atomic compression." The flash contains only a small part of all the explosion's energy, but the radiation spreads at the top speed possible in nature, moves faster than the explosion's fragments, and can do the necessary symmetrical work before the more material and less symmetrical fragments from fission arrive a millionth of a second later.

Sakharov would much later call the physics of this millionth of a second "paradise for a theorist." Each individual has his own idea of paradise, but there is little doubt that this objective—to understand what is occurring at temperatures in the tens of millions of degrees and use this understanding to produce a specific engineering construction—is a mighty one.

At such vastly high temperatures, both the energy and the pressure of light become enormous, but enormous to different degrees. Their relationship, which Lebedev measured in his most subtle experiments half a century earlier, is essential for the physics of the Third Idea. This relationship is easy to observe while simply sunbathing on the beach: a ray of sunlight, or in modern parlance, a stream of photons, carries fully visible and sometimes burning energy that exerts no perceivable pressure. You might think this pressure doesn't exist at all on the beach. Lebedev demonstrated that the pressure of light—even though it is very small—exists and has a quite definite relationship to its energy.

A stream of all kinds of particles bring both energy and pressure, whose magnitude is related to the speed of the particles:

$$\text{pressure} \sim \text{energy/speed}$$

Because the speed of photons is the largest possible in nature, their pressure is the smallest possible at the same amount of energy.

And now let's look again at the *ADS* diagram to understand why light is the best instrument for the compression of Sloyka. It's much easier for the partition-deflector D to withstand the force of the light's pressure than the force of the fission's material particles, even though the light has an enormous energy that can be used for compressing Sloyka.

Translated into simple language, what Sakharov's associate meant by "the high coefficient of reflection of impulse radiation from the walls made of heavy material" is that Sakharov discovered that the flash of light from the atomic explosion has sufficient time to "do quite a bit of work" inside the thermonuclear bomb casing and provide the energy for the symmetrical compression of the thermonuclear charge. How this "quite a bit of work" is done specifically, how the energy of the device is used to compress the superdevice, has not yet been officially declassified.[13]

To develop the initial idea into a theoretical model and then into an engineering embodiment took the physicists and engineers a year and a half of work. The work was so intensive that they didn't waste time writing intermediate reports. The only document extant from that period is a report about Sakharov's theoretical sector 1, dated August 6, 1954. It says that "theoretical research on atomic compression is being conducted jointly with the associates of sector 2 [Zeldovich]" and names two main topics: the "radiation yield from the A-bomb which produces the compression of the main [thermonuclear] object" and the "transformation of radiation energy into energy which compresses the main object."[14]

On December 24, 1954, the scientific-technological council chaired by Kurchatov decided to conduct a test of the new thermonuclear device RDS-37 in 1955. The final report from June 25, 1955, contains the names of thirty-one theorists. In the introduction, Zeldovich and Sakharov note that the development of the new construction "is one of the striking examples of collaborative creation. Some contributed ideas (many ideas were needed; several of the same ones were proposed by several authors independently of one another). Others stood out more through developing methods of calculation and clarifying the meaning of various physical processes."[15]

Sakharov talked about this in his *Memoirs* thirty years later:

> Apparently, several people in our theoretical departments came up with the "third idea" simultaneously. I was one of them. I think that I understood the basic physics and mathematical aspects of the "third idea" at a very early stage. As a result, and also due to the respect I had earned by then, my role in the acceptance and execution of the "third idea" was perhaps a decisive one. But the role of Zeldovich, Trutnev, and several others was undoubtedly very great and perhaps they understood and foresaw the prospects and difficulties of the "third idea" as well as I. At the same time, we (I at any rate) had not time to think about questions of author's priority, especially since that would be "dividing up the skin of a bear we had yet to kill," and restoring all the details of our discussion in hindsight is impossible, and perhaps unnecessary.

There was a critical moment in their collaboration at the earliest stage of the work. It was when Sakharov came up with a way to deal with very complex physical processes, which were seminal for the Third Idea, and mathematician Nikolai Dmitriyev then substantiated his approach: "I [Sakharov] remember to this day that initially Zeldovich did not appreciate that I was right and then believed in it only after Dmitriyev's work; this was unusual, for he's very sharp." Zeldovich considered Dmitriyev's mathematical talent to be very high. As to the complexity of the problem, one can sense it in the recently declassified diagram, Figure 11.2.[16]

After the physicists untangled the raveled skein depicted in the drawing with the help of the mathematicians, the Third Idea materialized into an engineering construction. Malyshev, the first Minister of MedMash, who took over Beria's position in the nuclear project, became the main obstacle on this path. Although not a reactionary, he was a government official who honored government discipline. He considered himself and the whole atomic project bound by

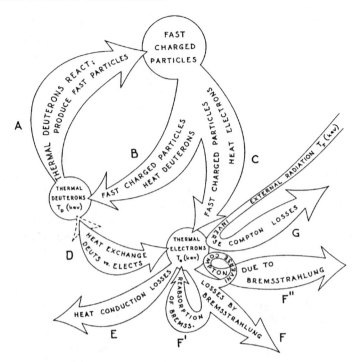

Figure 11.2 Thermonuclear diagram drawn in Los Alamos by George (formerly Georgi) Gamow, to whom FIAN owes its origin. Had he not left his native land in 1933, it is entirely possible that he would have been working on this same problem twenty years later (with his friend Landau) and drawing the same diagram—but in Russian. Physics is international, even top secret physics.

the government resolution drafted in November 1953 at Sakharov's own recommendation. The minister deemed it irresponsible that the theorists were changing course so abruptly. He flew out to the Installation and tried to reestablish order:

> He delivered a passionate speech that could be called brilliant if only we had not been right. Malyshev began losing self-control and started shouting that we were adventurers playing with the fate of the country and so on. He could not fully ban work on the "third idea" nor did he want to, but he could not control the amount of enthusiasm, or lack or it, with which we treated the classic device [Sloyka]. Similar meetings, dragging out for a half day, were repeated a few more times; they became ever more pointless and exhausting.

In a similar situation seven years later, Yefim Slavsky, a former Red Cavalry man who succeeded Malyshev in this ministerial position, crudely vented his indignation at these theorists, who dream up new devices while they're "sitting on the toilet and suggest the devices be tested before they've buttoned up their pants."

In 1954, Malyshev punished Kurchatov, who had sided with the Installation physicists, by giving him a party reprimand. All the same, the Third Idea came to occupy the most important place in the Installation's work:

> We were convinced that such a strategy would be ultimately justified, although we understood that we were entering a field full of danger and surprises. Working on the "classic" device [improving Sloyka] at full speed and simultaneously moving in a new direction was impossible, our forces were limited, and we also saw no point in pursuing our former direction.

At the end of June 1955 a commission chaired by Tamm approved the new thermonuclear design. The test took place on November 22, 1955, at the Semipalatinsk test site. A thermonuclear bomb was dropped from an airplane for the first time in history. Its projected force (150 Hiroshima-type bombs) was deliberately decreased by a factor of two by substituting inert matter for thermonuclear explosive in order to decrease the fallout zone. The yield of the explosion matched the calculations. The Soviet theorists could be proud that their predictions were so much more precise than those in the American test of Bravo.

Espionage and Physics

Having praised the glory of Soviet H-bomb physicists, it's time to ask how such a one-sided picture of Soviet thermonuclear success can be painted in the era of glasnost. Were these achievements really independent?

Indeed, both the vigor of the Soviet intelligence agents in the American atomic project and the enormous volume of spy information about the American A-bomb—thousands of pages—are now commonly accepted. In 1996 five veteran Soviet agents were given the title of Hero of Russia for their efforts that made possible "the liquidation of the U.S. nuclear weapons monopoly in the shortest possible time."[17]

Some people feel that these belated heroes' stars dimmed the Hero of Socialist Labor stars awarded to the physicists forty years ago. Didn't the stream of intelligence information about American nuclear arms yield anything of substance about the thermonuclear bomb? In 1997 American journalists introduced Theodore Hall, a former Soviet agent nicknamed Mlad, to the world.[18] He had worked in Los Alamos in the mid-1940s, and he worked for both countries. His biographers wrote that besides passing on information about the A-bomb in October 1947, he informed Soviet intelligence that Americans were using lithium in their work on the superbomb: "The Soviets were quick to seize the importance of the idea and perfect it on their own. In December of the following year, Soviet physicist Vitaly Ginzburg proposed the use of lithium-6 deuteride as a source of tritium in a Soviet H-bomb."[19] Here the American journalists referred to German Goncharov, the same Russian veteran physicist from Arzamas-16 whose publications are also key sources for this book.

There were other agents who have not been exposed yet—a certain Anta, someone named Aden, and an indeterminate number of others. So, some American thermonuclear veterans think that the second Soviet H-bomb (1955) was made with American help. The first bomb, Sloyka, did not arouse doubts simply because no American analog had been built. But it was the second bomb that embodied the basic principle of thermonuclear weaponry. Confronted with the puzzle—how their four-year lead on the A-bomb could have been reduced to a year and a half on the "real" H-bomb—the Americans saw the answer in the incident that occurred on January 7, 1953.

On that day John Wheeler, a prominent participant in the American thermonuclear project, was going to Washington by train carrying a secret document with important information about the thermonuclear bomb. He had to take it with him when he went to the toilet, but he forgot it there on his way out. When he remembered and returned for it, the document was gone. Wheeler immediately informed higher levels of authority about the loss, but the most careful searches yielded nothing. The situation was aggravated by the fact that the train was filled with leftist demonstrators traveling to the White House to demand mercy for Ethel and Julius Rosenberg, who were awaiting execution. It's entirely possible, as the then chairman of the U.S. Atomic Energy Commission, Lewis Strauss, thought, that there were foreign agents among the passengers who knew the nature of Wheeler's work and were following him.[20]

Where could this missing document have gone? For a long time the answer seemed almost obvious—into Beria's hands. This was the view of many American thermonuclear veterans. In particular, this was the understanding of Tom Reed, who in the 1950s and 1960s worked at the Livermore Laboratory, then was Air Force Secretary and presidential special assistant.[21]

However, archival evidence recently made available virtually rules out that a Soviet agent had gotten such a valuable catch. Moreover, it rules it out from both the American and Russian sides. It has become clear from the American side that whereas Wheeler was carrying two documents in a manila envelope, only one document disappeared from the envelope in the train toilet.[22] It's difficult to imagine such a modest spy, satisfied with only a part of an available trophy. It's easier to imagine a person peeking into a stranger's envelope out of curiosity and seeing the terrifying classification "Top Secret." He probably would have been stung by the realization that the fingerprints he left on this damned paper would lead him to the same electric chair that awaited the Rosenbergs and felt he had to destroy the paper at once.

But more important, there is no place in Russian thermonuclear history for the supposition that the document that vanished in January 1953 wound up in the USSR. Indeed, there's no mention of the key ("Third") idea yet in the Zeldovich-Sakharov report about the *ADS* diagram that came out a year later in January 1954. And why was Beria's successor, Malyshev, so opposed to the development of this idea? And why was Kurchatov reprimanded?

Thus, the only "material" evidence of the Soviets' thermonuclear dependence melted away. There remain some questions about the role of intelligence: why didn't atomic espionage go "hydrogen," and why did it drop off so drastically? The answer to both is that American counterintelligence went into full swing right at that point. Klaus Fuchs was arrested in January 1950, then a few months later, the Rosenbergs. Those who remained kept a low profile. Besides, the motives of the Soviet agents of that time were primarily ideological. These motives abruptly weakened after the 1949 Soviet A-bomb test—after the balance of power was reinstated.

But if there really wasn't any thermonuclear espionage, how was it possible that it took the Americans three years (1951–1954) to implement their "third idea," while their Soviet colleagues only needed a year and a half? To speak of the role espionage played here, we should distinguish between three stages of the Soviet nuclear project: the 1949 A-bomb, the first thermonuclear bomb of 1953 (Sloyka), and the full-fledged thermonuclear bomb of 1955.

That thousands of pages of American secret reports wound up at the disposal of Kurchatov, the director of the Soviet atomic project, is firmly established fact. That the first Soviet A-bomb, detonated in 1949, was a copy of the American plutonium bomb ("Fat Man") tested near Los Alamos and dropped on Nagasaki has also been confirmed.

At first glance, it seems impossible to reconcile these facts with the assessment made by Lev Artsimovich (one of Kurchatov's deputies) that espionage saved the Soviet Union a year or two in making the A-bomb. Although Western historians and the *Encyclopedia Britannica* agree with this assessment, the physicist Artsimovich can be too easily suspected of being biased in favor of the physicist's contribution .

But Hans Bethe, who wrote in a 1946 article published in the United States that "any one of several determined foreign nations [including Russia] could duplicate our work [to develop an A-bomb] in a period of about five years," is harder to mistrust.[23] This physicist not only received a Nobel Prize for the theory of thermonuclear energy of the stars but also directed all the theoretical work in quite a practical manner at Los Alamos. That is why Bethe knew very well exactly what scientific research was necessary to develop the A-bomb. He knew that this research did not require genius, just enough high-level physicists. And he was also familiar with the caliber of Soviet physicists.

Soviet physicists had had to do all the required calculations and experiments themselves. After all—they answered for the results with their own heads. That was why many veterans of the Soviet projects initially so distrusted the information about how much espionage had contributed—they were the ones who had measured and calculated everything. Kurchatov, who knew the American results, helped them avoid some of the blind alleys, which is what shortened the path by one-fifth, if one trusts Bethe's evaluation.

Does saving a year or two for the Soviet A-bomb merit the title of Hero of Russia given to the espionage agents? It probably does if an atomic attack on Russia would otherwise have occurred in that year or two.

The seeming disparity between thousands of pages of espionage materials and a savings of only 20 percent isn't the only paradox in the history of atomic espionage. The biographers of Hall/Mlad did an excellent journalistic investigation and gave us an opportunity to learn about the motives of people who betrayed the most precious Western secrets to the West's chief enemy. These people had, of their own free will, served what they thought was socialism when, in fact, it was Stalinism. And they served unselfishly, if one considers the satisfaction of personal heartfelt aspirations unselfish.

The American journalists who told Hall's story lacked knowledge of the history of physics and Soviet history to adequately fit his contribution to the history of the Soviet atomic project. The paradoxical conclusion one must reach from Hall's biography is that in passing on these most valuable secrets, Hall actually hindered the Soviet atomic project instead of helping it.

Amazing as it seems, Beria suspected "disinformation" in the espionage dispatches and considered "that the enemy is attempting to lure us into expending enormous resources and efforts on work which has no future promise," even after Hiroshima, when there was no doubt that the A-bomb was feasible.[24]

How could that be? After all, Beria had on his desk three independent espionage reports from the very center of the American atomic project. Hall was unaware that right there at his side was Klaus Fuchs, another Soviet agent, older and higher placed, who passed on more complete and accurate information about the A-bomb to Soviet intelligence. There was a third active agent—Sergeant David Greenglass.

Beria was anything but stupid. Two of his predecessors as Stalin's chief gendarme wound up in the meat grinder they themselves had operated. But Beria, after seven years of service, became a marshal and successfully directed the atomic project for the following seven years. This took more than mere cunning and wariness.

As the chief of intelligence he could distrust his own agents, considering their successes too good to be true. And really, as Hall's biographers write, "This was a rare phenomenon, unique in American history: Three individuals, unknown to one another, decided for reasons of political philosophy to commit espionage at the same time, in the same place, giving approximately the same kind of information to the same foreign government."[25]

The most valuable and reliable of the three agents was Fuchs, a mature, balanced thirty-three-year-old who held a high-level scientific position and had been an active antifascist in Germany. And the most suspicious character was the nineteen-year-old Hall, the son of a furrier who showed up on the doorstep of the New York Soviet intelligence office. This volunteer's youth and his petty bourgeois background—unfavorable by Soviet standards—aroused the fear that he was a tool of disinformation.

Unlike his comrades in the Politburo, Beria seemed free of prosocialist sentiments. And it was more difficult for him to believe that so many Americans were willing to risk their lives all at once for the sake of the socialist nation than to suppose that American secret services were trying to deceive him, and in a rather crude way. Consequently, not only did Mlad arouse his suspicions, so did Fuchs, since the content of their information coincided.

But if Hall did not help the Soviet A-bomb, then perhaps he did something at the next stage of developing a superweapon? He himself denies that he made any contribution to the Soviet H-bomb. Even if he did report that the Americans were thinking about using lithium in the superbomb, that doesn't constitute help.

As Goncharov discovered in the Russian archives, Soviet intelligence learned in 1947 that lithium was involved in American research. But the physicist Goncharov emphasized that intelligence did not report anything about the isotopic composition of lithium. Hall's biographers paid little attention to such "minor details"—the main thing was the word "lithium." And for greater effect, the journalists called lithium, along with hydrogen, helium, and beryllium, "the four mysterious light elements." But they didn't explain what was mysterious about them—they really are the lightest elements, but they had been discovered a long time ago. Thermonuclear explosive is made not just from lithium, but from its rare isotope lithium-6, combined with deuterium (LiDochka). This substance was used for the first time in Sloyka.

Hall's biographers apparently did not know that lithium is the first solid substance among the light elements and thus easier to deal with than deuterium and tritium, which are gases. Lithium was used in the very first nuclear reaction made by a particle accelerator in 1932. And the biographers most certainly didn't know that Ginzburg, the future father of LiDochka, talked about this reaction in his popular 1946 pamphlet *The Atomic Nucleus and Its Energy.* He illustrated nuclear energy's reserves "using lithium as the example" (rather than commonplace uranium): "You could use 100–200 grams of lithium instead of an entire train load of coal."[26]

That is why, when Ginzburg was included in the work on the hydrogen bomb two years later, he was just the person to start with lithium. Just going from simple lithium to the specific mechanism of utilizing LiDochka was already a true achievement, which Ginzburg reached in November 1948. Teller arrived at it a year and a half later in the United States.[27]

But even without this chronology, it's difficult to imagine that if Beria had in his hands information related to the construction of the superbomb, he would have given it to the just recently formed auxiliary FIAN group and not to Zeldovich, whose group that had been working inside the project from the start.

Let's remember that Beria allowed even purely scientific data to be given to Tamm only after Khariton's intercession.

Fathers and Grandfathers
of the H-Bomb

The main argument in favor of Sloyka's independent creation is the fact that Americans never tried to build a similar device. But espionage's most important, albeit purely administrative, contribution to the genesis of Sloyka was, as has been said, Fuchs's March 1948 report on the H-bomb. It was simultaneously overestimated and underestimated.

The directors overestimated the report, considering it evidence of intensive American work on the H-bomb, and for this reason created Tamm's auxiliary theoretical group "within a two-week period." Zeldovich, who was given access to these intelligence materials, underestimated Fuchs's opinion. So did—for five years—Fuchs's colleagues at Los Alamos, including Teller, who took part in a conference on the hydrogen bomb at Los Alamos in April 1946. Fuchs left the United States shortly thereafter, and two years later reproduced his two-year-old understanding in his report for Soviet intelligence.

The H-bomb project Fuchs reported on belonged to the "Classic Super" track. Two problems had to be solved in this track: (1) igniting the thermonuclear reaction at the tip of the "cigar," and (2) making the thermonuclear flame spread along the entire "cigar." The second problem turned out to be unsolvable. The American physicists found this in early 1950, while it took the Soviets until the end of 1953. From this chronology alone it is clear that the American "secret of the hydrogen bomb" did not cross Soviet borders. Indeed, the most important secret and the simplest to pass on was: *the Tube was a dead end.* Just a few words. And the Soviet physicists weren't aware of this secret, since they spent more than four extra years going down a blind alley.[28]

In 1948 Fuchs did not, of course, know that the second problem, that of making the thermonuclear flame spread along the entire cylinder, was unsolvable. His information was about the first problem—how to ignite the thermonuclear reaction. The design passed on by Fuchs contained an important idea—the compression of thermonuclear material using an atomic explosion.[29] This was later called "radiation implosion" in America, but Sakharov called it the "Third Idea" in his memoirs. Teller revived this idea in the dead-end situation of early 1951 after Stanislav Ulam proposed the idea of atomic compression. From all appearances, Sakharov discovered this very same idea in reflecting on the (im)possibility of atomic compression in the spring of 1954.

The Teller-Ulam design (and the Third Idea) differed from the design in Fuchs's report in that radiation compresses not only the tip of the thermonuclear "cigar" but the entire "cigar." In the United States, to make this shift they had to find themselves in a blind alley, then take a fresh look.

The course of events in the Soviet project was similar. In February 1950, after Truman's directive generated an analogous (but secret) directive from Stalin, Zeldovich reviewed various methods of igniting the Tube—different variations on an atomic match—in the report "The Hydrogen Deuterium Bomb." Although he mentioned Fuchs's scheme, he favored a different one using material fragments from the explosion.[30] He preferred the physics he knew well from previous experience, and held this opinion another four years, until the dead end became clearly evident. Such a lengthy period of misjudgment on Zeldovich's part might explain why he didn't immediately appreciate how correct Sakharov was in suggesting the Third Idea.

Taken as a whole, American and Russian history helps to resolve the most significant doubt about the originality of the Soviet Third Idea. The veteran-physicist Lev Feoktistov in the article "The Hydrogen Bomb: Who Betrayed Its Secret?" says that the Third Idea seemed to have dropped from the sky, implying that, in his opinion, someone had to have put this (American) idea into that Soviet sky:

> Several months after [the *ADS* atomic compression] new ideas suddenly appeared like light in a dark kingdom, and it became clear that the moment of "truth" had arrived. Talk ascribed these fundamental thoughts, in the manner of Teller, now to Zeldovich, now to Sakharov, or to both, now to someone else, etc. By that time I was well acquainted with Zeldovich. But never once did I hear a direct confirmation about this matter from him . . . I'm still left with the feeling that we weren't entirely independent at that time.[31]

Zeldovich never was excessively humble in his sense of authors' priority, and such restraint on his part is truly revealing. In accepting Sakharov's idea of using radiation for compression, Zeldovich should have recognized the content of Fuchs's 1948 report (to which he received access as far back as June 1948) and regretted that he had not appreciated it then.[32]

As for the suddenness with which the Third Idea appeared, it is useful to juxtapose it with the history of its American equivalent. During a discussion in 1952, while the trail was still warm, Bethe expressed the opinion that "The invention of [deleted (the Teller-Ulam design?)] in 1951 was largely accidental. It is unpredictable whether and when a similar invention was made or will be made by the Russian project."[33] However, Teller, who participated in this invention, noted, not without irony, "It is difficult to argue to what extent an invention is accidental: most difficult for someone who did not make the invention himself." Teller stated that the idea of the 1951 (Teller-Ulam) design was a minor modification of ideas generally known in 1946, and only two elements had to be added: to implode a larger volume, and to achieve greater compression by keeping the imploded material cool as long as possible. Teller also emphasized, "The main principle of radiation implosion was developed in connection with the thermonuclear program and was stated at a conference on the thermonuclear bomb, in the spring of 1946. Dr. Bethe did not attend this conference, but Dr. Fuchs did."[34]

Teller could be chagrined that (like Zeldovich) he hadn't discerned any sooner the potential of the idea that was "known in 1946." Fuchs filed an application for a patent for this idea on May 28, 1946, along with the mathematician von Neumann.[35] It was only in 1951 that this exceptional idea turned into a mechanism of radiation implosion in Teller's hands.[36]

The situation appears paradoxical. Bethe, who was not inclined to fall into raptures, called Teller's contribution "a very brilliant discovery" and "a stroke of genius."[37] Teller himself, not at all inclined to excessive modesty, did not agree with the high assessment of his contribution and asserted that his was but a small modification of previously known ideas.

Half a century later, "material evidence" from declassified Soviet archives was unexpectedly added to the opinions of American experts. This was the information atomic spy Klaus Fuchs, the "extremely brilliant" physicist, "one of the top men in the world on atomic energy," had passed on to Soviet intelligence in 1948.[38] (Bethe's assessment shortly after Fuchs's arrest in early 1950 is in quotes.) The information contained a diagram, probably the same one Fuchs patented in 1946, while still working at Los Alamos. After studying this diagram, Goncharov, a Russian veteran of the thermonuclear project, agreed with Teller's opinion that "the secret of the hydrogen bomb" was related to the ideas of 1946.[39]

It's difficult for a historian of science without access to secret atomic archives to take part in the disagreement between Bethe and Teller about whether Teller's 1951 invention was accidental and unpredictable—or a relatively small modification of previous ideas. It's safest to accept that the truth lies some-

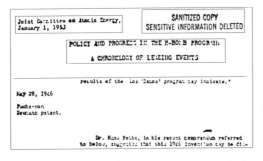 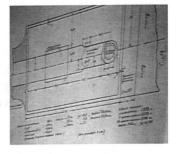

Figure 11.3 (Left) Sanitized and declassified excerpts from "Policy and Progress in the H-Bomb Program: A Chronology of Major Events," Joint Committee on Atomic Energy, January 1, 1953. The May 28, 1946, Fuchs-von Neumann patent is indicated; its description is deleted. (Right) A diagram from Fuchs's 1948 report to Soviet intelligence.

where between their positions. The closer this "golden mean" is to Teller's opinion, the more reason to call Klaus Fuchs the grandfather of all three hydrogen bombs—the American one, the Soviet one, and the British one—because he worked both as part of the establishment and as an agent.

The history of science shows that the genesis of a new idea is not infrequently surrounded by a thick shroud, and it's even difficult for people directly involved in events to separate what occurred into initial impulses, specific ideas, and critical remarks. And questions of who and what came first, when many outstanding minds are taking part in discussions, frequently do not have a simple, unambiguous answer. Teller is of the opinion that "The Work of Many People" (the title of his 1955 article), should be written in the hydrogen bomb's birth certificate.

However, it's important to remember that it took American physicists five years to realize the potential of the "1946 idea," and it took the Soviets six years to rediscover this idea anew. Just from this, Bethe's reasoning about the need for an additional "stroke of genius" is understandable. The disagreement between two prominent figures in American thermonuclear history helps us understand a similar disagreement in Russian thermonuclear history. If Bethe perceived the Teller-Ulam design as an unanticipated discovery, then Feoktistov's impression of the Third Idea's unexpectedness also becomes more comprehensible. And these two impressions, received on different sides of the Iron Curtain, reinforce one another and attest to how innovative the Third Idea was.

A whole cluster of problems had to be solved on the road from a key physics idea to an engineered construction. The Soviet team required a year and a half to complete it.

Wasn't this too fast compared to the all-star American team? In fact, it only took a bit longer in the United States. Mike was tested a year and nine months after the American version of the Third Idea emerged. If the Second Idea had appeared in the United States earlier, and had the production of lithium-6 been set up, Mike might have become the first hydrogen bomb, 16 months before Bravo.

But more important, there were good reasons why the road from the Third Idea to an engineered construction was shorter at Los Arzamas than at Los Alamos. One of the key elements of this construction is the spark plug, which was a plutonium rod located along the axis of the thermonuclear cylinder. This spark plug yields a "spark" in the form of an internal atomic explosion (as a result of the external atomic compression) and ignites the thermonuclear fusion along the entire length of the cylinder. There was nothing comparable in the American Classical Super, while Sloyka's core had an atomic charge from the outset. One had only to realize that the spherical Sloyka compresses into a cylinder on its own from the atomic compression coming from the sides, and that it is better to prepare this cylinder in advance. So just as important to the Third Idea's implementation was the experience Soviet physicists acquired in doing the calculations and tests on Sloyka.

To create the superbomb, the American physicists had to cross a chasm in one big leap, while the Russians took two smaller leaps, with Sloyka as an interim "support." The most important factor for success in these leaps was Sakharov and Zeldovich's collaboration at various stages—from the rapid acknowledgment of Sloyka to the implementation of the Third Idea. Their different styles made this collaboration especially productive. From an eyewitness:

> These two prominent theorists had very different "styles of thinking." Sakharov was characterized by inventiveness and great profundity while Zeldovich by very quick thinking and high erudition. These scientists created an extraordinarily creative climate; the Institute [Installation] became orphaned after their departure at the end of the 1960s.[40]

Another eyewitness recalls how interesting it was to follow the discussion of these outwardly opposite individuals:

> One was short in stature, bespectacled, rapid in his movements, and spoke clearly; the other was tall, languid, and spoke with a slight burr. But they were linked by sharp minds and enormous physical intuition. Mutual problems stimulated their thinking and they quickly grasped the crux of processes; hardly anyone managed to follow the course of their reasoning.[41]

For the thermonuclear steps of 1953 and 1955, Sakharov and Zeldovich received two stars of the Hero of Socialist Labor each.

Sakharov himself did not underestimate the heroism of what he had done. Twenty years later, when he received an invitation to come to the United States and lecture, his wife asked him what would interest him the most in America. By that time his imagination was already involved in cosmology and the physics of elementary particles, and he had an altogether different view of the government for which he had created thermonuclear weapons. However, he told his wife that he wanted very much to sit side by side with Ulam to compare the paths by which they had arrived at the same solution[42] (it was in the 1970s, when the roles played by Ulam and Teller in creating the H-bomb were not clear).

Zeldovich admired Sakharov's talent, treated him "extraordinarily carefully," "timidly," and said: "What am I? Now, Andrei, he's something else!"[43] According to another witness, Zeldovich said: "I can understand and take the measure of other physicists, but Andrei—he's something else, something special."[44]

His "understanding of incommensurability" was formed at the Installation, during the years of their greatest closeness, when they were creating Soviet thermonuclear weapons, and especially when they worked on the "Third Idea."

12

THEORETICAL PHYSICISTS
IN SOVIET PRACTICE

In the 1950s Sakharov and Zeldovich were connected by daily contact in the small world of the Installation: "We would visit one
another's offices several times a day to share a fresh idea or doubt, or just to
joke and talk." They first became acquainted in 1948 when Zeldovich "instantly" saw the value of Sakharov's Sloyka, a factor that helped determine the
project leadership's enthusiasm for the new recruit. Subsequently, Sakharov
never experienced anything resembling "unhealthy competition" on the part
of Zeldovich, who was seven years his senior. Both shared in equal measure
a chivalrous attitude toward science, maintaining that "truth is the highest
value." For a long time, Sakharov did not realize that for Zeldovich this truth
was qualified with the epithet "scientific."

Scientific truths are born, live, and sometimes die within the world of human
relationships. Delighting in the ideas of others or admitting one's own ideas to
be erroneous, cooperatively and unselfishly seeking the truth and courageously
defending it from ignorance and stupidity—all this was Zeldovich's way to do
physics. But he separated the pure, moral country of science from the rest of
the world by a strict boundary and guarded it behind especially prickly barbed
wire. Sakharov understood this moral geography much later on, but he never
accepted it.

Closeness with Zeldovich meant a great deal to Sakharov, not only because
their personal contact and Zeldovich's papers on cosmology "served as a stimulus and jumping off point" for Sakharov's return to pure physics in the 1960s,
but also because friendship in science—his main pursuit in life—created the
feeling of their general closeness. Sakharov still considered Zeldovich a friend
in the early 1970s, by which time he was already combining human rights activism with physics.[1]

Sakharov recalls their conversation in the 1950s: "Once Zeldovich remarked
to me: 'Do you know why it was Igor Tamm who proved so useful to the 'cause,'

and not Dau [Landau]? Tamm has higher moral standards.' Moral standards in this case meant a readiness to devote all one's efforts to 'the cause.'" Sakharov added a conclusion: "This statement, it seems to me, neither described Tamm's position accurately nor was it completely sincere on Zeldovich's part."[2]

This dialogue doesn't so much clarify the relationship between the four prominent theorists as raise questions about it.

Tamm, Landau, and the "Cause"

Sakharov and Zeldovich became three-time Heroes of Socialist Labor for their contributions to "the cause" of creating Soviet nuclear weapons, while their mentors were celebrated with Nobel Prizes for their achievements in pure science. Tamm and Landau also received the highest state awards—Heroes of Socialist Labor and Stalin Prizes for their contributions to the nuclear military "cause."[3] And they quit the project approximately at the same time, soon after Stalin's death and after the successful test of the first Soviet H-bomb.

Tamm and Landau stood side by side in the social life of Soviet science. Suffice to say, one of Tamm's closest students, Ginzburg, did his best-known work (on the theory of superconductivity) together with Landau in 1950. On the other hand, the Central Committee considered the "Tamm-Landau" group as resisting "Party influence" in Soviet science.[4]

Nonetheless, Tamm's and Landau's scientific styles were very different, and their attitudes toward being involved in the atomic project radically differed: the former experienced enthusiasm and satisfaction; the latter, fear and coercion.

In September of 1949, upon meeting an acquaintance, Tamm jubilantly asked: "Have you heard?! Our guys exploded the bomb! They built it so quickly!"[5] Tamm was referring to the A-bomb, in which he did not participate. Two decades later after the successful test of the thermonuclear bomb, Sakharov wrote: "To the end of his life, I think, Igor Tamm rightfully felt satisfaction in recalling those days [in the thermonuclear project]."[6] The overwhelming majority of Soviet physicists who took part in "the cause" felt a similar satisfaction. But not all of them.

Landau's attitude was entirely different. The Central Committee Archive contains an impressive collection of his opinions gathered by the KGB through eavesdropping.[7] This is what he said to physicists who were close to him in 1952–1953:

> One must use all one's strength not to get involved in the thick of atomic work. But one has to be very careful refusing it . . . If it weren't for Box Five [Jewish ethnicity[8]], I would not be doing special [nuclear-weaponry] work, but pure science, in which I now lag behind. The special work gives me a certain amount of personal security. But it's far from my serving "for the good of the Homeland" and the like, which emanated from your letters to me. You

can write letters like that to the Central Committee, but spare me. I don't care what place Soviet physics occupies—whether it's in first or tenth place. I have been reduced to the level of a "scientist slave" and this defines it all.

In the early 1930s he wasn't at all indifferent to the place Soviet physics occupied. The Great Terror of 1937 and the year Landau spent in the Lubyanka prison (1938–1939) changed his attitude toward the Soviet regime. Pro-Soviet fervor turned into anti-Soviet cynicism. The KGB tape recorder in 1957 provides evidence of this:

> Our regime, as I know it from 1937 on, is definitely a fascist regime, and it could not change by itself in any simple way . . . As long as this regime exists, it's even ludicrous to hope that it will develop into anything decent . . . If our regime is unable to collapse in a peaceful way, then a third World War with all its attendant horrors is inevitable. So the question of a peaceful liquidation of our regime is essentially a question about the future of humankind.

In addition to box 5, Landau's "criminal past" had to be neutralized—for although he was released from prison, the charges against him were not dropped. Landau limited his secret work to computational mathematics. As his associate put it, he pursued the work "the way he pursued science in general, that is, giving it his all and with great seriousness"—"everything he did, he could only do at a very high level."[9] But after Stalin's death, he quit the project when the first opportunity came along.

Zeldovich was close enough to Landau to know how he felt about this work. Zeldovich considered Landau his teacher, and it was on Landau's recommendation that Zeldovich was elected Corresponding Member of the Academy of Sciences. However, in the early 1950s, Landau berated Zeldovich with the foulest possible language when the latter attempted to drag him more deeply into secret work in spite of his unwillingness.[10]

It's now clear that Landau's case was exceptional. But it may be that Mikhail Leontovich's case was even more exceptional. When Sakharov and Tamm's work opened up the area of controlled thermonuclear fusion in 1951, Leontovich— on Tamm's recommendation—was appointed director of theoretical work, while the initiators (who were focused on "uncontrolled" fusion) were made permanent consultants. Shortly thereafter Leontovich pounced on a close student of Tamm's whom he encountered:

> Listen, what is your Igor Tamm up to! He's drowning in a swamp and pulling me in after him! You know, it's like when horrible corpses of the drowned, nearly decomposed and covered with seaweed, sitting at the bottom of a godforsaken pond, who suddenly see someone new floundering above them, drowning. And they beckon him with their bony hands, calling: "Come to u-us, to u-u-s, come down, down!"[11]

He didn't say this just for effect. Thirty years later, a few days before he died, Leontovich's fear returned in the form of a hallucination. He was agitated then and whispered to his son-in-law: "I'm very frightened. I'm completely helpless now and *they* will be able to force me . . . To do what *they* require. Now even Igor [Tamm] can't guarantee that at the first turn of events, they won't try to entice me into IT, too."[12]

Leontovich had not served a prison term, like Landau had. He, unlike Landau, had been involved (and enthusiastically) in physics for defense purposes—radar—during the war years. Now work in a new field of real physics lay ahead. But Leontovich already knew that *they* wanted to extract a weapon from this real physics. And most important, by that time something new about *them* was revealed that disgusted Leontovich.

Signal events took place after June 1948, when Tamm landed in "*their* swamp": the Lysenko pogrom in biology, intensified preparations for an analogous pogrom in physics at the end of the year (which, by a miracle, did not take place), and the state anti-Semitism campaign against "rootless cosmopolites" that began in January of 1949. Authorities began to take stock of the ethnicities of scientific research personnel and "improve" the proportion of Jews at FIAN.[13] How Leontovich felt about this is evident from how he distinguished a nationalist from a chauvinist: "His own shit smells good to the former, while to the latter—*only* his own shit smells good."[14]

Leontovich was soon to find out that the harshness of late Stalinism was mitigated in Beria's "godforsaken pond." Work on controlled thermonuclear reaction, the reason for his "invitation" into the project, soon lost its military significance and was declassified five years later through Kurchatov's efforts. Sakharov said about Leontovich's role in developing this area:

There was no better director for the theoretical work to be found. He was rather skeptical about its success, but he still did everything within his power to facilitate it. He was demanding, paternalistic and selfless towards his associates. The enormous achievements in the theoretical physics of plasma would have been impossible without him.

Nonetheless, all this did not cancel out the bitterness for Leontovich in "being dragged into Beria's pond." If he forgave Tamm for it, it was because he knew and loved him too much. And he understood that Tamm believed in the potential of the Soviet system, not out of conformity to the powers-that-be but rather in conformity to the socialist prejudices of his youth. "Leontovich would say with a friendly grin: 'In spite of everything, the Elizavetgrad Soviet member still lives in Tamm.'"

An excellent theoretical physicist, Tamm was also a dreamer in science. His scientific dream generated persistent enthusiasm sufficient for long years of seeking solutions to a scientific mystery and implementing his scientific guess. That was the case in the 1930s and 1940s with investigating the mystery of the forces binding the atomic nucleus, and in the 1950s and 1960s with surmising the existence of minimal length in the physics of the microworld. Neither decade-long epic brought victory. Less lofty, more quickly achieved results made a name for Tamm and earned him a Nobel Prize. But his inspired failures attest to who he was as a unique individual.

His lifelong socialist dream lived right alongside his unrealized scientific dreams. The latter nourished his hope of understanding the true design of Nature, and thereby solving the many tormenting contradictions of the physical world, while the former inspired the hope of resolving the contradictions of social life that the young Tamm took to heart in prerevolutionary tsarist Russia. After the lessons of Bolshevik revolution-making, he abandoned politics for science, but remained an internationalist Menshevik in his soul—and always belonged to the internationalist minority. Contemplating the laws of nature, he hoped in his heart that there was a regularity in history akin to natural laws, and that the events occurring in Russia—despite the terrible brutalities—were turning the wheel of history forward, toward a better future.

He did not submissively watch this wheel crush people. During the terror, "like a madman," he vouched for his arrested brother, who had already been stigmatized in the newspapers; he didn't renounce others who were crushed beneath that same bloody wheel. In the 1950s he defended genetics from Lysenko's "biology," contrary to the party line.

Why wasn't his experience enough to make him reexamine, as Landau did, his attitude toward the essential nature of the Soviet system? First and foremost, Tamm was not given a chance to personally see the secret workings of this system with such prison persuasiveness. Those who disappeared in 1937 did precisely that—they disappeared. The relatively few individuals who managed to return from beyond that horizon kept quiet out of fear that they would be sent back there. Public silence continued until 1956 when Khrushchev vociferously disturbed it at the Twentieth Party Congress. Until then, understanding what was going on largely required using one's own imagination. But Tamm, like many others, saw Khrushchev's "unmasking of Stalin's personality cult" as "socialism's self-purification" and a sign of its viability.

Worldview and general temperament played a key role in this. The terror of 1937 was known as Yezhovshchina, after the then People's Commissar of Internal Affairs. Yezhovshchina's final act was the disappearance of Yezhov himself. A social pessimist might see this as a generic characteristic of any terror, but an optimist like Tamm would think that "it can't get any worse." Tamm's worldview did not allow for the conclusion reached by Landau: the necessity for the separation of science and state.

Leontovich did not share Tamm's optimistic daydreaming at all, but knew him well: "After Mandelshtam's death, there was no physicist closer and more precious to Leontovich than Tamm, with whom he had been friends since the 1920s," when graduate student Leontovich had prepared problems for Tamm's book *The Theory of Electricity*.[15] It was through the prism of this experience that

Leontovich viewed his senior friend who had dragged him into Beria's nuclear swamp.

Let's emphasize once again that Landau and Leontovich were the rarest of exceptions. At that time the predominant mindset of the project participants was that of Tamm and Sakharov. Of course, for many of them a general attitude was not a vital issue; in practice they essentially had no choice. Graduates of universities were subject to "allocation": having provided them with a free education, the state directed the young specialists where it deemed necessary. Even the senior specialists received instructions that would be considered antisocial behavior to refuse.

A justification of one's part in classified work could even be coupled with nonacceptance of the regime. For example, one physicist, due to family circumstances, found out as a child that "we were governed by a gang of bandits."[16] This created a moral dilemma for him when he was confronted, as a university graduate, with becoming "allocated" to the project. But he concluded that he still had to work on this, because "the West would destroy us, as soon as they gained a real advantage." After all, for the West,

> we represented a fascism which—unlike German and Japanese fascism—was victorious, a fascism which dominated two hundred million people, and which would stop at nothing until it reached the Channel, and perhaps even crossed the Atlantic. That is why as soon as the Americans get the H-bomb they won't pass up the chance to get rid of the threat of fascism on the planet once and for all.[17]

However, the American bomb would destroy not only the leaders, but the people, too. That is why the Soviet bomb would protect the people, who were temporarily in the grip of Soviet fascism.

It was a question of psychological self-preservation, of the cover story behind which one could take cover and live peacefully, raising children and pursuing one's work. For the majority, the cover story was very close to the Soviet Union's official position that it needed a nuclear shield after Hiroshima and Nagasaki. The recent devastating war and the harsh Cold War that replaced it, the collapse of hopes for a peaceful and free postwar life—all this within a totally controlled society obscured the cancerous symptoms of Stalinism. And naturally there were apolitical physicists who simply took advantage of the opportunity afforded them to do interesting, prestigious and well-paid work.

"Ignorant Criticism of Modern Physics"

The physicists involved in the atomic project felt protected compared with those in the other sciences, primarily biology. They did not permit their science to be subjected to "Lysenko-ization"—the All-Union Conference of Physicists, for which elaborate preparations had been made, was canceled five months before the first Soviet A-bomb test in 1949. That was a purely defensive "use of nuclear weapons," and the stronghold of "university physics" did not even shudder. The ideological energy of the self-appointed patriots, not exhausted in 1949, emerged three years later, in 1952. Sakharov wrote only a few words in his memoirs about how this energy was rendered harmless: "When physics reached public acknowledgment Kurchatov managed to clear out all this mold." However, the dramatic events of 1952–1954 deserve more attention, especially since Sakharov also took part in them.

As before, Leonid Mandelshtam (deceased eight years) and his students were the main ideological target. The new occasion was the 1950 appearance of the fifth volume of Mandelshtam's *Works*, containing lectures he delivered at Moscow University. Mandelshtam's students, by then themselves prominent physicists, prepared their notes for publication, and Leontovich edited the volume.[18]

Naturally, the lectures were about physics, but the party supervisors, armed to the teeth with quotes, searched between the lines for philosophical foul play, and whoever seeks shall always find. The ideological authority operated not only with head-on directives, they also skillfully "worked with the cadres." In this case, they were able to enlist a man of science who was interested in the philosophy of science.

Aleksandr Aleksandrov gave a talk "On the Subjective and Idealistic Errors Committed by Certain Soviet Physicists" in FIAN on January 28, 1952.[19] The lecturer, a prominent Leningrad mathematician with a powerful public persona considered it his right to judge physics and its philosophy.[20] But at the beginning of his lecture he also indicated the mundane reason for his interference in matters of physics: "I was obligated by instructions given to me by a certain organization to investigate several physics books."[21]

Aleksandrov, a party member since 1951, was forced to submit to party discipline. And he wasn't the only one. After his lecture at FIAN, a commission was created at FIAN "to review the materials in Volume 5 of Academician L. Mandelshtam's Works"[22]—one year after the death of Sergei Vavilov, director of FIAN and president of the Academy of Sciences, who had written the preface to the volume.

The hawkish materialists were on the offensive, and half a year later the Academy of Sciences Publishing House came out with the collective work *Philosophical Problems of Contemporary Physics*. Its preface called for "doing the kind of work among Soviet physicists that had already produced significant results in agro-biology." All they needed to was to find an activist in physics comparable to the agrobiologist Lysenko. This collection was called "green poison" at FIAN because of its content and the color of its cover. The only physicist on the editorial board, Terletsky, combined work at Moscow University with work at the KGB. Maksimov, a Corresponding Member of the Academy of Sciences and a philosophical overseer of physics of many years standing, was head of the editorial board.

However, Maksimov was eager to speak his mind to a larger audience without waiting for the academic work to appear. On June 13, 1952, he published a large article that took up the lower half of two newspaper pages, "Against Reactionary Einsteinism in Physics," hammering Mandelshtam, the chief domestic reactionary Einsteinian. Why this article appeared in the newspaper *Krasnyi Flot* [Red Navy], a publication of the Naval Ministry, remains a mystery.

An even greater mystery until recently was the article written in response, which demolished Maksimov just by its title—"Against Ignorant Criticism of Modern Physics Theories." It is clear why the author of this article, Academician Vladimir Fock, one of the preeminent Soviet theoretical physicists, defended the honor of "the deceased Academician L. Mandelshtam, major Soviet scientist."[23] He had sent a telegram to Mandelshtam's widow eight years before: "Stunned by the death of my greatly beloved senior friend."[24] In addition to being a fearless individual, Fock was sincerely devoted to what he considered dialectical materialism, and he must have been especially offended that Maksimov, "incapable of understanding the subject, was indiscriminately criticizing our remarkable scientist" on behalf of this philosophy. And, finally, Fock must have felt additional responsibility because Aleksandrov considered himself his student.[25]

All this is perfectly clear. But what isn't clear was the timing of Fock's devastating article and where it was published. The place was *Problems of Philosophy*, the nation's main philosophical journal, where Maksimov served on the editorial board; the time was during the final months of the Stalin era when the "Doctors' Plot" was thundering across the land, the state anti-Semitism was peaking, and a Jewish-looking surname such as Mandelshtam seemed like an accusation.[26]

Documents from the Central Committee Archive relate how it happened.[27] Fock wrote his article in the fall of 1952. Then eleven prominent physicists who worked on the atomic project wrote a letter in support of its publication to the project director, Beria. Sakharov signed this letter, along with Tamm, Leontovich, Landau, and others. This was how the name of the thirty-one-year-old Candidate of Sciences, not yet a Hero or Nobel laureate, appeared in the political arena for the first time. The letter talks about the advisability, "in connection with the abnormal situation which has developed in Soviet physics," of publishing "an article by Academician Fock in the central press featuring criticism of Maksimov's article," especially after Maksimov's "incompetent and anti-scientific article." Kurchatov appended his own letter in which he stated that he shared Fock's views, and submitted the article and his letter to Beria.

On December 24, 1952, in accordance with the laws of party bureaucracy, Beria sent these materials to Malenkov, the Secretary of the Central Committee, who was in charge of ideology. In an accompanying letter, Beria informed his Politburo comrade: "Physicists whom you know, comrades Kurchatov,

Alikhanov, Landau, Tamm, Kikoin, [Anatoly] Aleksandrov, Artsimovich, Sakharov, Golovin, Meshcheryakov, Flerov, Leontovich, who share the views contained in the article by Academician Fock, appealed to us with a request to publish this article" since they consider Maksimov's article "anti-scientific and erroneous in how it orients our scientists and *engineers*." Beria added "*engineers*" on his own, implying that these physicists were working on the most important engineering project, and not just Einsteinism (be it reactionary or progressive). This is how the philosophical question was resolved.

And although the Scientific Council of FIAN adopted the resolution on February 9, 1953, "On the Philosophical Errors in the Works of Academician L. Mandelshtam," at a public discussion, Leontovich simply stated that he personally objected, while Fock refused to sign the commission's resolution.[28]

Cleansing university physics of "mold" began again in the fall of 1952 with the help of "nuclear weapons." In this case, Kurchatov came to the aide of the prominent mathematician Ivan Petrovsky, who became the rector of Moscow University in 1951. The new rector discovered that the physics department of the nation's premiere university was not permitting the country's leading physicists, including the authors of the best university textbooks, access to students. At the same time the department was supporting a ludicrous laboratory headed up by Znoyko, an individual who had no relationship to physics. Yet the rector's efforts were not enough to return university physics to the right track.

In December 1952 a commission directed by one of Kurchatov's deputies investigated Znoyko's laboratory. As the physics department party official complained to the Central Committee, this commission, "after a six-hour 'raid' on the laboratory, wrote a conclusion submitting that the objective of its work was anti-scientific."[29]

It took another year before two ministers (of MedMash and Culture) and the president of the Academy of Sciences sent their proposals to the nation's leaders. The letter of November 1953 began with a description of the general situation:

> The Physics Department of Moscow University has been governed for many years now by an unscrupulous group without, for the most part, any scientific or pedagogical value. Members of this group in their day drove out a whole host of major physicists from Moscow University. Under the pretext of struggle with idealistic views, the group discredited the leading scientists of our nation and at the same time has supported individuals with no knowledge of contemporary physics, for example, engineer A. Znoyko (the director of Laboratory No. 15 of the Physics Department). Attempts by Academician I. Petrovsky, the rector of Moscow University, to attract major scientists to professorial and teaching positions were met with extreme hostility by this group, and Academician Petrovsky was accused of preaching the cult of personality and of wanting to use the authority of prominent scientists to push aside and trample young scientists.

Strong measures are suggested in the letter; Sakharov's name is included among those with whom they were discussed.[30]

As a result, the dean of the physics department, Arseniy Sokolov, was dismissed by a Central Committee resolution, and the most militant defenders of "university physics" were forced to leave the university. The top secret Laboratory No. 15 and its chief Znoyko vanished without a trace.

Actually, a single trace did remain. In 1952 the Publishing House of Foreign Literature came out with a translation of Enrico Fermi's *Lectures on Atomic Physics*, and the "works of Znoyko" were mentioned in the anonymous editorial preface to the translation. Leontovich had been in charge of editing physics at this publishing house until 1951. After he left for the project, this position was passed on to the dean of the Moscow University physics department, which in itself says a great deal about how powerful the patrons of university physics were. The anonymous preface, which has preserved the name of the "empirical physicist" Znoyko for our times, shows how serious the threat of Lysenko-ization was in Soviet physics. If history had united the abilities of Terletsky, Maksimov, and Znoyko into one activist and had diminished the abilities with which Kurchatov was endowed, the fate of Soviet physics might have been much more dismal.

Dialectical materialism, the ruling philosophy of science in the USSR, made it possible for Maksimov to attack contemporary physics, and Fock to defend it. This in itself casts some doubts on the validity of Marxism-Leninism. As for its other component, historical materialism or the philosophy of social life, it is difficult to reconcile its founding principle, "social being determines consciousness," with common sense. The same Soviet "social being," the same way of life, generated too great a diversity of consciousness. A literate Marxist would say that in this case Marx was speaking about social consciousness. But what is most interesting is precisely the diversity of individual consciousnesses and how they coexisted with one another and the Soviet way of life. And how the consciousness of Andrei Sakharov, the main theorist of Soviet thermonuclear arms, coexisted with it and changed itself.

Family Life at the Installation

Andrei Sakharov's life and that of his colleagues was not exhausted by thermonuclear invention and struggling with the scientific mold described above. There was also ordinary family life, and everyday ordinariness in the unusual town that had been erased from geographical maps.

Even without geographic maps, the residents of the surrounding villages knew that since it disappeared behind the guarded barbed barrier, something unusual must have happened to it. Large Sunday markets took place in one of these villages. It appears that this was where the supposition arose that a trial model of communism was being built there—the shoppers from there didn't skimp and count their pennies.[31] Moreover, the university-educated Installation shoppers looked radically different from the local market folks and hinted at a future with universal higher education in the communist era.

The Sakharovs bought live chickens at these markets, then housed them in a barn, as was customary; the chickens laid eggs and roamed around their cottage yard to the great delight of their daughters. But the chickens were not for entertainment. During the first years of "trial communism" at the Installation, they were an important addition to their food supply. It was especially important because Luba, the younger daughter, was often ill as a child, and chicken broth was supposed to strengthen her health. That is why sometimes one of the chickens would disappear—much to the horror and indignation of the girls.[32]

The father of the Soviet H-bomb's role in the household consisted mainly of two things: chopping wood and killing chickens, with the same tool. Once a chicken without its head ran around the yard, and he tried to bring this amazing biological fact to the attention of his children, but they weren't up to biology. The girls were deeply put off by this treatment of live creatures, rejected any sort of explanations, and felt that their parents' actions were at odds with their moral admonitions. It was very difficult to reconcile it with the gentle, solicitous manner of their father, who had never once punished them.

Their father spent much more time at work than the official eight hours, but upon returning home, he belonged to his family completely; he devoted time to the kids, played and took walks with them, like all good fathers. He patiently answered all their children's questions, although he didn't especially ask the children many questions himself. The children liked it very much when the lights went out in the house (due to irregularities in the electricity supply), which happened frequently. He would get out the candles and sugar right away. The sugar melted on the flame, slowly dripped and hardened into candy balls. What could be more sweet than sucking on homemade candies?

In the summertime, when Sakharov came home early, the whole family would fetch the neighbors' children, get into his official car, a large seven-seater "ZiM," and go down to the river to swim. In the winter, they enjoyed skiing. In the evenings, the mother and father played their traditional game of chess. And when the children were in bed came the obligatory story, which their father made up as he went. Initially, the main characters were traditional: Baba Yaga and all sorts of forest creatures. When the children grew older, stories "about us" appeared in which the characters were members of the family. These were captivating and rather realistic adventures. A family verb evolved, "about-us-telling." There was only one peculiarity the children had to contend with: sometimes the storyteller would fall silent right at the most exciting point in the story, detach, and go off into his own reflections, and then it wasn't that easy to bring him back to the story.

The girls couldn't tell what he was thinking about. When they asked their father what he did at work, at least in general terms, he would invariably reply jokingly: "It's a ve-ry, ve-ry big secret!" All the children knew was that he was involved in something of extraordinary state importance. Seeing his constant companions, his "secretaries," who protected his body and soul, was proof enough.

Physicists proved to be such valuable state property that for a time they were guarded just like top government leaders. The "secretaries" were appointed to look after Sakharov from the summer of 1954 until November 1957:

> These bodyguards were from a special KGB detachment, and their job was to protect my life and—they made no secret of this—to prevent undesirable contacts. My "secretaries" lived next door to me, both at the Installation and in Moscow. I had to signal them whenever I went out by pressing a special buzzer, and I was supposed to summon them in the event of danger. One was a KGB colonel who had served with the border troops and, later, in Stalin's personal guard . . . Then he had worked "in the arrests" (as he put it) in the Baltic states, which was dangerous work then. He was very tactful, and even considerate. He was starting to think about retirement. My other "secretary," a lieutenant, was also diligent and considerate; he even made an effort to "educate" me politically, but without great success. He was taking a correspondence course in law. The "secretaries" carried Makarov pistols which they kept concealed and showed to me only at my request. They once told me that they could shoot without drawing them from their pockets.

On Sundays when the entire Sakharov family went out into the countryside—the girls would pick wild berries with their mother and he would simply go walking in the woods, lost in his own thoughts—one of the "secretaries" would keep him in his field of vision all the time. The children also noticed the difference between the "secretaries." The older man was gentler and readily played and talked with the children, while the younger one was cold and formal.

In addition to the bodyguards, there were many signs of the constant, unconcealed presence of the eyes and ears of strangers in their home. Once, as he was walking toward his house, Sakharov saw a stranger walk out of it and calmly leave; Sakharov called out to him, but the stranger ignored him. Another time in their Moscow apartment they discovered a cigarette butt in a conspicuous place; apparently the KGB people had come to test their equipment. Tolerating the presence of these strangers was especially painful for Klavdia, Sakharov's wife, whose life was restricted to the confines of her household (she never worked after the birth of their first child).

The Free Thinking of the Top-Secret Physicists

It seems astonishing that the physicists, working on such quintessentially state affairs in such a heavily guarded situation, could engage in free thinking, and that the Soviet regime would allow them to do so.

One of the reasons was that the regime had an overwhelming need for free thinking in their professional work. Another was that physicists almost to a man agreed about what they considered the essence of socialism, and consequently gave themselves permission to disagree with "separate" mistakes and to speak about their disagreements even during Stalinist times.

Soon after Sloyka became the FIAN group's acknowledged project, KGB General Fyodor Malyshev, Beria's representative at FIAN, invited Sakharov to join the Communist Party. Sakharov refused, saying that he would do everything in his power to make the project succeed as a nonparty person: "I couldn't join the Party because some of its actions in the past seem wrong to me, and I don't know if I would have additional doubts about it in the future." He explained that he considered wrong the arrests of innocent people and dispossession of the kulaks. The general stated that mistakes were made, but that the party had corrected them; finally he left the very promising young physicist alone.

Two years later, in 1951, by then at the Installation, and already as a "leading scientific worker," Sakharov again freely said what he believed. A commission "from above" arrived at the Installation to check up on political consciousness, and he was asked what he thought about genetics (destroyed by Lysenko with Stalin's blessing). Sakharov replied that he considered the genetic theory of hereditary correct: "The members of the commission exchanged glances without saying anything. There were no repercussions."

A few days later Sakharov got a chance to see that only his high position at the Installation protected him. When an associate of a lesser rank answered a similar question in a similar manner, he was immediately faced with the threat of dismissal. Sakharov's intercession helped to save him.[33]

The imprudent honesty of these responses meant that the respondents felt themselves within the Soviet system, rather than "internal émigrés." And the exposure of Stalinism's crimes in 1956 at the Twentieth Party Congress reinforced this free thinking even more. It was not at all automatically anti-Soviet, even for Landau, who had condemned Soviet socialism mercilessly. In 1956, two international crises involved the Soviet Union; Landau's responses were quite different—in the first case, he went along with the government, and in the other, he was completely against it.

The crisis which began with Egypt's nationalization of the Suez Canal in October resulted in the invasion of Egypt by Israel and then England and France a week later. The invasion ended when the Soviets threatened to intervene. Then in Hungary popular dissatisfaction grew into an uprising and ended with the proclamation of a multiparty system. In November, Soviet troops crushed this revolution.

Here are some of Landau's statements about these situations, obtained by the KGB:

The Egyptians are as admirable as the Israelis are vile and dishonorable toadies. All my compassion is entirely on the side of the Egyptians. The Israelis infuriate me. I, who am a rootless cosmopolitan, feel total disgust toward them . . . The Hungarian revolution means that virtually the entire Hungarian population, which rose up against their oppressors, that is, against a small Hungarian clique, but mainly against ours . . . Ours stands in blood literally up to their waists. I consider what the Hungarians did the greatest achievement. They were the first in our time to do severe damage, to deal a real stunning blow to the Jesuit idea of our time [that is, Soviet Communism]. A stunning blow![34]

At the same time, a person at the Installation of an entirely different social orientation, Nikolai Dmitriyev, openly expressed a negative attitude toward Soviet intervention in Hungary. He was expelled from the party for this (although he was later reinstated).[35]

This was that same mathematician Dmitriyev who helped Sakharov convince Zeldovich of the correctness of the Third Idea; both physicists valued his talent to an extraordinary degree.

"Kolya [Dmitriyev] was always interested in general topics—philosophical, social, political ones. Absolute intellectual honesty, and a sharp, paradoxical mind were always evident in his positions on these topics"—is what Sakharov wrote many years later, when he considered that Dmitriyev the nonconformist "opposes in equal measure" both official ideology and his—Sakharov's—position. He was the only brave one from the Installation who came to Sakharov's home, in his dissident years, to discuss and mainly to oppose his political views. Even after the disappearance of the "socialist camp" in the 1990s, Dmitriyev disagreed with Sakharov. He never doubted Marxist and Communist ideals, even as he saw that the Soviet implementation of these ideals had serious defects. He considered the division of the world into two opposing social systems fundamental, and viewed all political events through the prism of this confrontation.

In a certain sense the opposing side—America—was more clearly seen from the Installation than from other places in the nation. The Installation library subscribed to *The Bulletin of the Atomic Scientists*, founded by Sakharov's American counterparts who were concerned with the social consequences of nuclear weaponry. They let the nuclear genie out of the bottle, but they provided an example of social responsibility. Sakharov was especially attracted to Leo Szilard, one of the main instigators of the American A-bomb who became one of its chief opponents after its creation.[36]

He discussed this subject with Viktor Adamsky along with other topics going beyond framework of secret physics: "I loved to drop in on him in his little workroom near the staircase to chat about politics, science, literature and life." But it was impossible in the Installation library to become familiar with the most widely distributed book on earth—even though there was no end to "critical" literature about it. Adamsky gave Dmitriyev the primary sourcebook as a gift on his birthday in 1952, and the Bible made a profound impression on him without making him in the least bit religious at all.[37] This shows what the young physicists and mathematicians at the Installation were interested in and could permit themselves.

The end of the Stalinist era could be seen here even more visibly then in other towns in the nation. The signs of the Gulag vanished—no more gray columns of prisoners—and the signs of a normal town appeared: a theater with plays for children and adults, a movie theater, and a club.

Tamm and Sakharov

There were plenty of people around Sakharov who were singularly focused on physics, even just on H-bomb physics alone, for example, on how to use the laws of nature to create a "device" of such and such dimensions with a yield of N kilotons. Sakharov himself seemed completely focused on physics during the period of his own classified triumph. And this is how he seemed to the person closest to him in science.

In October 1953, several days after his election to the Academy, Tamm wrote a reply to a telegram of congratulations from the widow of his student Semyon Shubin, who perished in the 1937 Great Terror:

It was only just now, when I stopped home for a minute, that I received your telegram, sent two and a half hours ago and it's the first one I am answering—I haven't written back to anyone else yet.

Of all the congratulations I received, yours is the most precious to me.

Every person who has lived such a long, diverse and tough life as mine slowly creates his own personal invisible Pantheon. Semyon occupies a very special place in mine. In the first place, I always considered him the most talented, not only of my students (and I am spoiled by how many I had), but of all our physicists, who correspond in age to my students. Andrei Sakharov has only recently appeared—it's difficult to compare them, because much time has passed, and because their scientific profiles are different, and also because Sakharov completely focuses all of his spiritual energies on physics, while for Semyon physics was only *"prima inter pares"* [first among equals]—and this is why one can only say that they are comparable to one another in order of magnitude.

But besides all this, Semyon was one of the people who was closest to me in his emotional makeup—although we were very different people, and although I love many of my students, I have never established such emotional intimacy with any of them. And this is why, of all those who have departed about the same time, my brother and Semyon always remain most vivid of all in my memory.[38]

And so after eight years of close acquaintance with Sakharov—at first when he was a graduate student, then a close associate—Tamm considered that he "completely focuses all of his spiritual energies on physics."

Sakharov saw their relationship very differently. Discussing the physicists who had the most influence on him, he emphasized:

> Igor Tamm's role in my life was especially great, and he was the only one of the four to influence my opinions on—or more precisely, my fundamental approach to—social phenomena. Tamm worked at the Installation from April 1950 until August 1953. This was the period of our closest contact, and I grew to know him in ways that were unavailable to me in Moscow. We worked together throughout the day, had breakfast and lunch together in the cafeteria, and ate supper and spent our evenings and Sundays relaxing together.
>
> In 1950 Tamm was fifty-five, and of course, I was well aware of his brilliant achievements in science. I was aware that he came to his scientific career late in life; he had dedicated his youth to political struggles, impelled by his socialist convictions and his activist bent. During the Civil War, Tamm had some narrow escapes, crossing the front several times on dangerous assignments. He began to pursue science only after these adventures, influenced and greatly aided by Leonid Mandelshtam, whom he met in Odessa during the final stages of the Civil War. Tamm talked about his life and many other things during the evenings we spent one-on-one in the semi-darkness of his hotel room or quietly wandering among the deserted trails in the woods. We talked about the most sensitive topics: the repressions, the camps, anti-Semitism, collectivization, the ideal and real faces of Communism. Tamm primarily influenced my social principles: my own opinions, especially now [in the early 1980s] might differ markedly from his. The underlying principles guiding Tamm were absolute intellectual integrity and courage, a willingness to re-examine his ideas for the sake of truth, and a readiness to take action.
>
> In those early years, Tamm's every word seemed a revelation to me—he had clearly understood many things I was just beginning to notice, and he understood then more profoundly, keenly and energetically than almost anyone else with whom I could speak freely.

Why didn't Tamm notice any interest on the part of his student and close associate in anything else besides science? The simplest explanation lies in the differences in their emotional makeup and life experience—after all, a quarter of a century separated them. The age difference might have made the already reserved Sakharov, filled with love for his mentor in science and life, even more silent. Evidently, this was what made it difficult to notice that he was thinking intensely about what he was learning from Tamm.

The contrast in their temperaments, you might say, couldn't be greater: the slow-moving, unathletic, and quiet Sakharov and the emotional, energetic, fast-talking Tamm (who, in the letter quoted above, wrote the date exactly to the minute: "26.10.53, 17:53").

Of course, this incompatibility in their natures should not have prevented Tamm from noticing Sakharov's conduct, which seems almost heroic to an outsider—for instance when Sakharov offered his Moscow apartment to Agrest and his large family when Agrest was banished from the Installation. Since the apartment stood empty, Tamm, who had himself actively defended the banished man, might have considered this simply the normal conduct of a good person.

In due time, Tamm fully appreciated who his student Sakharov was. In 1968, having been awarded the highest award of the Academy of Sciences (the Lomonosov Gold Medal), Tamm prepared a speech, but couldn't read it himself due to his illness (which sent him to his grave three years later). He asked Sakharov, whose dissidence had already emerged, to read this speech at the General Meeting of the Academy of Sciences.[39] Sakharov reminisced with pride about the trust placed in him.

It is the opinion of David Kirzhnits, who knew both of them well, by that time "Tamm felt such power in Sakharov—its totality, the power of personality, power which neither he himself nor any other of his students had."[40]

But in the years of their joint work at the Installation, the power of Tamm's personality was formative for Sakharov. In a 1971 memorial article about Tamm, Sakharov mentioned Tamm's lessons of integrity at the meetings with the leadership.[41] It's possible that he was recalling the incident that another participant of these meetings related. Sakharov and Zeldovich had come forth in favor of a new radical decision, both with similar enthusiasm, but without sufficient grounds. Tamm spoke out against it and insisted on conducting special calculations that, as a result, refuted his student's prediction.[42]

Another instance of Tamm's integrity could have destroyed the student–teacher relationship if Tamm hadn't been Tamm and Sakharov hadn't been Sakharov. The evaluation that got the thirty-two-year-old Sakharov elected academician, skipping the level of Corresponding Member, was signed by Kurchatov, Khariton, and Zeldovich. The absence of Tamm, Sakharov's teacher and head of the department in which he worked at FIAN and at the Installation, could not be accidental.

His review of Sakharov's doctoral dissertation in early June, a few months before the Academy elections, clearly shows Tamm's evaluation of his student at the time:

A. Sakharov is one of our nation's major leading physicists.

It would be insufficient to say that he possesses broad erudition; his entire creative manner demonstrates that the laws of physics and connections between phenomena are directly visible and tangible in all their internal simplicity to him.

This gift, in combination with a rare originality of scientific thinking and intensity of scientific creativity, has made it possible for him in the past five years to advance three scientific-technological ideas of paramount importance. Each of them is based on the use of unexpected combinations of indisputable physical ideas, which makes it possible to indicate fundamentally new and, at the same time, exceptionally effective ways of solving the current problems of new technology.

The paramount importance of A. Sakharov's ideas for the state has led to the expenditure at present of very large human and material resources for their practical implementation. At the same time, A. Sakharov himself is successfully directing the general scientific research and development of this entire extensive activity.

There can be no doubt that A. Sakharov deserves not only the scholarly degree of Doctor of Physical Sciences but also election to the Academy of Sciences of the USSR.[43]

The last sentence presumed election to Corresponding Member of the Academy of Sciences.[44] Why didn't Tamm subscribe to the recommendation to bypass the first level and elect Sakharov academician, since the opportunity had presented itself?

Tamm could hardly add his signature to a review in which a field of "new technology" was called "the foremost field of physics" and in which it was claimed that Sakharov's scientific and technological ideas were "determining the course of the most important part of Soviet physics." This was no longer exaggeration—this was not true.

Tamm did not in the least confuse the "great importance of Sakharov's ideas for the state" with their scientific importance, and understood that the opportunity to elect Sakharov an academician directly was granted by the government. The Soviet government paid all the expenses of the Academy of Sciences and thus called the tune. However, it appears that Tamm considered it unacceptable for the government to interfere in the very process of composing the music, when a coarse hand from above indicated to the Academy whom it should or should not elect. This had already occurred in 1943, when Tamm as well as Kapitsa had opposed the government's will to directly elect Kurchatov academician, bypassing the Corresponding Member status. At the same time, both Tamm and Kapitsa treated Kurchatov favorably, and he evidently understood their pure motives, since this cast no shadows on their relations.

However, Kurchatov, as "a doer of the Stalinist era," apparently saw no harm in carrying out the instructions (possibly even prompting them) and securing the election of a young, talented physicist and a morally pure person, even if his talent for the time being was emerging mainly in the field of "new technology," not pure physics. In the society in which they live, science would win in general if people like that were in charge of it.

It's unlikely that logic like this would satisfy Tamm. A great deal in the society in which he lived needed to be corrected. And along with Kurchatov's reasons, Tamm knew that state considerations of a completely different nature played a role in Sakharov's supersonic academic advancement. Once, when he suggested to ministerial officials that some young specialists should be admitted into the Installation, what he heard in response was: "Why are all of them Jews!? Give us some Russians."

It was only when the atomic project began, when there was both a lack of time and a dearth of people to choose from, that a physicist's questionnaire data were not the deciding factors, and many Jews wound up in the top scientific positions. The situation required "normalization." Several years later the first specially trained cadres appeared, and along with them the opportunity to carry out the personnel policy, which, of course, reflected the general government policy. Given the situation, a Russian like Sakharov, and such a talented one at that, was a real find.

He himself knew what kind of ideas the leadership was entertaining regarding him. After talking about his refusal to join the party in 1948, he added: "I think that had I agreed, I would have probably been appointed to a major administrative post in the atomic system, scientific director of the Installation or a parallel position. But what good would that have been for the project—I'm no administrator!"

In Sakharov's case, Tamm was more worried about the harm to his student than about the good of the project. Of course, he knew that Sakharov was a real Russian *intelligent*, intolerant of Russian chauvinism. But leaping over an academic level in such troubled waters? What for? And, finally, such a rapid elevation—not on the basis of scientific merit done—was just no good.

In 1953 Tamm could not be sure that his student would survive honorably such an unearned honor. And this could have only come to the party authorities in a nightmare—that one day Sakharov in his Memoirs would list the Jews in the Installation leadership and add, "and I, though not a Jew, was maybe something even worse."

Sakharov said very little about his 1953 academic ascension—apparently, it had little significance for him. And most importantly, it had no bearing on his relationship with Tamm, about which he said, summing up the differences in their views about social phenomena at the end of the 1960s: "Our differences did not change that respect at all, and even, I would dare to say, love, which we felt for one another."

Sakharov wasn't the only one who felt respect and love for Tamm. Talking about his mentor, Sakharov simply cited (with a small change) the words of Evgeny Feinberg, another student of Tamm's:

> In late 19th-century Russia there existed something of fundamental importance—a solid, middle-class professional intelligentsia with firm principles based on spiritual values. That milieu produced committed revolutionaries, poets, and engineers, convinced that the most important thing is to build something, to do something useful. That was the milieu that produced Igor Tamm, and he shared its virtues, and its shortcomings. Perhaps most important of all was his independent spirit in matters large and small, in life and in science.[45]

And Sakharov summed it up: "Perhaps the primary good fortune of my youth was to have had my character developed by the Sakharov family, whose members embodied the 'generic traits' of the Russian intelligentsia which Feinberg described, and then to have come under the influence of Igor Tamm."

13

THE PHYSICS OF
SOCIAL RESPONSIBILITY

Igor Tamm's participation in the H-bomb project came to an end while he was at the test site, in the Kazakhstan steppes, in August 1953. Something happened there that had a greater impact on him than the success of the test of Sloyka. His young colleague Vladimir Ritus, who had not been at the test, became aware of this when he encountered Tamm after the test while both were on vacation at the Black Sea. After taking him to a place where a conversation about top secret matters could be held under sufficiently secure conditions, Tamm seemed less interested in reporting that the yield of the thermonuclear device confirmed their calculations than he was in telling Ritus about how the physicists had saved the people living near the test site from highly dangerous exposure to radiation produced when the bomb was detonated.[1] This close brush with disaster had a great effect on Sakharov, too. Just before he took Tamm's position at the Installation, he faced his first trial in social responsibility.

As Sakharov explained in his *Memoirs*,

> When we arrived at the test site, we learned of an unexpected and very complex situation. The test was planned to be ground level. The device at the moment of detonation was supposed to be in a special tower erected in the center of the site. We knew that ground-level detonations created radioactive "patterns" (bands of radioactive fallout), but no one had thought that with the powerful explosion we were expecting the band would go far beyond the test site and endanger the health and lives of many thousands of people who had nothing to do with our work and knew nothing about the threat hanging over them.

Victor Gavrilov had been aware of the possibility that people living within a certain radius from the test site could be in grave danger of exposure to radiation from fallout. He had specialized in meteorology at university and he knew more about prevailing winds than his colleagues. They worked nonstop for several days to arrive at a full appreciation of the danger. The Soviet scientists were helped by the work of their American counterparts, recounted in the book *The Effects of Atomic Weapons* (1950).[2] The calculations by the Soviet team made it clear that tens of thousands of people should be evacuated from the danger zone. The central government in Moscow took the threat to the population seriously.

> Each of the specialists, including Kurchatov, had to give personal confirmation of the need for evacuation. Malyshev called on us by name; the person had to stand and give his opinion. It was unanimous. Of course, our concerns were not only about the radioactivity, but the success of the test; however, for myself, those concerns took second place to my anxiety for human life. A glance in the mirror shocked me, I had changed so much in those days, my face grayer, so much older.

Hundreds of army trucks evacuated the populace. The residents of one village could not return until the spring of 1954, but at least they were spared the fate of the Japanese fishermen who suffered radioactive fallout from an American test that same spring.

Yet Sakharov shared the fate of the Japanese, to a lesser degree. Malyshev, the head of the testing program, invited him to go look "at what happened there." They stopped just twenty meters from the epicenter. The soil, covered with a black glassy crust, crunched under their feet. There were drops of thermonuclear rain on the ground—shiny black marbles. They had been formed by sand sucked into the air and melted by nuclear fire. They were later called "Kharitonki" after Yuli Khariton, the scientific head of the Installation, and collected as part of the analysis of the test results—surely they had been right in the nuclear inferno.[3]

A vivid image was burned into Sakharov's memory:

> The cars braked near an eagle with burned wings. It tried to fly, but couldn't. Its eyes were clouded, perhaps it was blind. One of the officers got out of the car and killed it with a strong kick, putting the poor bird out of its misery. I was told that thousands of birds died with each test—they fly up with the flash and then crash, burned and blinded.

Even though Sakharov spent no more than half a minute at the site—in protective gear—he felt that this visit to the site shortly after the detonation was the cause of a mysterious bout of severe tonsillitis accompanied by a fever of 106 degrees Fahrenheit, delirium, strong nosebleeds, and changes in his blood three months later. (Malyshev died in 1957 of leukemia.)

The Clean Bomb

The test helped Sakharov in his subsequent work on the weapon, but it also forced him to think about aspects of his work that had implications going far beyond physics.

At the test site, Marshal Vasilyevsky reassured the physicists by telling them that several dozen men die during every military exercise, and those losses are considered unavoidable—and because nuclear testing was much more important for the country's defense capabilities, more casualties were acceptable. That logic did not reassure Sakharov. He had to do everything in his power, using his professional knowledge, to prevent avoidable deaths.

He recalled Zeldovich's words on the eve of the test. "Don't worry, everything will be fine. Our worries about the Kazakh kids will turn out well, and recede into the past."

No amount of optimistic self-reassurance could cancel out the real consequences of technical and political decisions. One of the most obvious steps was to give up surface testing, which caused the greatest amount of radioactive pollution because it irradiated the soil, which in turn meant that the effects would affect the entire natural life cycle. After Sloyka in 1953, the USSR no longer conducted surface explosions of that magnitude.

The bomb testing issue took on enormous political significance after the Japanese fishermen aboard the *Lucky Dragon* felt the breath of the nuclear dragon during American testing in 1954.

A Ban-The-Bomb movement began in the West. In 1957 Albert Schweitzer appealed to all countries with a nuclear capability to call a halt to testing in his "Declaration of Conscience," and a few weeks later Linus Pauling drafted "An Appeal by American Scientists to the Governments and People of the World," which called for an international agreement to stop the testing of nuclear bombs.[4] The campaign had its opponents, who maintained that the danger of testing was being exaggerated, even on a purely biological level, and if one took the defense of Western democracy from communist dictatorship into account, the threat was negligible and the low level of risk justified.

Very different factors were at work, onstage and behind the scenes of these heated debates. The question of the technology of safe testing was linked to a serious scientific problem—how fallout radiation, dispersed throughout the atmosphere, would affect the entire population of Earth. The question at the heart of the issue was what effect small doses of radiation would have over an extended period of time, so it was difficult to get convincing experimental data. To this day—four decades later—this problem provokes heated scientific debates.[5]

The passion of discussion was captured in a book published in 1961 by Erwin Hiebert, an American historian of science who worked on the Manhattan Project in 1944–1945, and the reviews of his book written by two leading participants in the Manhattan Project, Arthur H. Compton and Klaus Fuchs.[6] The former reviewer, a famous American physicist and Nobel laureate, headed a laboratory that created the first chain reaction. The latter reviewer, a former Soviet agent released in 1959 from a British prison, was living in East Germany and at that time was deputy head of the Institute of Nuclear Research. The reviews, unsurprisingly, contradict each other, but both are very critical of the American historian—Compton accusing him of being anti-American, and Fuchs of being pro-American.

Behind the scenes of Western discussions there were powerful nonscientific factors. After the fall of U.S. Senator Joseph McCarthy in 1954, liberalism made a comeback. On the other hand, the military-industrial complex, although not yet called that, was a real force by then.[7] Probably no one could be considered more a part of the military-industrial complex than Edward Teller, "the father of the American hydrogen bomb." Teller confirmed his position as a hard-liner with the publication of his 1958 book, *Our Nuclear Future: Facts, Dangers, and Opportunities.*[8] The book was intended "for the layman who has no knowledge about atoms, bombs and radioactivity." Teller explained it all in a masterly way. His explanation had two clear goals: one short-term, the other long-term. The first was to defend the safety and necessity of nuclear testing, the second to persuade people that the peaceful use of atomic energy was inevitable and vital to the future of civilization. For Teller's arguments to prevail, antinuclear fears in the "Free World" had to be quelled.

Teller, numbers in hand, demonstrated that the danger from radioactive fallout was negligible. First he showed that the radiation from testing composed only a few percent of the natural radiation constantly coming to earth from space (cosmic rays) and from the naturally radioactive elements of the earth. Moreover, even a simple move from sea level to the mountains—say, from San Francisco to Denver—added five times more to the natural background than did all the tests.

He ridiculed the worries of antitesting proponents another way. He compared the danger of radioactive fallout with other, more familiar, dangers in a table that showed how much each reduces a life span.

He summed it up this way: "The reader will see that the world-wide fallout is as dangerous as being an ounce overweight or smoking one cigarette every

Various Factors Affecting Life Expectancy

Factor	Reduced Life Expectancy
Being 10 percent overweight	1.5 years
Smoking one pack of cigarettes a day	9 years
Living in the city instead of the country	5 years
Remaining unmarried	5 years
Having a sedentary job instead of one involving exercise	5 years
Being of the male sex	3 years
The worldwide fallout (lifetime dose at present)	1 to 2 days

two months." That gave average Americans a graphic picture of the threat facing them and was supposed to make them think, "so what's all the fuss?" To which Teller responded, in effect, that Albert Schweitzer and Linus Pauling were marvelous people but not professionals when it came to the bomb or its biological effects. The former was a musician and religious philosopher who became a physician in order to serve the people in the African jungle. The latter was an outstanding chemist. Their Nobel Prizes had no direct bearing on the issue, either. Schweitzer received the Nobel Peace Prize in 1952, and Pauling the Nobel Prize in Chemistry in 1954.

And here was Teller, a specialist in bombs and radiation explaining things with great scientific competence, making a clear and forceful statement. In conclusion, the thermonuclear specialist announced that America had very recently developed the *clean bomb*, which strikes with "heat and blast, but only a negligible amount of radioactivity. The energy of such a bomb is derived almost entirely from the fusion process." The testing of such clean bombs created even fewer reasons for worrying about radioactive fallout. Moreover, the testing was necessary to make nuclear weapons even cleaner.

And the clean thermonuclear weapons were needed not for "the killing of millions of civilians"—which was pointless from a military point of view—but "to stop the armed forces of an aggressor," to counter the attempts of "the Red bloc to take over one country after another, close to their borders, as opportunities arise." Such bombs were to be used to deter communist aggression, not "to kill people in the very country whose liberty we are trying to defend."

The average American did not know that Teller's attitude toward "the powerful communistic countries which strive for world domination" had a profoundly personal component. As he explained in 1996 to his Russian colleagues, "the events in the Soviet Union got an emotional emphasis when my good friend, the excellent physicist, Lev Landau, was jailed by Stalin. I had known him in Leipzig and Copenhagen as an ardent Communist. I was pushed to the conclusion that Stalin's Communism was not much better than the Nazi dictatorship of Hitler."[9]

Teller relied on the account of a friend of his youth (who had seen the destruction of Landau's school firsthand):[10]

> My second published paper in physics was a joint undertaking with my good Hungarian friend, Laszlo Tisza. Shortly after our collaboration in Leipzig he was arrested as a communist by the Hungarian fascist government. He had lost his chance of obtaining an academic position and I referred him, with my strong recommendation, to my friend Lev Landau in Kharkov, Ukraine. A few years later Tisza visited me in the United States. He no longer had any sympathy with Communism. Lev Landau had been arrested in the Soviet Union as a capitalist spy! The implication of this event was for me even more defining than the Hitler-Stalin Pact. By 1940, I had every reason to dislike and distrust the Soviets.[11]

In 1958 Soviet bomb physicists, with the exception of Landau, were completely unprepared to accept Teller's political picture of the world. However, almost all the physicists at the Installation agreed with the professional, biophysical picture that Teller painted in 1958 on the basis of measured and calculated magnitudes. Sakharov was the only one to see a serious flaw in the picture—one that could not be obscured by physics and mathematics.[12]

"Moral and Political Conclusions from Numbers"

The stimulus to more closely examine this seemingly benign biophysical picture was the "clean bomb," which, in Teller's opinion, obviated the problem. In his book, Teller mentions, favorably, the test of a "clean bomb" carried out on July 19, 1957.[13] That same summer, President Dwight Eisenhower supported the last word in bomb science, discrediting those who were arguing against nuclear testing.[14]

Physicists knew very well that even an "ideally clean bomb"—that is a purely thermonuclear bomb without a fission component—inevitably produced radioactive fallout. Thermonuclear neutrons create a radioactive isotope of carbon-14, right "out of the thin air," from the nitrogen in the air. This radioactive pollution quickly becomes the deadly inheritance of all humanity. Carbon is the basic element of living things and radioactive carbon is indistinguishable from ordinary carbon in all biological processes. Therefore, the level of radioactive carbon in living organisms quickly rises to that of the level of radioactive carbon in the atmosphere. Once in the organism, radioactive carbon lives and dies according to its laws, and in dying, disintegrates and emits radiation and damages the organism. For a physicist, the question of quantity is essential—how much radioactive carbon is formed and how harmful is its radiation. Teller does not mention radioactive carbon in his book at all.

To rebut Teller's assertion, Kurchatov asked Sakharov to write an article on the radioactive non-cleanliness of the "clean bomb."[15] "The primary goal of the article was to denounce the new American development without touching on 'ordinary' thermonuclear weapons. That is, the goal was blatantly political, and therefore there was an unseemly element of one-sidedness," Sakharov wrote. "But in the course of work on the article and consequent reading of the vast humanitarian, political, and scientific literature I went far beyond the original framework."

In Sakharov's article, dated July 8, 1958, and published in the scientific journal *Atomnaya energiya*, the physics and mathematical calculations (based on biological data) lead to very definite numbers: every megaton of the "cleanest" thermonuclear blast in the atmosphere creates an amount of radioactive carbon that will kill 6,600 people in the course of 8,000 years all over the planet.[16]

From the statistical point of view those figures match Teller's assessment— one day of life or one cigarette every two months. Why then did Sakharov choose another form of expressing it? To better serve his propaganda goal? No, he felt that his conclusion responded to the *physical* aspect of the social phenomenon under study. Humanity menaced by radioactive carbon is like a street crowd in which a madman starts shooting randomly. A smaller level of radioactive carbon in the atmosphere does not mean that stray bullets will become softer or lighter, only that the shots will be less frequent. Therefore, it is wrong to say that everyone alive will have a day less to live. What will happen is that those unlucky enough to get hit by a stray bullet will die. You cannot average those deaths over the life span of everyone in the crowd.

This danger is fundamentally different from the ones Teller used in his table. Overweight, smoking, and other old-fashioned dangers depend on a person's behavior. But no one can hide from the effects of radioactivity.

A villainous madman with a radioactive carbon gun in his hand is invisible to passersby, but Sakharov sees him through the eyes of a theoretical physicist. For him the death of people from the fallout of testing, no matter how small compared to death from other causes, is a proven scientific fact. As was the fact that if the tests were stopped the madman's arsenal would be cut in half only in 5,570 years, the half-life of radioactive carbon. But if they are not stopped, every new megaton of testing adds another 6,600 deadly rounds to his arsenal.

Sakharov was a theoretical physicist who had full confidence in his figures. But he was a practical humanitarian, and he felt his personal responsibility for the actions of the invisible killer. That is why in his article he raised the question that goes beyond physics: "What moral and political conclusions must be made from these numbers?" And so in 1958, he used the words "moral and political"—not part of the physics vocabulary—for the first time in his writing. The natural conclusion that followed from his numbers is that in terms of dangers to the atmosphere, testing the "clean" bomb was no different from the "dirty" one, and that continued testing will kill innocent people.

When Sakharov wrote a popular version of his article (May 24, 1958) at Kurchatov's request, he used phraseology that corresponded to the clichés of Soviet newspapers. According to Sakharov, that wording reflected his position at the time, "which was only beginning to deviate slightly from the official one." But just the order of words "moral and political" shows that Sakharov's political rhetoric, which was not notable for its originality, had a secondary significance. The moral theme predominated, and that is doubly noteworthy. First, because it deviated from the relationship of morality and politics in Soviet ideology— everything that promoted the victory of Communism was deemed moral. And second, Sakharov's moral conclusions distinguished him from his colleagues in the military-industrial complexes on both sides of the Iron Curtain.

In 1958 Teller concluded his defense of testing with words on political morality, too. "It has been claimed that it is wrong to endanger any human life. Is it not more realistic and in fact more in keeping with the ideals of humanitarianism to strive toward a better life for all mankind?"[17] That rhetorical question can be translated into the Russian proverb, "When you chop wood, the chips fly" or—in English—"You have to break eggs to make an omelet." Sakharov wrote that in Stalinist times he had accepted that popular "law" of Soviet history. In 1958 he would have been ready to accept the American version if it included "the ideas of peaceful coexistence, noninterference, disarmament, and most importantly, a halt to nuclear testing rather than the risky ideas of military parity (that is, the arms race), which are only one step away from preventive war."

And he added quite an official stance of Soviet politics:

> In order to guarantee security in the face of the US and England armed with nuclear weapons, the Soviet state was forced to develop and test its own nuclear weapons. However, the goal of the policy of the USSR and other countries in the socialist camp is not an arms race, but peaceful coexistence, disarmament, and a ban on nuclear weapons—weapons of mass destruction. An important step in the direction was made on March 31, 1958 [when the Soviet Union declared a unilateral halt in nuclear testing]. The position of Soviet scientists is clear. This is unfaltering support of the historic and humanitarian decisions of the Supreme Soviet of the USSR. We firmly believe that this is the position of the great majority of scientists abroad.[18]

Obviously, the father of the Soviet hydrogen bomb had his version of the political world as did the father of the American hydrogen bomb—where "the Red bloc was striving for world domination." Looking from our vantage point now at the two competing pictures, separated by the Iron Curtain, Rudyard Kipling's line comes to mind: "Oh, East is East, and West is West, and never the twain shall meet." And if they had? It would have been very interesting to watch the two physicists compare their black-and-white—or red-and-white—political pictures. Perhaps they would have become aware of the depth of the mutual distrust separating the two blocs and that might have affected their political views. But they did not meet until thirty years later, and Sakharov had to develop his political views in the narrow space allowed for such thoughts in his country and without much knowledge of the range of such thoughts in the American scientific community.

But even if Teller and Sakharov had met in 1958, the differences in their understanding of the problem of testing would have been harder to reconcile than their politics. Nor did Sakharov find sympathy for his moral position among the majority of his Soviet colleagues. "My opinion formed back in the 1950s that nuclear testing in the atmosphere was a direct crime against humanity, in no way different from, say, secretly pouring deadly germs into a city water reservoir, found no support among the people around me."

Sakharov mentions only one exception—Viktor Adamsky. "He was sympathetic to my views on the harmful effects of testing, which was support for me in the general atmosphere of lack of understanding or what seemed to be cynicism to me." Other colleagues thought he was inflating or exaggerating the seriousness of the problem. He seemed, to them, to resemble Ivan Karamazov, who refused to accept higher world harmony if it required the tears of just one tortured child.

However, quite unlike Ivan Karamazov, Sakharov was moved by a feeling of personal and professional responsibility, rather than by general contemplation of world harmony. He felt *he* was responsible for thousands of defenseless victims—and he refused to evade that responsibility by extending the guilt to the many participating politicians and physicists or by diminishing it on the grounds of the small scale of sacrifice. In presenting his conclusions in the articles written in 1958, his primary audience may have been his Soviet colleagues, who did not understand him.

He came up with various ways of explaining the moral situation he saw so clearly. Why not regard these "human costs" as payment for technological progress, like the casualties of automobile culture? Because car accidents are caused by the carelessness of specific people who bear criminal responsibility, while the victims of nuclear testing are entirely anonymous and therefore their "unpremeditated" killers—including himself—are not liable. "The suffering and death of hundreds of thousands of victims, including people in neutral countries and in future generations" is a crime, and an unpunished one, "since it cannot be proved that radiation was the cause of death and also because future generations have no way to defend themselves from our actions."

And then he musters a very obvious argument: "A judge examines a murder charge independently of thousands of other deaths and accidents in a big city and reaches a verdict without regard for mitigating circumstances, whatever small percentage the particular tragedy may be of the entire mass of tragedies."[19] But it did not work. His colleagues refused to regard themselves as defendants in a murder trial.[20] In any case, they did not speak out since questions like this were not discussed publicly in the USSR.

Therefore, Sakharov was writing about his own moral accountability when he declared,

> The remote consequences of radioactive carbon do not mitigate the moral responsibility for future victims. Only an extreme lack of imagination can let one ignore suffering that occurs out of sight. The conscience of the modern scientist must not make distinctions between the suffering of his contemporaries and the suffering of generations yet unborn.[21]

Sakharov's voice of conscience corresponded with the goals of Soviet policy then. Khrushchev personally approved the publication of both articles. The popular article was translated into several languages and was used in Soviet propaganda abroad.

But it was not published in Russia, most likely out of fear of alarming the Soviet people. People might not understand that he was talking about something that would happen thousands of years later. Sakharov apparently did not care about the nonpublication of his popular article—he was addressing physicists and politicians, not the "broad masses." He offered to send the text of his article to Senator Clinton Anderson of New Mexico, who had participated in a discussion of the problem of nuclear testing.[22] This offer was not approved by the Soviet government.

What effect did Sakharov's political debut have in the West? None at all.[23] Sakharov's name was not known in the West, and no one paid attention to his articles, even though they were translated into English. Soviet propaganda took care of the popular article, and the Americans translated the scientific one themselves. Consultants Bureau, Inc., was an agency that translated Soviet scientific literature, including the journal *Atomnaya energiya*. Interest in Soviet science had risen sharply after the launch of Sputnik in 1957. Yet neither the Western foes of testing nor its supporters heard that single quiet voice from the USSR. They were too busy arguing with each other. But both sides could have used the Soviet article for their own means.

The pro-testing people could have reproached their opponents for being Soviet stooges. Those who opposed testing also had good material.[24] They could also have pointed out that it was a physicist working in thermonuclear weapons, perhaps even the "father" of the Soviet hydrogen bomb, writing about the moral criminality of testing—to the shame of his American bomb colleagues.

Could Sakharov's status have been figured out without the help of the CIA? One should only pick up the Soviet *Biographical Dictionary of Natural and Technological Science* (1958–1959) and find the seven-line entry for "Sakharov, Andrei." It reported that he had been working in the Physical Institute of the Academy of Sciences of the USSR since 1945, that in 1953 (soon after the testing of the Soviet hydrogen bomb) he became an academician at the age of thirty-two, and that in 1950 he worked jointly with I. Tamm on an important project in thermonuclear physics.[25] But there are no publications on this project listed; only three articles in other fields of physics published by Sakharov in 1947 and 1948 are mentioned. Taking into account that Tamm achieved international fame with his Nobel Prize in 1958, then it would not have been difficult to make the obvious connection—that Sakharov was no mere propagandist but a physicist of the highest caliber in the Soviet nuclear project.

However, Pushkin's observation that "we are lazy and not curious" apparently applies outside Russia, too. Sakharov's identity remained undiscovered, so Western scientific and political activists were not forced to wonder why the situation in the Soviet Union was so different from the United States that one of the key developers of its nuclear weapons arsenal would feel compelled to speak out against testing. But now we know that things were not at all different in the Soviet Union, and that Sakharov was the black sheep in the family of nuclear arms designers.

But why was he so deeply troubled and not the others? Sakharov asked himself the same question and tried to answer it.

> A large psychological role here (and later) was played by a certain abstract quality of my thought and characteristics of the emotional sphere. I'm talking about it here without self-praise or self-condemnation—I'm merely stating a fact. The peculiarity of long-term biological consequences of nuclear explosions is that they can be calculated and a more or less precise number of victims can be determined, but it is almost impossible to ascertain who exactly those victims are, to find them in the human sea. And on the other hand, seeing a person who died of cancer, say, or a child born with congenital deformities, we can never maintain that this particular death or defect was caused by nuclear testing. This anonymity or statistical nature of nuclear testing's tragic consequences creates a psychological situation to which people respond in varying ways. I, for one, could never understand those for whom this problem simply does not exist.

Sakharov's thinking had the rare ability to combine the very abstract with the completely concrete. As Tamm said about him in a 1953 letter of recommendation: "the laws of physics and connections between phenomena are directly visible and tangible in all their internal simplicity to him." The success of his bomb inventions was determined by a fluency in abstract theoretical physics and an ability to deal with the "device" in which he forced the physics to work. In the same way, he moved quite naturally and easily from a device he invented at the Ulyanovsk cartridge factory to abstract questions of theoretical physics.

Thinking about the biological consequences of nuclear testing, he could clearly see the concrete result—death and congenital deformities, however anonymous they would be. He considered the theoretical averaging of those actual tragedies by dividing them up over all of humanity—shortening the life of every person on the planet by one day—as simply incorrect theorizing, if not political trickery. This manner of thinking strengthened the *emotion* of social responsibility, but it could not determine it. Life experience and a moral legacy were the other components that formed his worldview.

Sakharov took away a lesson from the testing of his Sloyka in 1953. Only a stroke of luck prevented a catastrophe like the one that befell the crew of the Japanese *Lucky Dragon* in March 1954, but on a much larger scale. The stroke of luck came from Viktor Gavrilov, a man who was not directly responsible for the course of the test. We can imagine how Sakharov would have suffered if due to his lack of forethought radioactive rain had fallen onto the heads of Kazakh children who had not been evacuated.

One of Sakharov's colleagues recalled his response during his later human rights activism—"If not me, who?"—and perceived it as a manifestation of fearlessness.[26] It is just as plausible to see Sakharov's sober comprehension of his social position, as "father of the Soviet hydrogen bomb." That empirical fact strengthened his feeling of personal responsibility, whose roots naturally go back to his birth family and his "scientific family" in which he matured, and thus all the way to the legacy of the Russian intelligentsia.

Khrushchev's Moratorium Declared, Violated, and Revoked

Sakharov took the responsibility for radioactive poisoning of the planet's future, but he also accepted responsibility for its present. There was only one way to prevent nuclear war, Sakharov and his colleagues believed in those days— by maintaining parity of nuclear arms with the United States. Soviet nuclear power was supposed to force the West to seek political solutions for all problems, including the most important one, reliable guarantees of peace. And at the time, an agreement on a total ban on nuclear testing seemed to him, as it did to the Soviet leadership, a fully viable interim step toward total peace. Maintaining a balance of nuclear forces meant work—research and development of ever more sophisticated weapons. For many of Sakharov's colleagues this was interesting, prestigious, and well-paying work.

And then on March 31, 1958, the Soviet government proclaimed a unilateral moratorium on nuclear testing. In his May article on the dangers of testing, Sakharov mentioned the unfaltering support of Soviet scientists for this "historic and humanitarian decision."

But the scientists he saw every day at the Installation were hardly quite so unfaltering in their support. After all, they did not share Sakharov's opinion of the dangers of testing. And a halt on testing put their professional futures in jeopardy. But even the most conscientious—to use Soviet phraseology for those who put society before the individual—were stunned to learn of the government decision. It was as much of a surprise to them as it was to the U.S. government.

Sakharov had learned about the decision accidentally, when he came to the Central Committee on another matter. He had wanted an end to testing, but even he was taken aback by such an autocratic way—without consultation with specialists—of handling the issue that involved so much high technology. But he placed too much trust in Khrushchev. People who maintained a socialist orientation forgave Khrushchev many things because of his exposé of Stalin, the mass rehabilitation of the victims of the Terror, the general cultural "thaw," and the call for peaceful coexistence at the Twentieth Party Congress in 1956.

In the fall of 1956 Sakharov asked Tamm if he liked Khrushchev. "I added that I liked him very much, for he was so different from Stalin. Tamm replied to my ardor without a shadow of a smile, saying that yes, he liked Khrushchev and of course he wasn't Stalin, but it would have been better if he differed from Stalin even more." In hindsight, it's easy to say that Sakharov had overestimated Khrushchev's differences from Stalin, while Western leaders underestimated them. However, from our point of view, enriched with information from the archives, it is easier to appreciate the Western leaders. When you do not believe in the historical inevitability of the total victory of Communism, it is much easier to notice that the actual structure of the Soviet government had not changed with the advent of Khrushchev. The pyramid of power still had very steep walls—too much depended on the very few at the top, especially the one at the very top. As before, the pyramid protected itself from erosion by external winds of change, forbidding its citizens contact with the outside world. As before, the press was under the total control of the government. It isn't so important what it was called—autocracy, dictatorship, or Soviet rule. What is important is that it is difficult to deal with such a regime, hard to predict its actions, hard to trust, and even harder to verify its actions. And the fact that the Soviet leaders really did believe in their country's historical mission to liberate the world made them even less reliable partners in international politics.

The exposé of Stalin's crimes appeared to Sakharov and many of the Soviet intelligentsia as a self-purging and "restoration of true socialism." The Western pragmatic politician saw the exposé as nothing more than an instrument in the struggle for personal power. Even now, when it is clear that both factors were at work, it is difficult to assess their relative proportions.

Both factors worked in "peaceful coexistence" entering the arsenal of Soviet politics. This important shift in ideology occurred during 1954–1956, after the long years of Stalin's strong belief in the inevitability of world war, which would lead to the global victory of socialism. The birth of the hydrogen bomb changed the situation.[27]

In 1953 two American presidents highlighted the change. Two months after the first thermonuclear explosion (the "Mike" test), Truman in his January valedictory, without disclosing the actual power magnitude of the new weapon, proclaimed that the thermonuclear war would "*destroy the very structure of a civilization.*" In December Eisenhower, in his "Atoms for Peace" speech to the United Nations, talked about the "*the probability of civilization destroyed.*" Four months after the test of Sloyka, he explained the brand-new situation created by the new weapon: even a significant superiority in the number of nuclear arms does not make the price a country must pay worthwhile in event of a nuclear war.

Physicists—Einstein, Bohr, Szilard—realized this right after the birth of the first generation of nuclear weapons in 1945. Before coming to the same conclusion, politicians had to see that the power of nuclear weapons had grown thousands-fold in a few years and had still not reached its limit.

The first Soviet leader to recognize the new reality was Malenkov, who announced publicly on March 12, 1954, that a new world war would mean the end of world civilization. He was better prepared than the other Soviet leaders for this assessment—he had a scientific education (he graduated from the Moscow Higher Technical School) and he was well informed. "His man" Vyacheslav Malyshev (also a graduate of the same university) headed the atomic project after Beria. On March 1 the Americans tested Bravo, a bomb forty times more powerful than Sloyka. Malenkov received an analysis of the resulting situation prepared by Kurchatov and three other outstanding scientists.

So on March 12 Malenkov was responding to the December speech by the American president and to the March demonstration of American thermonuclear might. However, his realism in foreign policy contributed to his downfall on the domestic front. Khrushchev took the opportunity to rid himself of his main political rival and criticized his deviation from the party line and his nuclear panic. But just two years later, Khrushchev, having elbowed aside his recent comrade-in-arms, proclaimed that the party had taken "peaceful coexistence" into its arsenal, so to speak.

What could a Western observer think about this new Communist slogan? Was it a fundamental revision of the political world map in a nuclear age, or pure demagoguery and propaganda?

Khrushchev's subsequent actions and his *Memoirs* suggest that it was not just political sham. He was no fanatic prepared to sacrifice his people for the triumph of an idea. His simple common sense was against war. However, there are no signs that he appreciated the fundamentally different political reality created by science. His level of scientific comprehension is evident from his continued support of the quack Lysenko to the end of his political career.

In January 1958 *Time* magazine proclaimed Khrushchev Man of the Year.[28] The symbol of the year was Sputnik—actually, the first two satellites launched by the USSR (four months before the United States). The article described other achievements, primarily political, of "Russia's stubby and bald, garrulous and brilliant ruler," who in 1957 "outran, outfoxed, outbragged, outworked and outdrank them all."

In its conclusion, the article gave him some advice: "At 63 Nikita himself does not yet have absolute power, is still best described as chairman of the gang. And to control such a gang, as Nikita well knows, takes far more political skill than Stalin ever required. Khrushchev's Russia needs its thinking men—its scientists and its technicians—and Khrushchev must allow them to think. They demand respect. They can do without Khrushchev, but Khrushchev cannot do without them."

It looks like his staff didn't show Khrushchev the advice. In any case, he made the decision to halt nuclear testing in March 1958 without any consultation with scientists. American politicians couldn't even imagine such a thing.[29] They saw it as yet another Oriental—communist—trick: the Russians must have tested everything they needed to test, and now they can afford to stop testing for a long time, earning political points and spurring on the leftwing opposition in the West. And so the Americans responded that, before joining the moratorium, they had to complete their planned tests. Over the summer of

1958 they detonated some thirty nuclear explosions, and Khrushchev lost his nerve—he ordered a renewal of Soviet testing.

In those years there is no evidence of political interest on Sakharov's part that went beyond the limits of his professional competence. Like a true professional, he treated other professionals with respect. As a man of honesty, it was easier for him to assume that he did not understand something important in foreign policy than to question either Khrushchev's ability to rule or his country's social system.

So Sakharov tried to save the moratorium and the face of the USSR with his professional strengths. He offered to change the construction of weapons earmarked for testing, to make them so reliable that they could be certified without full-scale testing. He persuaded Kurchatov of the technical viability of this path, and Kurchatov flew to the Crimea to see the vacationing Khrushchev.

The country's leader apparently knew better than the scientists and he rejected the recommendations made by the physicists. A series of twenty tests took place in October, and when Kurchatov spoke a few months later at a party congress, he reported that "they showed the high effectiveness of some new principles developed by Soviet scientists and engineers. As a result the Soviet Army has been given even more powerful, more perfected, more reliable, more compact, and more inexpensive atomic and hydrogen weapons."

In citing these words in his memoirs, Sakharov agreed with him. And the two loyal physicists Kurchatov and Sakharov were given reason to think that perhaps the country's leader did know better than they what to do: after the Soviet test of November 3, 1958, nuclear calm prevailed on the superpowers' test sites for almost three years.[30]

But outside nuclear policy, those three years were far from calm. To the surprise of both superpowers there was a revolution in Cuba in early 1959. The prosocialist words and deeds of the young bearded revolutionaries—*los barbudos*—elicited antipathy in the United States and equally fervent support in the USSR. A Soviet person did not need to be interested in politics to share these feelings. Especially after April 1961, when the *barbudos* fought off, without outside help, a thousand-man army of counterrevolutionaries trained and armed by the United States.

Another major anti-American event occurred on May 1, 1960, a day that Soviet tradition celebrated as a holiday of peace and labor (even though it was called the Day of International Solidarity of the Workers). That day an American U-2 spy plane was shot down over USSR territory. Thinking that the pilot was killed, the American government first tried to deny the espionage nature of the flight. However, the pilot had been captured alive and presented to the public along with his espionage equipment. This international scandal led to a cancellation of a May 16 summit of the four nuclear powers.

Now, with Soviet–American tensions running high, after two-and-a-half years of nuclear moratorium, Khrushchev decided to revoke it. He announced his unexpected decision to nuclear weapons designers at a specially convened meeting at the Kremlin in July 1961. He cited changes in the international situation and the fact that the USSR was lagging severely behind the United States in the number of tests. At the time, the score was 83 to 194, but it had not changed since the start of the moratorium.

Sakharov felt that a renewal of testing could yield very little technically—there were no developments that needed to be checked, and he said so in a speech. Since there was no reaction to his statement, he wrote a note to Khrushchev, expanding his argument.

Khrushchev responded to the note a bit later, during the banquet for the participants of the meeting. According to Sakharov, the head of state started out calmly but got wound up and shouted that Sakharov was sticking his nose where it didn't belong, that he knew nothing about politics, and that he had many illusions. He promised to take him along the next time he went to negotiate with the capitalists. "Let him see them and their world with his own eyes, maybe then he'll understand a few things." No one present supported Sakharov.

A few weeks later, during a meeting to review preparations for testing, Khrushchev asked Sakharov whether he had understood his error. He replied in an unusual way: "My point of view is unchanged. I do my work, I fulfill the order."

How is that to be understood? Was he prepared to fulfill whatever order came from the government? Or that he had not learned anything to contradict his beliefs but accepted that Khrushchev had reasons that could not be revealed to him? After all, he never did negotiate with the capitalists and did not know what arguments worked on them.

Sakharov trusted the country's leader—trust fed not only by the unseemly behavior of the Americans but also by the fact that Khrushchev had been right in 1958—that the Soviet autumn tests had not interfered with the establishment of the moratorium. The greatest reason for trust on the part of Sakharov and many others was the process of de-Stalinization under way in the country; its greatest guarantor was Khrushchev personally, against the silent resistance of the majority of the *nomenklatura*—influential officials who are party appointees.

Let us not forget that 1961 was the year of the Twenty-second Party Congress, where Khrushchev passed the resolution to remove Stalin's mummy from the mausoleum on Red Square. It was the year when Solzhenitsyn finally came out of his literary underground, and Khrushchev himself made the decision to publish his groundbreaking novella *One Day in the Life of Ivan Denisovich* about life in the Stalinist Gulag camps.[31]

Lydia Chukovskaya's words in her diary, dedicated to her friendship with the great poet Anna Akhmatova, convey the feelings of the times: "A new time, which we have lived, lived, lived to see."[32] These two women lived for poetry, without any pro- or anticommunist feelings. Their social ideal was being able to speak freely about what one thought and felt. The "new time" allowed Akhmatova to read her poems aloud and even to dictate them to Chukovskaya, who wrote them down and took them with her. In the "old time" Akhmatova wrote her poems down on paper for Chukovskaya to read when she visited and then burned them on the spot.

That may not seem like much evidence for the coming of a new era. That is, if you look at it from a few decades in the future or several thousand miles to the West. But Chukovskaya and her peers used another measuring stick—one with Stalin's stamp on it.

The Tsar Bomb

Viktor Adamsky recalled their mood when scientists from the Installation were preparing for the tests of 1961. "All of us, including Sakharov, held a naively patriotic point of view that we had to have the most powerful and most effective weapons, and that this must be known to 'potential enemies' as well as 'people of good will,' who would influence their governments to agree to the ban."[33] Sakharov participated in the development of two devices for that series of tests. They had no reasonable military application (which is what he had referred to when he spoke out at the Kremlin).

One was designed for 100 megatons but was being tested in a "clean" version, which reduced its power by half. But its yield was still more than three times greater than the American record.

This device is sometimes called the Tsar Bomb. Whoever came up with the name, a parallel with the Tsar Cannon and the Tsar Bell, obviously had not paid attention to the tour guides on the Kremlin grounds. The 40-ton Tsar Cannon never shot a round and the 200-ton Tsar Bell never tolled—they are symbols of meaningless grandeur—whereas the Tsar Bomb exploded and resounded throughout the world. The energy of the blast was several times greater than the total of all the bombs of World War II, including the two A-bombs.

American arms experts saw no military point in such a blast. But the political point was made, particularly inside the USSR. Khrushchev announced the upcoming record-setting blast on the first day of the Twenty-second Party Congress and on the final day the delegates learned of the test's success.

The "potential enemies" learned about it, too. Their physicists studied air samples and concluded that the Russian bomb was fifty-eight megatons and 98 percent clean. The Americans had overestimated the Russian device a bit, but its actual fifty megatons and 97 percent purity remains a record. However, even a 100 percent clean fifty megaton bomb detonated in the atmosphere equals, according to Sakharov's calculations, $50 \times 6,600 = 330,000$ victims over 8,000 years. What did he think of those figures? Could he have lacked the imagination to see the suffering those numbers represented? The infants born with genetic defects, the adults dying of cancer at their peak. How did he control his imagination?

He "felt that we had to squeeze everything out of this session so that it would be the last one." And so he was preparing one more experimental blast:

> At the same time as the "big" one I worked assiduously on a device that I called "the initiative" to myself. In one of its parameters it was an absolute record. For the time being it was being made without a "request" from the military, but I assumed that sooner or later such an "request" would come, and when it did, it would be a rush. And then, a situation could arise analogous to the one in 1958 that led to a resumption of testing. And that I wanted to avoid at any cost!

The record-breaking aspect of the "Initiative" was not as understandable as number of megatons. And that is why Sakharov acted on his own initiative beyond the official plan. "I managed miracles of pull (the only time in my life), borrowing pieces of plutonium to create the parts."

He did all this just to make this series of tests the last. That seems like the truth. But it also seems like illusory self-justification.

What had Academician Sakharov and his associates worked on during the three years of the moratorium from 1958 to 1961? At the Kremlin meeting in July 1961 he spoke of several "science fiction" developments being done in his section, for instance, a nuclear engine for a spaceship—a "blastopter." However, they were not paid for science fiction at the Installation. Sakharov was surrounded by young physicists who were looking for applications for their talents. And Sakharov himself was young and a creative and active physicist and designer. Thus, "during the three years of the moratorium we collected a large 'backup' of ideas, calculations, and preliminary elaborations."

One of these "elaborations" was inspired by reports in the foreign press that a superbomb with a 1,000 megaton yield might be feasible. Sakharov and his two associates proposed a specific scheme for such a "device." This became the basis for the Tsar Bomb of 1961.[34]

But back in 1958, in his popular article Sakharov mentioned that the power of a typical hydrogen bomb was five megatons and wrote confidently, "in fact hydrogen bombs can be built 10 and even 100 times bigger." That was Sakharov the physicist writing—for a physicist there were no fundamental problems with that. For Sakharov the designer there was the problem of embodying the idea in bomb hardware. That was another level of the problem, but ever since his days at the munitions factory, it, too, was capable of capturing his imagination.

The most humanitarian of Sakharov's colleagues, looking back a half century, wrote that the task of creating a superpowerful bomb had been solved "elegantly."[35] To assess the rightness of his judgment, one needs access to top secret technical information. But to be appalled by this description, one need only think what task was being solved. How could they step back from the human aspect of this "elegant" task?

In Kurt Vonnegut's novel *Cat's Cradle*, written around that time (1963), the character Dr. Hoenikker was one of the fathers of the A-bomb and a Nobel laureate. He enthusiastically took up the creation of Ice-9, which freezes whatever it touches. He was so interested in the physics question that the danger of freezing everything on Earth to death was overlooked. This character is a caricature, but it is a caricature of reality. Yes, there is room for creative imagination even in the terrible sphere of nuclear weapons, and people do get used to the daily routine of this terrible physics. Sakharov gave a bitter description of this habitual psychological attitude in his *Memoirs*, when he tried to find a military application for the Tsar Bomb:

> After testing the "big" device I was worried that it didn't have a good carrier (bombers didn't count because they're easy to shoot down), in other words, in a military sense we were working in vain. I decided that an effective carrier could be a big torpedo fired from a submarine. I imagined that a nuclear jet engine that converted water to steam could be created for the torpedo. An enemy port, several hundred kilometers away, would be the target. A war at sea is lost when the ports are destroyed—so the navy tells us. The body of such a torpedo could have been made of very strong stuff, so that mines and protective nets would pose no threat. Of course the destruction of ports—either with an above-water blast from a surfacing torpedo with a 100-megaton charge or from a similar bomb—inevitably would entail very large numbers of casualties.
>
> I discussed this project with Rear Admiral Fomin. He was shocked by the "atrocious" character of the project and noted in a conversation with me that military seamen were used to fighting armed adversaries in open battle and that the very idea of such mass killing [the population of the port] was repugnant to him. I felt ashamed and never discussed my project with anyone else.

We know this story only from Sakharov's *Memoirs*. All recollection is a collaboration of memory, honesty, reason, and conscience. To give all coauthors their due, we must know the dry facts. Today, when much that was secret is now known, declassified, Sakharov's tale tells us more about his conscience than about historical reality.

The reality is this. The idea of a gigantic nuclear torpedo for attacking shore sites came up long before the Tsar Bomb, and even before Sloyka was tested, and it wasn't Sakharov's idea—the corresponding resolution was signed by Stalin on September 9, 1952.[36] The torpedo named T-15 was designed to be more than twenty meters long and to weigh forty tons.

As to Sakharov's Tsar Torpedo, there is evidence now that its idea came from overseas. A commander of an American submarine, observing the Soviet Tsar blast, expressed in an American magazine article his interest in having a similar weapon in the Navy's arsenal. Khrushchev was given a clipping from that magazine and he charged the Ministers of MedMash and Defense to "elaborate this question" with the help of Academician M. Lavrentiev.[37]

Admiral Pyotr Fomin (with whom Sakharov discussed the project) headed the elaboration of the question. He was in charge of the nuclear test site on Novaya Zemlya, that worked on all kinds of "atrocious" thermonuclear weapons—bombs, missiles, and torpedoes. The newest kind, the supertorpedo, was expected to create an underwater 100-megaton explosion with a resulting ocean wave 500 meters high, which could "wash away" imperialism from the face of the Earth. Fortunately for America, research disproved that theory.[38]

But even if the research results had been "positive," it is hard to say that the new weapon would be more atrocious than any other strategic nuclear weapon. What is more humane? To burn up a city's population with a thermonuclear blast or to drown it in a giant wave? No one considered that kind of dilemma.

The real dilemma was this: Would humanity commit nuclear suicide or not? Soon after the H-bomb was invented, it became clear that strategic weapons were created not to be used but to threaten. The ancient formula of war and peace had remained intact: If you want peace, put fear into the heart of your potential aggressor.

Frightening a potential aggressor is serious business. Sakharov worked at it conscientiously but also from inertia, without thinking through the new strategic threat equilibrium—an unstable equilibrium. It was to illustrate his mindset that he told the story of the Tsar Torpedo that never was. Unhappy with his own intellectual inertia, he shifted the blame on himself. Sakharov "imagined" only a new engine for the torpedo—the nuclear jet engine. Though his words "I decided" create the impression that he had been the initiator of the whole "atrocious" torpedo, in fact this wording allowed him to keep the secret that the Soviets worked to have a gigantic nuclear torpedo back in the 1950s.

Potential aggression, threats, political illusion, and weapons fantasies were the reality of the world in which Sakharov lived in his "heroic period." This illusory world created "the sense of the exceptional and crucial importance of [his] work in preserving world balance through mutual deterrence (later they began speaking of the concept of Mutual Assured Destruction)."

Khrushchev lived in that same illusory world. Judging by his memoirs, he did not fully understand Sakharov's objections to testing, but realized what forces moved the physicist who defied him. And therefore he felt something like awe for him and called him—after another two confrontations—"a moral crystal among scientists."[39] And when, after the tests in 1961, Khrushchev was given the honors list to endorse and did not see Sakharov's name there—because he had been against the testing—he was furious. On his order Sakharov was given his third star of Hero of Socialist Labor.[40]

Sakharov's illusory world cracked the next year.

"The Most Terrible Lesson"

Sakharov called 1962 "one of the most difficult" years in his life. First of all, his hope—"extremely naive," as he would later write—that the superpowerful blast in 1961 would stop testing throughout the world was dashed. Quite the opposite. The United States resumed testing two weeks after the USSR. That meant that they had been ready and just awaiting a signal. It was as if both countries had broken loose: in one year about two hundred Soviet and American explosions poisoned the earth's atmosphere.

A milestone in Sakharov's life was the detonation set for September 1962. Two blasts, actually, were scheduled. They were planning to test two versions of a warhead—very similar ones, according to Sakharov. By then he was used to measuring the power of a bomb by the number of victims of its fallout. In this case, it was a six-figure number for each device. But most important, Sakharov was certain that "two parallel tests were unjustifiable excess" and that "without any loss to the country's defense capabilities one of the tests could be cancelled." He made that his goal. He was the deputy to Yuli Khariton, the scientific director of the Installation. Khariton was no hawk, and in 1961 he had joined Sakharov's effort to cancel scheduled tests, in vain. A year later he did not even try to help Sakharov.

The twin versions were developed by two Installations. Following the U.S. model, in 1955 the USSR created a second nuclear weapons center, Chelyabinsk-70. And just as in the capitalist world, the two centers turned into competitors.

Jealous of the excitement generated by the 100-megaton device, which was "useless" in a military sense, the people at Chelyabinsk came up with a way of reducing power to a still-huge ten megatons and of fitting the warhead into a missile. An analogous idea had been developed at Arzamas, both devices were very similar.[41]

In order to avoid having two tests conducted and causing twice the damage to humanity, Sakharov took unprecedented action. He made a special trip to the other installation (this was his only visit there). He offered to give up further development of the version that his associates were preparing in favor of their alternative version. He tried to persuade Minister Slavsky. He even called Khrushchev.

But he lost—both bombs were tested. "A terrible crime had been committed, and I couldn't prevent it! A feeling of impotence, unbearable bitterness, shame and humiliation overcame me. I dropped my face on the table and wept. This was probably the most terrible lesson of my life: you can't sit on two chairs!" And what won? The single-minded military thinking: the more tests the better. Bureaucratic ambition. The inertia and self-perpetuation of the military-industrial complex.

In early 1961 President Eisenhower warned Americans in his farewell address that they "must guard against the acquisition of unwarranted influence, whether sought or unsought, by the military-industrial complex."[42] In September 1962 Sakharov discovered that the American president's warning extended to his socialist country, too. There were no corporations under socialism, but there was definitely profit in the form of social position and the ensuing commensurate benefits.

The Soviet military-industrial complex leadership beat Sakharov in September 1962. He could no longer think that there was something important he did not know—about foreign policy or how capitalists conduct negotiations. This was all domestic, internal, technical, all under his nose.

His illusory world cracked and his understanding of his personal responsibility became even stronger.

The 1963 Moscow Ban on Testing

In a desperate attempt to stop the double test, Sakharov threatened Minister Slavsky. "If you do not reverse the decision, I will not be able to continue working with you. You have deceived me." But he did not follow up on that threat. He had an important task he wanted to bring to a successful conclusion, and he could not count on anyone else to do it.

It was the nuclear test ban. That expression entered the political lexicon in May 1955, when the USSR introduced a motion in the United Nations calling for a total ban on testing. The question of control arose immediately, and in view of the mutual distrust, it was insurmountable, although negotiations continued and many speeches were made.

A hint at the solution to the fallout problem could be found in Teller's 1958 book, *Our Nuclear Future*. It not only praised clean bombs as "almost solving the problem of radioactive fallout," but also offers a total solution: "Deep underground tests will eliminate fallout altogether."[43] There was little written about it, perhaps because the United States did not conduct its first underground test until 1957, and it was still not clear whether underground testing could replace the terrible mushroom clouds.

By April 1959 things were clearer and Eisenhower proposed banning only above-ground tests. Khrushchev immediately rejected this proposal, probably because he did not appreciate its main diplomatic superiority: it did away with the issue of control, which was the most difficult part because it involved inspection at test sites. By then both sides had admitted that there were reliable methods for detecting atmospheric detonations without the need for on-site inspection. The main virtue of such an agreement was that it put an end to polluting the atmosphere and restrained the spread of nuclear weapons.

The negative Soviet reaction to Eisenhower's proposal might have been related to the fact that the USSR did not yet know in 1959 what an underground test was like. The USSR conducted its first underground detonation in October 1961 and the second in February 1962.

In the summer of 1962 Adamsky, who shared Sakharov's attitude toward radioactive fallout, got the idea that the physicists at the Installation could help the diplomats. A man with wide-ranging interests, he followed developments through the American *Bulletin of the Atomic Scientists*, which they received at the Installation. He prepared a draft letter to Khrushchev from the designers of nuclear weapons, in which he laid out concrete and professional conclusions why the Soviet government should propose banning only above-ground detonations.[44]

The new is merely the old forgotten, as they say. In diplomacy, a good new proposal is a not-too-forgotten old proposal from the other side. Adamsky showed the letter to Sakharov, who approved it. But he added that it would be better to go through the minister. He headed to Moscow the very next day.[45]

Sakharov describes what happened next:

> I told [Minister] Slavsky the idea of a partial ban, without mentioning either Eisenhower or Adamsky; I only said that this was a way out of the dead end of the Geneva talks, a way that could be very timely politically. If we were to propose this, the US would definitely support it. Slavsky listened very attentively and sympathetically. At the end of our conversation he said, "Malik (the Deputy Foreign Minister) is here now. I'll talk to him today and pass along your idea. Of course, the decision will be made by 'Himself' [Khrushchev]."
>
> A few months after our conflict over the double testing, Slavsky called me at work. He spoke in a conciliatory tone. "Whatever has happened between us in the past, life goes on, and we must restore our former good relations. I'm calling you to tell you that your proposal created great interest upstairs and most likely steps from our side will be taken very soon." I told him that this was very important news for me.

On July 2, 1963, Khrushchev made the proposal to ban above-ground testing, and just ten days later the agreement was ready. The official signing took place in Moscow in August.

Khrushchev's respect for the father of the Soviet hydrogen bomb must have made Sakharov's opinion a deciding factor. On the other hand, this was just months after the Cuban crisis of October 1962, when humanity came perilously close to the brink of nuclear disaster. Had Sakharov considered his role in the Cuban crisis, which had almost put an end to civilization? What role had the fifty-megaton detonation of the previous October played? Did it bolster Khrushchev's confidence when he sent Soviet nuclear missiles to Cuba? Or did it add to Kennedy's caution when he rejected the opinion of many of his advisors about a military strike? Did it keep both of them from playing for keeps?

In any case, by the summer of 1963 the autumn crisis of 1962 and Sakharov's recommendation had worked in tandem. Sakharov felt the historic significance of the agreement was that "it saved hundreds of thousands, and perhaps, millions of human lives"—those who would have perished if above-ground testing continued. No less importantly, it was the first international agreement on a nuclear "code of conduct" agreed to by the opposing superpowers.

Sakharov had reason to be proud of his part in the 1963 accords.

A Rejection of Lysenkoism

In Sakharov's 1958 article on radiocarbon the bibliography included two strange references to manuscripts on the effects of radiation on heredity. Why manuscripts? And how did the physicist learn of them?

Soviet readers of the time understood that the tragedy of Soviet genetics lay behind those cryptic references. Destroyed with Stalin's blessing, it had to remain in the underground for many years—Lysenko's political talent guaranteed that he and his successors would remain in Khrushchev's good graces for a long time. This despite the fact that in the early 1950s molecular genetics made a major breakthrough after the work of Crick and Watson in Britain.

The reason for Lysenko's dominance was simple—in the Stalin years he secured influential posts for his people in biology and agronomy, in the Academy of Sciences in particular. So even the removal of the bans of the Stalin period could not improve the situation without outside help. Soviet physicists, with their secure social position, tried to help the biologists.

Tamm was most active in this area. In 1955 he was one of the initiators of a letter to the Central Committee in defense of genetics.[46] In February 1956, a week before the Twentieth Party Congress, he gave a lecture on the molecular mechanisms of heredity at Kapitsa's seminar at the Institute of Physics Problems. This was the first public discussion of genetics since Lysenko's pogrom in 1948.[47] At about the same time, in the mid-1950s Zeldovich brought Sakharov to the house of geneticist Nikolai Dubinin, where Dubinin carried out his experiments on the classic subjects of genetics—fruit flies, often mocked in Soviet newspapers. (It is Dubinin's manuscript that is cited in Sakharov's article.) When in 1958 Kurchatov created a haven for geneticists at his institute—the department of radiobiology—it was headed by the physicist Tamm.[48]

So Sakharov was prepared when the time came for him to study the genetic effects of radiocarbon. Work in that area could only strengthen the astonishment that anyone could believe in Lysenko's "theory."

We can imagine Sakharov's feelings when he learned at a General Meeting of the Academy of Sciences in June 1964 that N. Nuzhdin, one of Lysenko's closest colleagues, was being considered for election to the academy. "All my anti-Lysenko passions flared up again; I remembered what I knew of the tragedy of Soviet genetics and its martyrs. I thought that the confirmation of Nuzhdin's candidacy by the General Meeting could not be permitted at any cost."

The candidates for election to the Academy of Sciences are nominated by the divisions in their appropriate field, and usually the General Meeting of

Academicians in all fields ratifies by secret ballot the departmental choices, trusting the opinion of their colleagues. Sakharov decided to speak out against the biologist's candidacy, unaware that the biochemist Vladimir Engelgardt and Igor Tamm had already made plans to do so. In his memoirs, Sakharov wrote that he spoke first. The archival transcript shows that his memory was faulty—Engelgardt spoke before him.[49] However, the transcript also shows that there was a good reason for his memory's trick. Engelgardt chose a very academic method of argument—he said that he did not know of any "contributions of a practical nature" made by Nuzhdin and that he had not found a single mention of his name in monographs or leading journals in recent years.

Sakharov called things by their real name. He called on "all Academicians present to vote so that the only ballots 'for' would belong to those people who together with Nuzhdin, together with Lysenko bear responsibility for those shameful, painful pages in the development of Soviet science, that are fortunately coming to an end."

Tamm's speech finished the job. Nuzhdin was not elected.

Twenty years later, in exile in Gorky, as Sakharov thought about the chain of events that led him there, he noted the turning point, the "fateful" significance of two very different factors: his planned struggle of many years for an end to above-ground testing and his impulsive—three-minute—intervention in the academic career of a biologist he did not know personally. The testing issues revealed to him the world of high state and international policy. The brief speech at the Academy of Sciences revealed him as a public figure when he had no intention of being one.

Lysenkoism was not only a sore on the body of biology, it was a blatant vestige of Stalinism. And for that reason, Sakharov's speech was perceived as a public action. People outside nuclear physics learned of the secret academician for the first time. They learned that he was not just a secret physicist.

14

FROM MILITARY PHYSICS
TO PEACEFUL COSMOLOGY

Inventor or Theorist?

In explaining why he did not leave the Installation immediately in 1962 (because of the disputed double test), Sakharov named his work on banning testing as the most important reason, but not the only one. Another reason was
that he had nowhere to go. Not in the sense that there wouldn't be a place found
for him somewhere in one of the institutes of the Academy of Sciences. But what
would he do there? Theoretical physics?

It was a difficult issue for him, with his honesty, sense of personal dignity, and
his manner of behaving "outwardly modestly, but in fact quite the opposite."

After Sakharov's election to the Academy of Sciences in 1953, Lev Landau
was asked his opinion of the young theorist. Landau asked, "Whom do you
mean?" When he was told it was Sakharov, he replied, "But he's no theorist!
He's an inventor physicist."[1] Landau knew him as an inventor. And Sakharov
was working then, and continued for another decade, as an inventor. True, he
had started out as a theorist. But there are many Russian businessmen and
politicians of today who have written dissertations in physics or mathematics.

Sakharov remembered how he had come to Moscow after a few years at the
Installation and ran into Ginzburg. He told him about some purely scientific
idea. "He snickered and said, 'So, you expect to do physics as well as your little
bomb?!'" In hindsight, Sakharov agreed that combining such things "turned
out to be very difficult, basically impossible."

He also remembered his father's sorrow, a few weeks before his death in late
1961. "When you were studying at the university, you once said that discovering the mysteries of nature was what could make you happy. We do not
choose our destiny. But I am sad that your destiny turned out to be different. I
think that you could have been happier." The development of nuclear weapons
is not simply far from discovering the mysteries of nature—it is contraindicated.

Sakharov watched the mathematical talent of Nikolai Dmitriev fade away. Zeldovich had said that Dmitriev was "perhaps the only one among us with that spark from God. You could think that Kolya is just this quiet, modest boy. But in fact, we all tremble before him, as if he were the highest judge." His lapidary talent for unique masterpieces was no longer needed once the Installation began mass production. And if a talent is unneeded, it is doomed to extinction.

Sakharov's unusual—and saving—grace was his double talent. He was recognized as a theorist in the mid-1960s, when he began attending seminars at FIAN and the Institute of Theoretical and Experimental Physics. His colleagues were impressed by the juxtaposition of his two gifts as a theorist and as an inventor and designer.[2] They are as different from each other as the talents of theoretical physicist and writer. That is why such combinations are rarely encountered.[3]

Ginzburg, whose comment had hurt Sakharov's feelings, expanded on his remark. "I can say this about him: He was undoubtedly a very talented man, he was made of the material of which great physicists are made. It's just that . . . He always had this inventive spirit."[4]

Sakharov had a good idea of the stuff he was made of, even as a graduate student. But material alone is not enough. He describes with sober objectivity his early attempts to give the material a working shape—his successes and failures in that regard. In 1947 he lacked the spirit, intuition, and boldness to follow the path that would have led him to a major problem in theoretical physics of the time.[5] He did manage to take the first step down that road. He is being severe to himself when he sums up this way: "Everyone does the work he deserves." However, this personal feeling does not block the general picture for him:

> Recalling that summer of 1947, I feel that I had never—before or since— come so close to Grand science and its forefront. Of course, I regret a bit that I did not end up at the top (no objective circumstances are an excuse). But from a broader point of view I have to take joy in the progress of science— and if I had not been privy to it, I would not be able to feel it so acutely.

His passionate preference is evident when he talks about theoretical physics. In describing his graduate work in his *Memoirs*, he lets fall mention about "pi-mesons" and their "isovector nature." Hastily, he apologizes in parentheses: "I do not explain certain terms in this book—let the non-physicist reader forgive me and simply regard them as misty, beautiful images." Is there a non-physicist who would find anything but cold mist in the word "isovector"? But we can all feel the beauty of the narrator's emotions over such intellectual matters. Especially given that he was in his sixties, writing in exile under the vigilant eye of the KGB, and that the KGB had stolen the manuscript of his book several times already.

In recounting the great honor (and at first great difficulty) of an assignment—Tamm's students had to report on the latest scientific articles at their
seminars—he recalls describing the work of an American physicist and feeling "like a messenger of the gods." When he finished, "Pomeranchuk rushed
up to the board, terribly agitated, hair on end, and said something like, 'If this
is true, it's exceptionally important; if it isn't true, it's also exceptionally
important . . .'"

Pomeranchuk would later review Sakharov's dissertation. He used the term
"bubbles" for tasks that were not related to grand science and not "exceptionally important." Sakharov used this term when describing his first successful
small theory, which he created in the spring of 1945, on the distribution of
sound in water filled with air bubbles. He noted bitterly, "I often dealt with such
unserious things, and in fact what I worked on between 1948 and 1968 was a
very big bubble."

We must remember that this was written in the early 1980s, in exile in
Gorky, after he had switched to theoretical physics. And after he had shed his
illusions about the Soviet state and anxiously wondered what use the country's
leaders might find for his inventions.

His penchant for inventing is obvious in his story of the first successful test
at the cartridge factory in Ulyanovsk. In the *Memoirs* he relishes that period
forty years earlier. He draws schematics in which "the core of a bullet slides,
with light friction down the inclined copper pipe through a magnetizing coil,"
and explains how another, "demagnetizing coil helps determine whether the
core is sufficiently tempered, consisting of steel with a reduced coercive force,"
and so on.

It is doubtful that more than one reader in a thousand would understand
his explanation. But the remaining 999 will believe that he was "very proud"
of the construction of a device to test the armor-piercing steel cores and that
he was "rather sorry to leave my designing, which was beginning to come
along." It is even easier to believe that the opportunity to invent in thermonuclear physics that would come up five years later attracted him just as much,
and that if confidentiality were not required, he would have enjoyed describing his designs with the same pleasure. That was a fascinating concept—to re-
create the star's source of energy on earth. He had complete freedom in selecting
physics ideas in the construction. And he had the opportunity and even the
requirement to embody the invention and test its viability. When Sakharov
called the physics of a thermonuclear blast "paradise for a theorist," it was not
just the theorist talking, but the inventor as well. Even when the duplicate test
of 1962 cast a black shadow on his inventor side, Sakharov continued weapons development for another six years, and not because he was forced.

It is amazing that after almost twenty years Sakharov was able to return to
creative work in theoretical physics, which had changed so much in those

decades. Besides, physics is youth's game, especially theoretical physics. People make their major contributions in theoretical physics when they are in their thirties. Sakharov returned to the field when he was in his forties.

Having two talents can be a help, and it can be a hindrance. Rarely, and this was the case in Sakharov, one can feed the other, keeping it from drying up. In hindsight, it is easy to speak of the benefits of switching creative energy from one field to the other. But it was hard for him to think in those terms in the early 1960s, when his inventiveness was confined to the military, and therefore harder to justify morally, and also becoming more boring.[6] Sakharov knew how quickly the train of scientific progress pulled away from the station, and he wasn't sure that a forty-year-old could catch up and jump aboard.

He felt the gap back in the time of his first visit to the Installation, when Zeldovich asked him to give a lecture on quantum field theory. "Unfortunately I had fallen behind (in two years), and a great leap forward had taken place in that period. I did not know the new methods and results obtained by Schwinger, Feynman, and Dyson; my talk was on the level of the rather antiquated books of Heitler and Wentzel."

Looking back at his scientific career, sixty-year-old Sakharov saw how lucky he had been to be able to return to theoretical physics. In a "written conversation" with his wife—hiding from the KGB bugging of their Gorky apartment—he wrote of "four years of my scientific maximum, late by usual standards. In fact, it was a gift of fate that I could do anything after special work [bomb making]. No one besides Zeldovich and myself ever managed. In the U.S., neither Teller nor Oppenheimer could return to pure science. The exception there was Fermi. But he died soon after, and he was a genius."[7]

Zeldovich's name is there for a reason. He played an important role in Sakharov's return to pure science—one could say that Zeldovich dragged him in with him.

The Physics of the Universe

Busy with his top secret duties, Sakharov followed from afar what was happening in the discovery of nature's mysteries. After every trip to Moscow, Tamm and Zeldovich returned with scientific news, which they related to the theorists at the Installation. But that wasn't the same as taking part in the work.

After Tamm left the Installation in 1953, two outstanding theorists remained by Sakharov's side—Zeldovich and Frank-Kamenetsky. They were seven and eleven years older than Sakharov, had obtained major results before the war, and while working on the nuclear project continued doing pure physics and publishing articles.

A piece of not-very-serious evidence survives about what Sakharov thought then of pure theory. It is a sheet of paper recording in his hand the bet he made in 1956 with Frank-Kamenetsky:

17 February 1973

The Problem of Quantum Determinism:
Is there a singular solution of the Schrodinger equation that describes the universe for all degrees of freedom for all times?

(17 years ago D.A.F[rank]-K[amenetskii] answered "No", A.D.S[akharov] answered "Yes")[8]

This was more than pure physics, almost metaphysics or even theology. Let us leave the question of how much truth there was in this joke to the last chapter. But apparently, Sakharov's part in the bet was greater than his calligraphy. Frank-Kamenetsky devoted a lot to the study of stars and wrote a hefty tome in that field, but it was "earthly" astrophysics, based on astronomical measurements, and somewhat close to the "astrophysical technology" that was the work of the Installation.[9] The universe as physical object—the subject of cosmology that awaited Sakharov—was far from Frank-Kamenetsky's interests.

It is a very special subject. You can't experiment with stars, either, but at least there are very many stars and you can observe and compare them. But the universe as a whole is by definition unique and even to declare that you see the object and not just some tiny random portion of it required exceptional intellectual daring. Or brazenness or irrationality. That was the opinion of Vladimir Fock, who helped his teacher, Aleksandr Freidmann, translate his famous 1922 article on the expansion of the universe into German and who wrote a fundamental monograph on the theory of gravity.[10] The same attitude toward cosmology in the United States in the 1950s is described by Steven Weinberg, later a Nobel Prize winner. "Everyone thought that the study of the early universe was not something to which a self-respecting scientist should

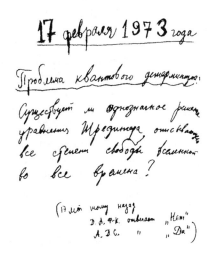

Figure 14.1 Cosmological bet of 1956, between Sakharov and Frank-Kamenetsky.

be devoting himself."[11] Cosmology was at a far remove from what was important in physics.

Einstein had made it possible in 1917 to speak of the universe as a physical object on the basis of his theory of gravity, which united Newton's law of gravity and the theory of relativity. But in the subsequent four decades cosmology essentially only gave the opportunity to speak in a mathematical language, not to take physical measurements and compare them with the predictions of theories, as it should be in the physical sciences. In the course of those decades, cosmology received only one measurable fact, albeit an important one. The story of that fact reveals the uniqueness of the greatest natural object, the universe.

The fact had been predicted in 1922, by Aleksandr Freidmann (1888–1925), a Russian mathematician who closely followed the revolutionary renewal of physics. Regarding Einstein's cosmological theory with the eyes of a mathematician, he realized that the great physicist had found only one—very particular—solution for his equations. If we were speaking of a pendulum, we could say that Einstein found the tension when the pendulum is motionless. But a pendulum also can be in motion. Freidmann, basing his thinking on Einstein's equations, described the "motion" of the cosmological pendulum—the universe. It turned out that the universe could expand—that is, its component galaxies could move away from one another.

Freidmann sent the article about his discovery, called "On the Curvature of Space," to a German physics journal in the spring of 1922 from a Petrograd (not yet renamed Leningrad) ravaged by the civil war. The unknown Russian's results were so unusual that it was easier for Einstein to suspect that the author had made an error in his calculations, which is what he wrote in his commentary published in the next issue of the journal. This is the famous error of Einstein. He soon understood it and published another note, in which he said Freidmann's results were "correct and shed new light."

It was not that theoretical light that helped cosmology take the next step, but the very dim light from distant foggy patches. They were studied by the American astronomer Edwin Hubble with a telescope. He was not studying gravity or the curvature of space. He concentrated on nebulae, which he recognized to be distant galaxies, and discovered that they were receding from our own galaxy, the Milky Way.

The changes in the tone of a locomotive's whistle as the train rushes past an observer allow us to judge its speed. Similarly, a sophisticated observer such as Hubble can measure the speed of distant galaxies by the changes in their weak light and determine an amazing fact: the farther away the galaxy, the greater its velocity of recession. This observation, made in 1929, is known as the Hubble Law.

Theorists following both astronomy and physics realized that this was the expanding universe predicted by Freidmann. It was the cosmological triumph of theoretical physics. No other triumphs followed, however, for another three decades. Astronomers merely made Hubble's measurements more precise.

No one doubted Hubble's law, but some physicists were uncomfortable with the grandeur of an expanding universe and sought a simpler explanation for his observations. They found one in the murky waters of microphysics in the making. It looked like the aging of light particles, photons, over the enormous time of their travel from galaxies to Earth. The small effect of the disintegration of photons replaced the grandiose picture of the universe flying apart in all directions.

However, this cozy hypothesis was elegantly disproved in 1936 by the Russian physicist Matvei Bronstein, who had a profound understanding of both microphysics and cosmology. As a result, the empirical basis of cosmology was strengthened. But a single point of support is not enough for stable development. It was quite different from the other areas of theoretical physics, which were based on thousands of measurements.

Neither cosmology nor gravity was needed then to study the composition of matter. The forces in microphysics are greater than gravity by an unimaginable number that is forty digits long. Only if an equally astronomical number of particles are gathered in one place would gravity be needed. But by then we move from physics into astronomy. And the theory of gravity and cosmology require a special mathematical language that was not needed in other areas of physics.

These circumstances made cosmology at best a respected but eccentric and distant cousin of the rest of the physics family. Very few physicists had the time and curiosity to maintain professional relations both with cosmology and microphysics. Among those few was Landau, who included the theory of gravity in his famous *Course on Theoretical Physics*.

This book might have helped Sakharov, back in the late 1940s, to keep both edges of the physics world in his field of vision. He kept a journal of articles that interested him, and next to the news of microphysics we can find a notation about the expanding universe from a 1949 *Physical Review*, the leading American journal.[12] The move to the Installation and H-bomb physics kept him from pursuing this for years.

In the early 1960s cosmology unexpectedly moved from dotty spinster aunt to intriguing debutante. By 1967 Zeldovich published a book with a coauthor summarizing the early years of cosmology.[13] The book recounted Bronstein's work on physical cosmology of the 1930s, even though no one doubted the expanding universe after the discovery in 1965 of cosmic background radiation homogeneously filling the universe.

This phenomenon, like Hubble's expansion, had been predicted (by Gamow in 1948) but discovered accidentally. Cosmic radiation was of the same nature as heat coming from a stove—one "heated" to minus 270 degrees Celsius, just three degrees above absolute zero. The discoverers won a Nobel Prize.

Theoretical physicists saw in this radiation not only expansion of the universe but also a sign of something about the early phase of the expansion.

If the galaxies are moving apart, they must have been closer to one another in the past, and therefore once formed a solid mass that was not separated by cosmic space. It was heated to huge temperatures, and consequently permeated by intense radiation. The unexplained event that took place billions of years ago, in the very beginning of the cosmological expansion, was called the Big Bang, or the birth of the universe. As the universe expanded, the radiation cooled down. Over the billions of years, it cooled down billions of times. But a sensitive apparatus picked up the relict of the Big Bang, and hence the name relict (or background) radiation.

Besides this, the most impressive cosmological discovery, in the 1960s astronomers discovered a few other amazing phenomena. New concepts entered the science vocabulary: quasar, pulsar, and black hole.

Zeldovich burst into the area, where the latest discoveries combined with theoretical puzzles of the distant past, with his first work on cosmology in 1961. "After him, I began thinking about 'Great Cosmology' myself," wrote Sakharov. By then Zeldovich had written several dozen works on fundamental physics; he never gave up his connection to pure science. But Sakharov was concentrating on H-bomb physics.

From the Atomic Problem to the
Problems of the Universe

Zeldovich and Sakharov were radically different even in the external style of their scientific lives. It is said that the predilection for polygamy or monogamy lies deep in the structure of the personality. Zeldovich had quick affairs with various scientific ideas, which he carried out to the birth of a publication. In his lifetime he published about three hundred works in pure science with several dozen coauthors.

Sakharov has only two dozen pure science works and no coauthors, except Zeldovich. This exception shows us that Zeldovich knew what he was doing when he tried to bring Sakharov into pure science. For physicists who lived outside the Installation and had to judge by his publications, Sakharov was a dark horse. Zeldovich knew, without any publications but from personal experience, that Sakharov was rather a "talking horse," as he put it.[14] Their first joint article of 1957 went back to Sakharov's 1948 secret report at FIAN on muon catalysis.[15] Zeldovich's work notebooks for 1957, dealing with this work, have a notation: "Sakharov's most profound idea."[16] Profound ideas don't come frequently. A physicist from the Installation recalls Zeldovich saying, "Andrei, it's two years since you've had a knock-out idea."[17]

For the self-contained Sakharov, Zeldovich was the best window into science. He could replace several seminars and the usual postseminar scientific chat. With his acute perception, fast thinking, and erudition, Zeldovich knew about everything and was interested in everything even if he wasn't working on it himself at the moment. In the prewar years, for instance, his work had nothing to do with cosmology, but when he heard Bronstein's elegant theoretical construction, it stayed with him until he used it thirty years later in the first Soviet book on cosmology. History of science per se did not interest him: "The past of the universe is infinitely more interesting that the past of science about the universe."[18] It may have been because science alone is not enough to understand the history of science, even the history of such a pure science as cosmology.

Explaining the sharp turn in his scientific life, in 1984 Zeldovich, then seventy, referred tactfully to the "atomic problem" that had "captured me completely."

> In those very difficult years, the country spared nothing to create the best working conditions. For me they were happy years. Big new technology was being created in the best traditions of big science . . . By the mid-fifties, some priority tasks were already resolved . . . Work in the field of the theory of explosion was psychological preparation for the biggest explosion—of the universe as a whole . . . Work with Kurchatov and Khariton gave me a lot. The most important is the inner sense that I did my duty for country and people. This gave me a certain moral right to subsequently work on such issues as elementary particles and astronomy, regardless of their practical value.[19]

There wasn't a syllable about bombs here, but the tactful Soviet reader understood. And almost all the necessary elements for explaining the turn in Zeldovich's scientific career are here, although some of the elements are turned around or need translation from Aesopian Soviet to ordinary language.

In the United States, a similar transformation took place with John Archibald Wheeler (b. 1911), the first American involved in the theoretical development of the atomic problem (through his and Bohr's paper of 1939). He also took part in the development of the H-bomb as director of Project Matterhorn at Princeton (1951–1953).[20] It was mentioned earlier that the secret document he lost on a train in January 1953 was suspected of reaching the Soviet Union. There is more reason to suspect that Wheeler had seduced his thermonuclear counterpart Zeldovich into pure gravity.

A few years before Zeldovich became the principal cosmologist in the Soviet Union, Wheeler became the principal U.S. theorist in gravity. One did not need to steal documents to learn that the prominent American nuclear physicist had changed profession. One merely had to read the *Physical Review*. But a physicist as talented as Zeldovich does not follow examples; he is motivated from within.[21]

An aptitude for leadership can explain why the two former weapons physicists became national leaders in gravitation and cosmology. But the switch in scientific orientation is tied to something else, which is the same despite the differences in socialism and capitalism.

If we spell out everything that is between the lines in what Zeldovich wrote, we get the following picture. In the late 1950s (a bit earlier in the United States), theoretical physics of thermonuclear weapons had exhausted itself and was replaced by engineering physics. The priority of new technology had been solved: American and Soviet physicists had created the greatest scourge for their politicians. It was called MAD, for mutual assured destruction, the ability of each superpower to destroy the other even after a sudden enemy attack.

As a result, the authorities recognized the connection between new technology and science, and out of respect for those who made the connection possible, permitted them to work on whatever they wanted (probably with the thought that their impractical interests might nevertheless give rise to the next major technological breakthrough). Theoretical research did not require a lot of funding. Incomparably greater sums were spent on experimental science—particle accelerators and spaceships.

"Work in the field of the theory of explosion" could be psychological preparation for cosmology only by accustoming scientists to the distance between theory and testing, and therefore to boldness in science. At the Installation, the leading physicists had to create a theory of the thermonuclear bomb without being able to test their calculations on small trial explosions in the lab. First, the complete theory, and only then a full-scale megaton blast . . . or a dud. This is comparable with cosmology in psychology rather than scale. You have to dare to create a theory on such a non-observable object as the universe billions of years ago.

And then there is what Zeldovich called the moral right to take up areas "regardless of their practical value." We can easily imagine how the physicists who created the horrifying thermonuclear mushrooms while doing their "duty for country and people" were ready to escape from their practical application.

The escape route might be prompted by the Soviet newspapers of the early 1960s. There were articles about new subjects of study—in space. The Installation's theorists knew better than anyone else that the signals from the first sputnik and Yuri Gagarin's smile were not really from the realm of science but from the next "big new technology," which would deliver the technology they had invented for distances of thousands of kilometers. But they also knew that moving away from the planet's surface by just a hundred kilometers vastly expanded the horizon—literally. Astronomical observations made without atmospheric interference promised great discoveries. This was confirmed very quickly. The discovery of background radiation—the remains of the hot time of the birth of the universe—came by accident in 1965, but it was no accident that the discovery was made while working on radio communication with satellites.

These factors came into play only because the theory of gravity and cosmology offered intriguing questions—real mysteries of nature in pure form. And only pure science could find the answers.

In describing his return to pure science, Zeldovich does not mention Sakharov's name in his 1984 autobiography. It was the fifth year of Sakharov's exile, and the Soviet censors kept a sharp eye for the unacceptable name.[22] But Sakharov and Zeldovich came to cosmology together twenty years earlier.

Zeldovich called what he was doing "relativistic astrophysics," that is, the physics of cosmic phenomena that required the theory of relativity for an explanation. Relativistic astrophysics covers the physics of exotic objects in space, and the physics of the exotically single object—the universe as a whole.

We can imagine quasars, pulsars, and black holes among the luminous stars in the sky. And it is easy to imagine how telescopes show these objects with greater magnification and in greater detail. But no telescope can show the entire universe. Here only the eye of the intellect can help.

If we judge by their publications, Zeldovich began before Sakharov. By the time Sakharov published his first work on cosmology (1965), Zeldovich had written more than two dozen. But if we look into their informal communication, a different picture arises.

Think of the bet Sakharov made with Frank-Kamenetsky in 1956, five years before Zeldovich's first publication on cosmology. The brief text of the bet makes clear that Sakharov already saw "the universe for all degrees of freedom for all times." Sakharov's serious style excludes the possibility that he would put together such seemingly incongruous words without consideration. Therefore, he was thinking of the universe as a physical object as early as 1956. In that era, it was an extremely exotic object for a working physicist. And no one in Sakharov's circle seems likely to have helped him develop that viewpoint.

Vladimir Ritus wrote of Tamm in his memoirs (without mentioning Sakharov's name): "When one of his senior students got interested in cosmology and put forward several rather abstract ideas, Tamm shared his surprise and regret with me, saying that these hypotheses were impossible to prove or disprove in the visible future."[23]

Knowing Zeldovich's attitude toward Sakharov's "profound" and "knock-out" ideas, we can assume that Sakharov's confidence in treating the universe as a physical object strengthened Zeldovich's resolve to take that object into his arsenal.

Ten years after the cosmological bet made by his friends at the Installation, Zeldovich had a wonderful idea: to fill all the space in the universe with vacuum. But when he brought it up at a seminar, he was met with ruthless criticism. He called Sakharov, who liked the new idea so much that he took the next step. According to Wheeler, it was an incredibly bold one.

Sakharov was emboldened by the pure science work he took up after a long hiatus in 1963. He worked following the direction of Zeldovich's thoughts, and he thanked him for "numerous discussions that led to the problem's formulation as a whole and enriched the work with many idea."[24] Following Zeldovich, in that work Sakharov assumed the hypothesis of the "cold" early universe. The discovery of the cosmological background radiation proved, however, that the early universe was hot. And so Sakharov's first work on cosmology went to the archives of "unemployed ideas" created in theoretical physics (perhaps up to 95 percent of new theoretical ideas end up in these archives).

Nevertheless, the work was a turning point for Sakharov, and he even remembered the day he found the solution to a difficult question—April 22, 1964. "I believed once again in my powers as a theoretical physicist. It was a kind of psychological warm up, that made possible my subsequent work of those years."

His new confidence shored up his "program for 16 years," which he wrote for himself on a single sheet of paper in 1966. Why for sixteen years? Perhaps because he had spent the previous sixteen at the Installation, removed from grand science. Apparently for the same reason, the program included sixteen topics, beginning with the solemn "photon + gravitation" and ending with the mysterious "megabittron."

Of particular interest is topic 14. Apparently, after deciding to find sixteen tasks, he pondered on this point, put a question mark, remembered how resistant science is to planning, and added: "This is just what I'll probably be doing." He turned out to be right, and he took up "just that" soon after, and even wrote two of his brightest theoretical works on that point.

First he found an explanation for why there are many more elementary particles in the universe than antiparticles—in the language of physics, he proposed a way to explain the baryon asymmetry of the universe. That was the most successful of his ideas in pure physics.

Rivaling it in beauty and unexpectedness was his new approach toward gravitation. In the old universal gravity he discerned a manifestation of ultra-microscopic qualities of space-time itself.

The two ideas that Sakharov envisioned in 1966–1967 deserve further discussion.

Symmetries in the Asymmetrical Universe

Finding a definition for "beauty" is a thankless task. Of its synonyms, the one most appropriate in the sciences is "symmetry." The concept can be expressed with mathematical precision and it is visible. The symmetry of a butterfly's wings is the simplest (and most appealing) example.

This simple quality from the everyday world moved into geometry, where it received a mathematical description. Symmetry is the rule of form that keeps the form unchanged under specific circumstances. If you reflect a butterfly's right wing in a mirror and put it in place of the left, no entomologist would be able to tell the difference.

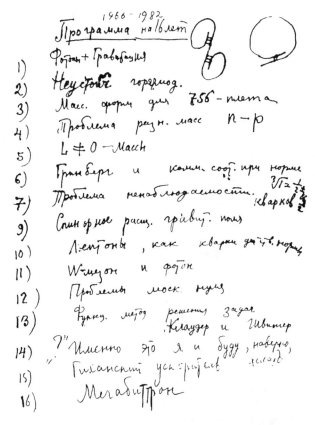

Figure 14.2 "Program for 16 Years," which Sakharov created for himself in 1966. In thinking about advanced physics and math, the academician skipped point 8. Otherwise, he would have had to come up with one more task. But that turns point 14 into 13, which might explain its special nature.

Armed with the power of mathematics, the concept of symmetry became an instrument of theoretical physics in the study of the deep structure of nature. Physics traveled a long way before finding in its laws manifestations of the universe's symmetries. Everyone knew that a spinning top will remain upright in one spot and not fall. It does not fall because, we could say, it does not know where to fall: all directions perpendicular to its axis are equal—that is, all directions in space are symmetrical relative to that axis. In the language of physics, that sort of symmetry determines the law of conservation of angular momentum, the main law for the spinning top.

Symmetry is one of the most workable concepts in physics. The behavior of a top or a single atom or a thermonuclear blast is determined by symmetry. A theorist always begins with the most symmetrical simplification of his problem. Every fundamental law of physics reveals a symmetry in nature. If an asymmetry is found in natural phenomena, the theorist sees a difficult but fascinating problem: to find a place for that asymmetry in the harmony of the universe.

"It is known that Maxwell's electrodynamics—as usually understood at the present time—when applied to moving bodies, leads to asymmetries which do not appear to be inherent in the phenomena." This is the beginning of Einstein's first article on the theory of relativity. With that theory he overcame the asymmetry that is not inherent to the phenomena themselves—he created a description in which that asymmetry is merely one facet of the profound symmetry of nature.

Another triumph of symmetry in physics was achieved by Paul Dirac. In the late 1920s he took on a purely theoretical problem. There were two fundamental theories at work at that time: the theory of relativity and quantum mechanics. The first provided description of phenomena with velocities up to the speed of light. The second described the behavior of microscopic particles. But nature does not keep its phenomena is separate compartments, and Dirac wanted to find out what directed the motion of the electron when both theories were necessary at the same time. He managed to unite the theory of relativity and quantum mechanics in one elegant, albeit unusual-looking, equation for the electron.

There was only one impediment. Dirac's equation required the existence of another particle, in some way quite similar to the electron, in others completely the opposite. The particle had to be exactly like the electron in mass but the opposite in charge. And a meeting of the two would lead to mutual destruction—annihilation.

Even though no other particles were known to physics then but the electron and proton, Dirac believed in the symmetry of his equation and, in 1931, predicted a new particle, which he called the "anti-electron." A few months later, it was found in studying cosmic radiation. In honor of its positive charge the experimenter discoverer, Carl Anderson, called it the "positron," but this name does not reflect the particle's main quality—being the anticopy of the electron. Later other elementary particles and their anticopies would be named properly: antiproton, antineutron, anti-sigma-hyperon, and so on (or anti-so-on).

The main relationship between a particle and its antiparticle is that they annihilate each other when they meet, giving birth to particles of light, or photons, which have no charge and inherit the energy of the parent pair. Conversely, if a photon has enough energy, it can turn into a particle and its antiparticle.

The power of the symmetry of equations in explaining the real world prompted Dirac to persuade many of his colleagues that "physical laws should have mathematical beauty." The story of his success is a favorite among theoretical physicist. In any case, Sakharov kept the story of the "anti-electron" close at hand. When he once was showing his ability to do mirror writing, he wrote "electron + positron = 2 photons."

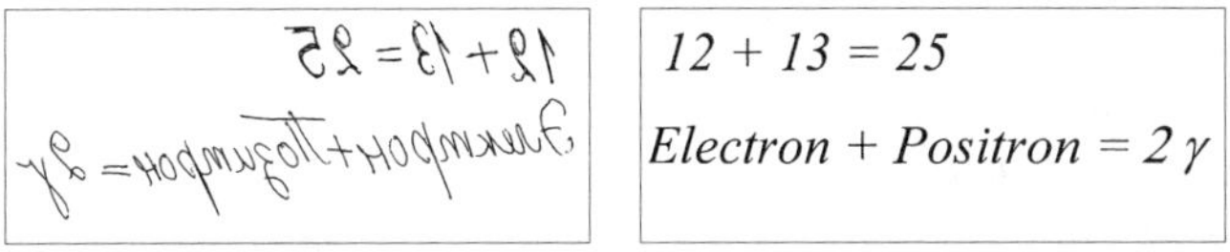

$12 + 13 = 25$
Electron + Positron = 2 γ

Figure 14.3 A summary of Dirac's success story in Sakharov's mirror writing and in English translation. The Greek letter gamma stands for "photon."

He also demonstrated his ability to write in different directions simultaneously with both hands, writing his hostess's name. When Lydia Chukovskaya, who kept examples of his writing, tried to repeat his trick, she failed (Figure 14.4).

Mirror symmetry—the symmetry of the butterfly—like mirror asymmetry, embodied in the double autograph shown below, has something to do with Sakharov's most significant idea in cosmology.

Matter and Antimatter in the Universe

In 1966, soon after Sakharov made himself a scientific plan for the next sixteen years, he noticed an asymmetry in nature that was becoming evident in those years: there were many more particles in the universe than antiparticles.

Ever since Dirac predicted the existence of antiparticles in 1931, matter and antimatter had an equal right to exist as far as physicists were concerned—in theory. In practice, after Carl Anderson discovered the first antiparticle, the antielectron, aka positron, in 1932, it took another three decades to observe the next antiparticle, the antiproton. And it was only a few years ago that physicists managed to create the first simple anti-atoms—antihydrogen atoms—out of antiprotons and antielectrons. They made nine. The first anti-atoms

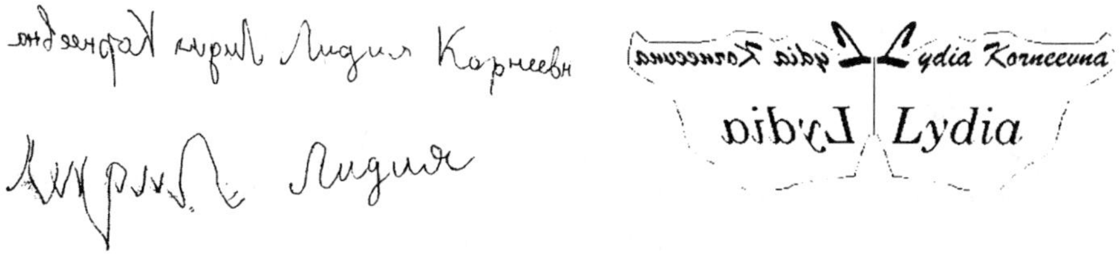

Figure 14.4 Sakharov demonstrated his ability to write simultaneously with both hands, writing his hostess's name and patronymic. Lydia Korneevna Chukovskaya, who kept this example of Sakharov's writing, tried to repeat his trick of mirror writing and failed. If she had been ambidextrous like Sakharov and if they both had written in English, they would have come up with something resembling the butterfly shown on the right.

existed for only billionths of a second before they met with normal matter and were annihilated.

In a popular article on antimatter, Sakharov gave this example: "The annihilation of 0.3 g of antimatter with 0.3 g of matter will give the effect of the blast of an A-bomb."[25] That was his second profession speaking. The contact of two small tablets would cause the same blast as 20,000 tons of ordinary explosives. That could make one lose any sympathy for the experimental physicists creating anti-atoms. Just imagine if antimatter were easy to create.

But sympathy for the theoretical physicists could increase. All the experiments with antiparticles did not change a thing in the theoretical equality of matter and antimatter that the theorists understood back in the 1930s. How could the empirical and theoretical results be made to match? What was the explanation for the fact that matter and antimatter were so inequitably represented in the universe?

The most tangible part of matter are nuclear particles—protons, neutrons, and their close relatives. Physicists gave them a family name—baryons. And the apparent absence of antibaryons is called the baryon asymmetry of the universe.

As long as physicists regarded the universe merely as a collection of various astronomical objects, they could think that matter predominated in the cosmic neighborhood of Earth and that farther away there were stars and planets of antimatter. Astrophysicists looked for signs of antimatter in space. Science fiction writers created dramatic encounters between spaceships from earth with extraterrestrial ones made of antimatter. There was even a joke that the best way to tell if a spaceship had come from an antiworld was if most of the physicists on board were anti-Semites.

The situation changed dramatically after the 1965 discovery of cosmological background radiation. Even skeptics had to believe the universe could be treated as a single physical object with its history determined by the laws of physics. It was clear that the universe had once been very hot. The remaining background radiation had cooled to a temperature only three degrees above absolute zero, but there was a lot of it—it filled the entire space of the universe. Ordinary matter was concentrated in the stars and planets, separated by enormous distances.

If radiation and matter were recalculated in particles—photons and baryons—then we would see that now for each baryon there about one billion photons, today's "barely warm" photons. But what had been the case yesterday? When the universe was smaller in size, the photons, according to the laws to radiation, were warmer. And if we go back far enough, there was a time when the energy of the average photon was enough to give birth to a baryon and antibaryon pair. Until that moment, photons easily turned into such pairs, and the pairs upon meeting just as easily turned into photons, by annihilation. In that hot period, there were approximately as many such pairs as there were photons. And that means that there were a billion times more baryon-antibaryon pairs than the excess of baryons over antibaryons that is observable today. These are the baryons that remained after all the baryon-antibaryon pairs were annihilated into photons, which in the process of expansion cooled so much that they lacked the energy to create new pairs.

That means that in the very young and hot universe, there was only one billionth part more baryons than antibaryons. Thus the asymmetry in nature is not simply small but challengingly small.

Sakharov had "trouble imagining" that originally, in the nature of things, for every 1,000,000,000 photons there were just as many antibaryons, 1,000,000,000, and only one more baryon—1,000,000,001. These startling numbers, in Sakharov's view, "offend the eye: 'it couldn't have happened that way.' It was this circumstance (as the reader sees, intuitive rather than deductive) that was the stimulus for a lot of studies on baryon asymmetry, including my own."

It was a stimulus for Steven Weinberg, 1979 Nobel laureate and author of a best-seller on the first three minutes of the universe. In 1977 he wrote:

The baryon number per photon might have started at some reasonable value, perhaps around one, and then dropped to its present low value as more photons were produced. The trouble is that no one has been able to suggest any detailed mechanism for producing these extra photons. I tried to find one some years ago, with utter lack of success.[26]

Therefore, Weinberg decided to ignore all "nonstandard possibilities" and accepted baryon asymmetry as an inexplicable fact. By the time Weinberg's book came out in Russian in 1981, it was clear that he should not have ignored the nonstandard possibility indicated by Sakharov in his paper of 1967[27] (and presented in the next paragraphs). Zeldovich, who was editor of the Russian translation, devoted a special appendix to that possibility.

But even Zeldovich, who was the first to learn of Sakharov's work, long considered it too strange to be correct. Sakharov recalls a conversation in 1967:

Zeldovich asked which of my purely theoretical works I liked best. I said: "Baryon Asymmetry of the universe." He winced and said, "Is that the work where the baryon number is not conserved and time flows backward?" "Yes, that's the one." Zeldovich said nothing, but it was clear that he had great doubts about the value of my ideas.[28]

Sakharov's new idea seemed "fantastic and crazy" even to Feinberg. When he received a copy of the article with the author's inscription, he thought, "Well, sure, Sakharov can indulge himself in anything, even fantasy like this."[29]

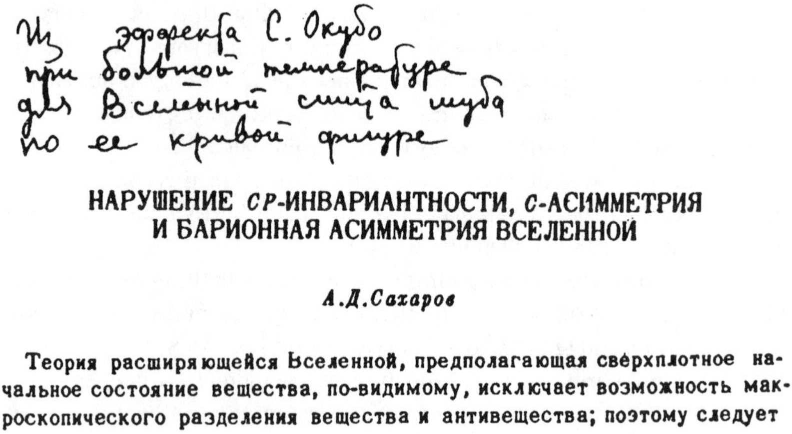

Из эффекта С. Окубо
при большой температуре
для Вселенной сшита шуба
по ее кривой фигуре

НАРУШЕНИЕ СР-ИНВАРИАНТНОСТИ, С-АСИММЕТРИЯ И БАРИОННАЯ АСИММЕТРИЯ ВСЕЛЕННОЙ

А.Д.Сахаров

Теория расширяющейся Вселенной, предполагающая сверхплотное на-чальное состояние вещества, по-видимому, исключает возможность мак-роскопического разделения вещества и антивещества; поэтому следует

Figure 14.5 Poem expressing the idea of baryon asymmetry of the universe (on a copy of the 1967 article Sakharov presented to Evgeny Feinberg).

The inscription was in verse:

> With S. Okubo's effect
> And at high heat,
> A skewed coat was made to fit
> And on the universe look neat.

What fantastic craziness lay behind the verse?

We know that the figure of the universe is (baryonically) skewed and we know that theoretical physicists were bothered by that fact it was only slightly skewed, by one billionth part. No tailor would alter a suit to accommodate a difference of one millimeter between right and left shoulders. Cosmologists were worried by a difference a million times smaller—but only because that differ-ence seemed to be related to the very genesis of the universe.

Sakharov's Three Conditions for the Universe

The American theorist Susumu Okubo was not thinking about cosmology. In the mid-1950s he was working on the physics of elementary particles when mysterious asymmetries surfaced. Before then the unspoken understanding was that everything in the world of elementary particles had to be extremely symmetrical. These minimal components of matter seemed to have nothing like right and left hands. And therefore there could be nothing like the asymmetry of the right-handed and left-handed in the world of people. In the world of elementary particles, it was assumed that mirror symmetry ruled—total parity of right and left, or *P*-symmetry, like the wings of an ideal butterfly (see Figure 14.6, left panel).

To be more accurate, it was believed that if a phenomenon was possible in the world of elementary particles, then if that phenomenon were reflected in the mirror—switching right and left—we would get a phenomenon that was just as possible.

In 1956 a momentous event occurred: it was discovered that the world of elementary particles is not fully *P*-symmetrical. That is, there were phenomena whose mirror image was not as likely. This observed asymmetry in the microworld disconcerted physicists. They began looking at two other symmetries, which until then had been considered as certain. Operation *C* replaces any elementary particle with its antiparticle, that is, any charge with its opposite. Operation *T* turns time backward, that is, replaces any movement with movement in the opposite direction. Let us imagine all particles as white billiard balls and their antiparticles as black ones. Then operation *C* changes the color of the balls to the opposite and operation *T* changes the collision of the balls, as recorded on video tape, to what it would be if the tape were played in reverse.

Each one, *P*, *C*, and *T*, is like the wave of a magic wand. So, in 1956 physicists learned to their surprise that a wave of *P*-wand changes the microworld. What did follow from the basic principles was that waving all three wands at the same time changed nothing in physics. They called that *CPT*-symmetry.

But how about separately? For a few decades, physicists had been certain that the microworld was symmetrical for any of the *C*, *P*, *T* magic wands. It might have been easier being a theorist in such a world—if people's right and left hands were the same, making gloves would be twice as simple. However, it is probably impossible to understand humans if you ignore the differences between the right and left hemispheres of the brain, the imaginative and logical locations. Simplicity can be worse than theft, as the Russian saying goes. An oversimplification of the world steals the possibility to comprehend it.

The problem is that there is no formula for avoiding oversimplification. The only reliable instructor in the matter is the experiment or the wise question skillfully posed. The mirror unevenness of the microworld, shown by experiments, pushed physicists to build castles in the air, in which the asymmetrical outbuilding was part of a symmetrical universe. Within a year, one castle was completed. Landau found that all then-known *P*-asymmetrical phenomena

Figure 14.6 (Left) *P*-symmetrical butterfly. (Center) *CP*-symmetrical butterfly. (Right) *CPT*-symmetrical butterfly.

were subordinated to combined CP-symmetry, and he proclaimed that symmetry a new law of nature: The simultaneous waving of the C- and P-wands does not change the world.[30] In other words, Landau proposed that the butterfly of the microworld had a form that did not change if you simultaneously switched the right and left and black to white—that is, the particles switch places with the antiparticles and the right switches places with the left (see Figure 14.6, center panel).

Landau's cutting-edge work drew a lot of attention. It was his first work sent to *Nuclear Physics*, and according to Okubo, the editor of the journal, Leon Rosenfeld, "was so pleased to have received a paper from such an eminent physicist as Landau directly from Russia, that he published it immediately without any delay instead of its being sent for refereeing as was normally done."[31]

A scientific paper's importance can be measured by how much it helps pose new questions to nature, and—if the answers are negative—helps to disprove itself. Landau's work helped Okubo ask: What if even CP-symmetry is not all-powerful in the microworld? And he figured out how to pose that question to nature. In his two-page article in 1958 he pointed out that if CP-symmetry is not valid, then a particle and an antiparticle with the same lifetimes could end their lives differently—breaking down differently into other particles.[32]

This remained a purely theoretical possibility until 1964, when it was discovered that CP-symmetry was not absolute. It, too, was violated, albeit rarely.

Strangely enough, the main Soviet theorist in thermonuclear weapons followed these pure subtleties unrelated to his work. This is evinced in Sakharov's evaluation of Landau's works on CP-symmetry, by request of the Lenin Prize committee (and sent on December 18, 1958). He summed up by saying, "In his influence on the development of science in our country and the entire world, Landau is one of the top figures."[33]

Landau was not then awarded the Lenin Prize, and probably not because the Central Committee of the party had a different view of the CP-symmetry problem. They had in their possession a lengthy KGB report that Landau had dared to name Lenin "the first fascist."[34] How could they give him the Lenin Prize?!

For all his disagreement then with Landau's anti-Lenin formulation, Sakharov was unlikely to have considered it relevant to an assessment of Landau's work. And it did not matter that Landau's 1957 hypothesis on combined CP-symmetry was disproved by experiments in 1964. The main thing was that his work had moved forward the search for scientific truth.

In 1966 it was Sakharov's turn to move the search forward. He began thinking of the experiments on violations of CP-symmetry and the Okubo effect in combination with the fact of the baryon asymmetry of the universe. And he came to the idea of a microphysical genesis of that asymmetry—the skewed figure of the universe. He began with the fact that in the microworld only the most general *CPT*-symmetry worked and the butterfly of the microworld looked like the right panel of Figure 14.6.

It does not change only if all three switches are done simultaneously: left and right, particle and its antiparticle, past and future.

Next to this butterfly of the microworld Sakharov apparently placed the butterfly of the expanding hot universe (see Figure 14.7).

He used the *CPT*-symmetry of microphysics to explain the asymmetry of the universe. In the era of the Big Bang, when matter was so compressed that the elementary particles felt one another, the universe sensed the laws of the microworld directly. It was then that the asymmetry of the universe, according to Sakharov, formed in the processes that were taking place in every micropoint all over the cosmic space. *T*-asymmetry allowed the birth of now-observable *C*-asymmetry—the difference in number of particles and antiparticles.

Besides the wing of the universe's butterfly that is visible to astronomers, Sakharov could visualize the other wing that opened before the Big Bang. The cosmological butterfly is *CPT*-symmetrical, but it cannot be seen whole because of the brevity of human life compared to the age of the universe.

In inventing the mechanism that creates an excess of baryons over antibaryons from the original symmetrical state, Sakharov used three gears:

Figure 14.7 Butterfly of the expanding universe.

(1) with "the effect of S. Okubo"—the difference in decay of particles and anti-particles; (2) "at high heat for the universe"—this condition creates the necessary cosmological effect for an ultrabrief period, while the universe is hot enough, and then the result freezes; and (3) "a coat is made" with a needle that was a new instrument for physics. Sakharov proposed that the baryon number is not conserved. This meant that the proton, the "brick of the universe" that was considered to be completely stable, had to decay on its own.

At the end of the paper, Sakharov thanks six physicists for discussion and advice, starting with Zeldovich.[35] Zeldovich had brought Sakharov to the Institute of Theoretical and Experimental Physics, to the students of Pomeranchuk—Boris Ioffe, Igor Kobzarev, and Lev Okun. They knew everything about C, P, and T and so on (it was in discussions with them that Landau came up with his CP idea).

Okun, who helped Sakharov with his suggestions, considers his paper on baryon asymmetry of the universe "one of the most profound and bold articles of the twentieth century."[36] No one doubted its boldness at the moment of its birth. Sakharov took aim at the seemingly untouchable law of conservation of baryon number.

In school we study only electric charges, whose conservation is one of the most basic properties of the electromagnetic field. Conservation of baryon charge was not a result of some profound theory of baryon field, it was based only on the fact that other behavior had not been observed—the decay of the proton. This is a fact worthy of respect, and Sakharov showed his respect by estimating the rate of decay of the proton in the theory he proposed. The decay was so "astronomically" slow that it explained why it had not been observed—it required an unthinkable precision of measurement.

Whether to respect a fact, bear it in mind, or obey it unquestioningly is a decision left to the researcher. In the late 1960s, the great majority of physicists, among them, Zeldovich, chose absolute submission to baryon symmetry. History held a mini poll on this issue among the theorist fathers of American nuclear weapons. In 1966 articles, both Robert Oppenheimer and Edward Teller expressed full confidence in conservation of baryon charge.[37] Teller was confident enough to suggest an explanation of quasars as a collision between a galaxy and an antigalaxy, of which there should be equal numbers in the universe. As the Russian bard Okudjava sang in those days, "everything is equal, everything is fair—for every wise man there is a fool, for every flow there is an ebb." For every proton there is an antiproton and for every galaxy there is an antigalaxy.

Then why did Sakharov decide to leave this unanimous chorus in 1966? Perhaps he had understood better than the rest the lesson of CP-symmetry according to which in physics, as in a society ruled by law, everything was permitted that was not forbidden by law. Or perhaps he had a deeper understanding of the fact of cosmological asymmetry of matter and antimatter and did not try to convince himself that the asymmetry observed on Earth, locally, could somehow be made compatible with the overall symmetry of the universe.

But what we are actually talking about is the depth of scientific intuition, which is based on facts and theories but is not reduced to them.

When Sakharov realized in 1948 that the H-bomb design he received from Zeldovich was leading nowhere and found a completely new path, this was his scientific intuition at work. Zeldovich "instantly appreciated the seriousness" of his discovery. In 1966, the path proposed by Sakharov diverged too sharply from the well-worn tracks, and it took Zeldovich, who saw Sakharov's intuition work "before his very eyes," years to fully evaluate its seriousness. It happened when the development of the theory of elementary particles—for the so-called Grand Unified Theory—also questioned the stability of proton. It was only then that Sakharov's explanation of the baryon asymmetry of the universe took its proper place at last in the arsenal of contemporary physics.

In Okubo's words, "Although the idea appears now to be so simple, it is due to the genius of Prof. Sakharov [that he was able] to combine many different aspects of theory into a coherent picture."[38]

It is too soon to put this picture in a gilt frame. Physicists in many countries are checking the elements of the picture experimentally. That is the usual way of development of ideas in physics, a development in which the international physics community extracts knowledge for humanity through cooperation and competition. We will certainly learn in this millennium where the experimental testing and development of the theory that explains the asymmetry of matter and antimatter will lead. But its more immediate prospects are described in *Scientific American*:

> It is imaginable that the universe was born skewed—that is, having unequal numbers of particles and antiparticles to begin with . . . Theorists prefer the alternative scenario, in which particles and antiparticles were equally numerous in the early universe, but the former came to dominate as the universe expanded and cooled. Soviet physicist (and dissident) Andrei Sakharov pointed out three conditions necessary for this asymmetry to develop.[39]

In proposing the nonconservation of baryon charge—the instability of the proton—as one of those conditions, Sakharov was being a dissident in physics. He was not afraid to speak about what he saw with his own eyes when others had not yet seen it or were afraid to look. It is not yet known whether he had discovered a new law of nature. But we do know that the mysteries of nature reveal themselves only to such dissidents.

Sakharov's work was not at the very center of theoretical physics or even of the part that Sakharov considered his own—"elementary particles, gravitation, and cosmology."[40] Hundreds of theorists were working in each of the three fields. What distinguished Sakharov was that he united them. His work was the first to define a concrete characteristic of the universe as a whole by the characteristics of the microworld.

Sakharov's explanation of baryon asymmetry of the universe in 1967 merely opened up a new direction for scientific research; it did not close it by being an exhaustive and complete theory. This direction is sometimes called cosmomicrophysics—the combination of the physics of the microworld and the megaworld.

To this day, revealing the decay of the proton is a goal of experimental physicists, and this goal also affects the Grand Unified Theory, which is intended to unify all fundamental forces with the exception of gravitation.

The Elasticity of a Vacuum

It was gravitation that the second of Sakharov's exceptional ideas was aimed at. Fate, as he used to say, gifted him with this idea in 1967, and Zeldovich was involved in that gift.

It started with Zeldovich's idea to fill emptiness with a vacuum. He filled the empty space-time of Einstein's theory of gravity with the quantum vacuum of microphysics. By that time the empty space-time no longer resembled a box without walls filled with the ticking of an invisible clock. In the late 1940s experimenters confirmed what the theorists had been saying since the early 1930s: if you remove all the contents of a vessel, what would be left was not lifeless emptiness. Life seethed quietly, with particles continually being born and dying—fluctuating—and that nonstop simmer is able to change the color of a flame. The change is slight, but the experimenters noticed it. In order to get away from the old-fashioned emptiness, they used the Latin word—vacuum.

Physicists were looking at vacuums through a microscope, so to speak, but Zeldovich suggested using a telescope. He proposed that the living vacuum revealed in microphysics could have a gravitational effect on the megaworld— on the rate of expansion of the universe. In this way he proposed to explain the new astronomical data on the strange distribution of quasars.[41]

Zeldovich spoke of his idea at a FIAN seminar and found no support. The idea contradicted the habitual viewpoint that the vacuum acted only on elementary particles and that for greater bodies, macroscopic ones, the vacuum was merely emptiness. Besides, physicists did not accept the reason that prompted Zeldovich to make such a bold statement. And in fact that "observable fact" that had aroused Zeldovich's imagination (a peculiarity in quasars distribution) soon vanished like a mirage amid new observations.

Anna Akhmatova's lines apply to more than poetry:

> If you could but know the trash from which poems grow,
> Without the slightest shame—
> Like the lowly weeds you can no longer name,
> Sprouting by the fence from seeds you did not sow.[42]

Scientific ideas sometimes also sprout by the fence.

Sakharov missed Zeldovich's talk, but he learned from him that the FIAN theorists were "sharply negative" about the idea. "After the seminar Zeldovich called me and told me the content of his work, which I liked instantly. A few days later I called him with my own idea that was a further development of his approach." Life had prepared Sakharov to accept Zeldovich's idea regardless of what had prompted its genesis. He had been thinking about the microphysical vacuum in 1948, before his "exile from Grand science." Twenty years later he did not simply support Zeldovich. He foresaw how the vacuum in microphysics could be united with gravitation at the most profound level—where gravitation might be rooted.

Zeldovich looked at the quantum fluctuation of the vacuum through a cosmological telescope, characterizing all vacuum life with a single number—its energy density. The "astronomically small" density of vacuum energy would have a noticeable effect only on astronomically large distances. But Sakharov tried to explain gravity as a characteristic of the continual seething going on in a quantum vacuum.[43] His paradoxical idea was that gravitation—Newton's gravity known to all schoolchildren—does not exist. Then what does? There is the "elasticity" of the vacuum, which leads to all the known manifestations of gravity—from apples falling to the collapse of stars and the formation of black holes.

But if Sakharov's paper "did away" with gravitation, then why did John Wheeler, one of the most prominent figures in the theory of gravitation, like it so much? He explained it enthusiastically in his books and articles.[44] Wheeler was interested not in preserving the classical Newton-Einstein theory of gravity at any cost, but in understanding it fully—that is, in solving the difficult questions that arose from that theory but had no answers. The most important of them was quantum gravity.

Sakharov's hypothesis opened an unexpectedly new view of the fortress long besieged by theorists. While his colleagues, settled in a military camp around the fortress, considered which catapults and battering rams to use on its thick walls, Sakharov seemed to have discovered an underground passage leading inside the fortress.

He proposed seriously considering the fact that in all points of space-time the vacuum life is pulsing and to take into consideration the effect of that life on the behavior of ordinary, macroscopic Newtonian bodies. The hope was that Einstein's theory of gravity with its curved space-time continuum, its collapse of stars and expansion of the universe, would become a consequence of quantum theory. And then out of Einstein's theory, when gravity is not very strong, Newton's law of gravity would follow.

The reader, recalling the formulation of that law in high-school textbooks,

$$F = GmM/r^2,$$

may ask, "Well, but where will the value of the gravitational constant G come from?" Sakharov based his approach on a new constant arising in the complete theory of the microworld, length *l*, which corresponds to the limit of applicability of geometric concepts known since Euclid. For distances less than *l*, the usual concepts of space and time must be replaced by some much more profound and less obvious concepts. Sakharov's theory does not require elucidation of those concepts. It permits theorists to continue their search for the complete theory of elementary particles. However, they are now given an architectural plan of how to unite their search with the search of the complete theory of gravity. And if the searches are successful, then length *l* will determine the constant *G*, which governs the fall of an apple and the movement of the planets.

According to Sakharov's idea, the gravitational constant is the result of the microscopic structure of the vacuum. He called his approach "gravitation as the elasticity of the vacuum." What does it have in common with the ordinary, familiar elasticity?

Man dealt with elasticity when he made his first bows and arrows. Intuitively, he took into account the coefficients of wood's elasticity. In more enlightened times elasticities were measured for various materials and encoded in tables. For the preparation of a good bow, however, it is enough to select a material with the appropriate coefficient, without thinking about how elasticity is determined by forces holding the atoms and molecules of the material. The maker of the bow needs to study the molecular structure of matter if he is not satisfied with trial and error—going through all materials—and if he wants to know how the bow will behave on the extreme edge of elasticity, before it breaks.

Just as for calculations on how an object moves in the gravitational field of Earth or the Sun, it is enough to take the value of *G*. But in order to learn what will happen to a star as a result of its unlimited compression in its own gravitational field, or how the expansion of the universe began, knowing the "molecular" structure of the vacuum is indispensable.

Theorist-Inventor

The mechanism of the formation of baryon asymmetry, invented by Sakharov in 1967, is still the only working hypothesis to explain the observable asymmetry of matter and antimatter. The mechanism he invented to explain the "formation" of gravitation from the characteristics of the microworld is still only an architectural idea. Therefore Sakharov's colleagues in theoretical physics have to rely on a combination of reason and feeling called intuition in their assessment of his ideas. A variety of intuitions is vitally important for scientific research. But the same variety leads to differences in opinion.

For instance, Sakharov considered Zeldovich's idea (born by the fence of astronomy), his own starting point, to be one of his best. Zeldovich, apparently, did not think so—in his 1984 scientific autobiography, he does not mention it.

Some sober-minded theorists do not give serious weight to Sakharov's hypothesis on gravitation as the elasticity of the vacuum. They do not wish to count their chicks before they are hatched. Others consider the idea one of the most significant of Sakharov's contributions to pure science and the next step after Einstein in discovering the nature of gravitation.[45]

Let us leave the final decision to history. But even before this decision, one can say that the theoretical physicist who in the course of a single year published two such "knockout" ideas as Sakharov did in 1967 has the right to feel proud, especially if that physicist is also burdened with being the leading designer of thermonuclear weapons.

This astonishing combination leads us to the question that began this chapter: was Sakharov a theorist or inventor? We can replace the "or" with a hyphen. He used the term "theorist-inventor" when talking of his military technology work,[46] but it applies to his theoretical physics as well. Theorists differ not only in their strengths of intuition, but in their methods of work. Some start with a general alluring idea and seek a way of concrete formulation. Others begin with a simplified theory of a concrete phenomenon. Still others, with the most general physical theory that they try to apply to the given problem.

In Sakharov's theoretical physics the inventor is visible. He devises a mechanism that nature could use in controlling its mysteries. An inventive engineer starts with scientifically known elements that he can combine. But an inventive theorist must first invent the very elements he will need to combine into a theoretical mechanism. The inventiveness can be measured by the level of unusualness of the elements.

We can imagine how the theorist and the inventor collaborated in Sakharov: the theorist saw unusual elements that do not contradict the fundamental laws of nature. The inventor, not abashed by their unusualness, used these elements to construct a working mechanism: a magnetic field as incorporeal walls of a vessel to hold the blazing lightning of fusion, the flush of radiation from an A-bomb to compress fuel to produce an H-bomb, and the superweak instability of the proton in the superhot early universe. The theorist tells the inventor that all these elements are acceptable in fundamental science. And the inventor figures out how to turn them into a working mechanism.

That may be the way the theorist and inventor in Sakharov collaborated in his creative laboratory. But how did these collaborators regard the completely nonscientific concerns of this lab's chief?

15

WORLD PEACE AND
WORLD SCIENCE

Sakharov was reflecting on the symmetries of the universe and on the nature of gravity at the Installation while he continued working conscientiously on the development of "big new technology," in the tactful terms of Zeldovich. Sakharov's professional work as a weapons designer ended only with his dismissal in 1968.

Taking a closer look at the events that led to that dismissal, it is particularly interesting to see what kept him at the Installation. Why had he not followed Zeldovich who left for the Academy of Sciences in 1964?[1] Sakharov felt that his "presence at the Installation could at some crisis point be decisively important." But what prompted him to think so? Was it only his successful intervention in the Test Ban Treaty of 1963?

Physicists Fortify, Marshals Guide

His loyalty seems even harder to understand in view of an event that occurred in 1955 and scarred him for life. They were celebrating the success of a test (the Third Idea) at a banquet given by Marshal Nedelin, the military head of the test. He asked Sakharov to make the first toast.

> I picked up my glass, rose, and said something like: "Let us drink to all our devices exploding as successfully as today, but over test sites and never over cities." A silence fell at the table, as if I had said something improper. Everyone froze. Nedelin chuckled, also stood with his glass, and said, "Let me tell you a parable. An old man in his nightshirt is praying before an icon illuminated by a votive light: 'Guide and fortify me, guide and fortify me.' His wife, in bed on top of their stove, said: 'Listen, old man, you just pray to be fortified, I'll do the guiding myself!' Let's drink to fortifying!"

I winced and I think turned white (I usually blush). For a few seconds no one said anything and then all spoke in unnaturally loud voices. I drank my brandy in silence and never opened my mouth to the end of the evening. Many years have passed, but I still feel as if I'd been lashed with a whip.

Now this story is seen as a symbol of Sakharov's situation in the Soviet military-industrial complex. But this was just one episode in his twenty-year career as a weapons designer, and Sakharov told it years after his dismissal—he already looked at many things with different eyes. And he knew that history would expand the marshal's joke into an edifying parable.

In 1960 Marshal Nedelin was in charge of testing of a new rocket. There was a problem just before the countdown, and the technical experts—the ones who "fortify"—recommended putting off the launch and calmly find out what was causing the problem. But the marshal—who "guided"—wanted the test to go off on time. He decided single-handedly to correct the problem during the test. To instill confidence in the engineers, he remained close to the rocket. The rocket's engine obeyed the laws of physics rather than the will of the boss, and when it started, it burned seventy-four people alive, including the marshal.[2]

Of course, Sakharov had other facts at his disposal, which permitted him to think up until a certain moment that Nedelin's joke did not represent the entire government policy and that the government needed to hear the opinion of technical experts and was capable of listening to them.

The character of the technology Sakharov was working on and its state significance demanded a broad view. One reason was that the nuclear weapons had to have appropriate means of delivery. Soon after the 1953 tests, Sakharov and his colleagues were informed about the space missile program. Their guide was its chief designer, Sergey Korolyov (1907–1966), who reminded Sakharov of Kurchatov.

Korolyov was not only a brilliant engineer and organizer, but a colorful personality. He had a lot in common with Kurchatov. A very important trait of Kurchatov was a love of science. Korolyov had his dream of space travel, ever since he worked for the Jet Propulsion Research Group in his youth. I don't think he considered Tsiolkovsky a dreamer, as some did. Like Kurchatov, Korolyov had a rather crude sense of humor, concern for his subordinates and comrades at work, and an enormous grasp of the practical, but perhaps more cunning, ruthlessness and cynicism. They were both major figures in the military-industrial complex, but they were enthusiastic about science and space travel, too.

Scientific enthusiasm and powerful personalities protected Kurchatov and Korolyov from being enslaved by the military-industrial complex, and that in turn made this complex look less evil in Sakharov's eyes. They could believe that the scientific and engineering elite were able to influence the life of the country, turning Stalinist socialism into a "scientific" socialism, true socialism.

Kurchatov actually succeeded in a few things. He provided "political asylum" for geneticists in his nuclear empire—which was against the official line. His sensational lecture at Harwell, the British atomic center, in 1956 broke the wall of secrecy in controlled thermonuclear reactions, which had important consequences for scientific and political contacts with the West.[3]

Kurchatov had Khrushchev's trust, and he could have used it even more for the good of science and the country if he had not died at such an early age—at fifty-seven in 1960.[4] He accomplished a lot because, besides his enthusiasm, he had other qualities; the KGB called him "a secretive man, careful, shrewd, and a great diplomat."[5] Sakharov had none of these qualities, nor did he seem to notice them in Kurchatov. But he had reason to think that simple scientific logic, without cunning diplomacy, was enough to influence government decisions.

He had that impression from his meeting with Leonid Brezhnev, the future leader of the country. In the late 1950s, when Brezhnev was in charge of military technology at the Central Committee, a resolution was being prepared that the Installation physicists considered wrong. It would have siphoned intellectual and material resources "from more important things (we meant in the military-industrial sphere: it was not a question of redistribution with civilian affairs)." Khariton decided to explain this to the Central Committee and took Sakharov with him as support to talk to Brezhnev. He heard them out "very politely, taking notes on a pad." As a result, the resolution was not passed.

In the early 1960s the social status of physicists was at its highest point in Russian history. The esteem was so high that when Khrushchev was being removed from office in 1964, his party comrades listed disrespect for physicists as one of his faults.[6]

The State Committee on Science and Technology was founded in 1965, and its chairman, Academician Vladimir Kirillin, held another post simultaneously: deputy head of government (under Aleksei Kosygin). In the spring of 1966, Kirillin invited Sakharov, Ginzburg, Zeldovich and several other major scientists and engineers to see him.

> Kirillin said that the U.S. is doing a lot of scientific and technological futurology, writing some lightweight and trivial things, but on the whole this work had its uses, giving us a long perspective which is very important for planning. He proposed that each of us write in free form how we see the development of our fields of science and technology in the coming decades and also, if we wished, to touch on more general questions. We left. In the next few weeks I worked enthusiastically and wrote a small article with a great flight of the imagination . . . For me work on that article had great psychological significance, once again concentrating on general questions of the fate of humanity.

So Sakharov once again addressed broader issues. His 1958 article was based on a very long-range social forecast: over the course of the half-life of radio-carbon—around six thousand years—the population will grow to approximately thirty billion. And as Sakharov noted, "this number is not contradicted by the food capabilities of the planet in light of scientific progress."[7]

The 1966 article "Science of the Future. A Forecast of the Prospects of the Development of Science" begins with the statement that many of its ideas and evaluations are the fruit of collective work and names eight of his colleagues from the Installation. The first section is narrowly professional—"the use of nuclear detonations for scientific and technological goals"—and even here, after listing several scientific and industrial applications, his imagination soars when he writes of directing the movement of asteroids through nuclear explosions.

The article exudes unbridled scientific gusto. "This goal justifies all expenses," he writes about the search for "laws of high energy dynamics that enhance our concepts of space and causality."

Only one paragraph is devoted to social forecasting:

> The progress of cybernetics will lead to profound changes in ideology and philosophy and will have great social consequences. I would think that the progress of technical and social cybernetics, along with achievements in biology, physics, and astrophysics, as well as the political creativity of democracy will bring the greatest and quite unexpected corrections into predictions about the social life and political structure of the society of the future.[8]

On the whole, Sakharov sounds like a technocrat, or rather, science-crat, unable to restrain his science-fiction dreams and unaware of the reality around him. But we should remember the assignment: to forecast the development of science. The readership was limited, too. The article was published in a collection with just 120 copies stamped "for official use," that is, only for the leadership elite.

Thus, in the summer of 1966 Sakharov was thinking about the progress of science separately from progress in any other area. In less than two years, he would be writing another article in which he combines progress of science, peaceful coexistence, and individual freedom—the article that would be the turning point in his life.

The Letter to the Politburo
on the Dangers of Defense

In recalling events that occurred in those two years, Sakharov described yet another article, which remained in manuscript form. But the story leaves a sense of doubt.

As he related, journalist Ernst Henri suggested he write an article, in the form of a dialogue with the journalist, "on the role and responsibility of the intelligentsia in the modern world," for *Literaturnaya Gazeta*. Sakharov accepted, but what he wrote was so radical that it scared off the editors. They wanted to get approval from higher up. Sakharov sent the manuscript to the Politburo and got a negative response. And so what did he do? He gave the manuscript to the journalist and "forgot about the whole business."

That certainly does not resemble Sakharov's persistence, despite his modest demeanor. And it doesn't fit at all with the beginning of the chapter "The Turning Point" in his *Memoirs*, where he writes of professional weapons concerns, when work brought him into strategic military discussions:

> What I learned was more than enough to have a particularly acute sense of the horror and reality of a major thermonuclear war, the madness and danger threatening all of us on our planet. On the pages of reports, at meetings on problems of operations research, including operations of a strategic thermonuclear strike on an assumed enemy, on schematics and maps the unthinkable and monstrous turned into an object of detailed examination and calculation, turned into practical life—as yet only imagined, but already regarded as a possibility. I could not help thinking about it—with the ever greater understanding that we were talking about not only and not so much technical questions (military and economic), as primarily questions that were political and moral.

The new hot issue of those discussions was antiballistic missile (ABM) defense, missile systems intended to destroy the enemy's attacking missiles. Despite the defensive intentions of the new weapons system, Sakharov came to the conclusion that striving for such a defense was extremely dangerous. A new race of defensive and offensive weapons would undermine the viability of the former guardian of peace—mutual assured destruction, or MAD. It is better to have such a mad guardian than none at all.

Are we to believe that with these tragically serious concerns in his professional field, Sakharov found the leisure time to entertain himself with an optional article on the role of the intelligentsia and then, when he was shushed by the authorities, immediately forgot about it? It doesn't seem likely.

The answer was hidden for a long time in the archives of the Central Committee. After the Central Committee was shut down (1991), the archives were opened, and—among other Soviet secrets—Sakharov's letter to the Soviet leadership dated July 21, 1967, became available for study.[9] It is a large missive, containing a nine-page letter marked "Secret" and a ten-page manuscript of the article prepared "jointly with the famous columnist E. Henri" for publication in *Literaturnaya Gazeta*.

The letter's topic is a moratorium on developing ABMs proposed not long before that (March 1967) by U.S. President Lyndon B. Johnson and Secretary of Defense Robert S. McNamara. The point was a "bilateral rejection by the USA and the USSR of the development of an antiballistic missile defense (ABM)

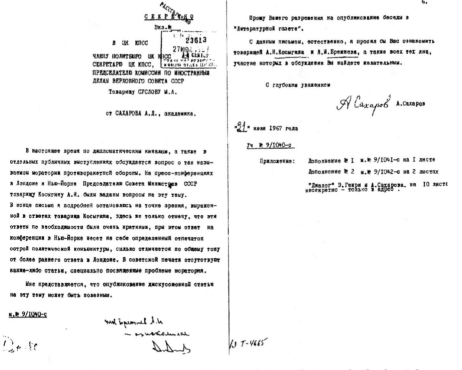

Figure 15.1 The first and last page of the secret letter to the Soviet leadership, July 21, 1967.

against a mass attack of a powerful enemy, while retaining those works that are necessary for defense against small-scale missile aggression."

The threat of small-scale missile aggression—that is, launched by a rogue state or terrorists—became a reality in the 1990s when the appropriate technology was available. But in 1967 Sakharov saw the problem of defense against "a small number of missiles from an aggressor or provocateur" and separated it from problems of "mass attack." He devoted a special addendum of the letter to small-scale aggression, reaching a practical conclusion: "creating ABM systems of defense from mass attacks is not realistic, while for individual missiles it is difficult but possible." This brings up the question of whether Sakharov was privy to antiballistic affairs. After all, his work for the military involved creating offensive weapons, not defensive ones.

In fact, ABM defense developed hand in hand with the means for nuclear missile attack. As early as in the summer of 1956, the ABM designers made a visit to the nation's main nuclear center, the Installation, where Sakharov and his colleagues were designing nuclear warheads. The point of the visit was to run experiments on shooting at one of the warheads—on the ground.[10] The ABM designers needed to learn the construction of the warheads in order to counter them. And the Installation designers had to consider how to make the warheads impervious to antiballistic missiles.

Sakharov, in his high position, was aware of all aspects of the ABM work. The correlation of methods of attack and defense were the topic of the most heated and top secret discussions in 1967.

Sakharov begins his secret letter to the Politburo with a polite but firm disagreement with Kosygin, one of its most important members and the head of the government. Not long before that, at a press conference in New York, Kosygin said that an ABM moratorium was possible only with a general disarmament treaty and also stated that means of defense were always moral, unlike means of attack. That "simple and understandable" point of view, a legacy of the prenuclear era, was popular in the West as well. Both the U.S. Congress and the Soviet Central Committee had members who were certain that they could handle any problem using plain common sense. In both military-industrial complexes there were experts with a personal interest in developing ABM systems who wanted to continue their interesting work. American ABM experts had persuaded themselves that they were twenty years ahead of the Soviets, while their Soviet colleagues were just as certain that they were "ahead of the entire planet."[11] Each side tried to convince its politicians that it was right.

But there was also a difference between the two sides. Robert McNamara came to politics from the world of free competition. He made sure that issues were studied through a competition of expert opinions. The U.S. scientific experts came to the same conclusion as Sakharov: It was impossible to create an AMB system invincible to mass attack, and a race in that area would significantly increase the danger of nuclear war and at best would lead to a pointless expenditure of enormous amounts of money.

The USSR also had knowledgeable experts, but they were dealing with political leaders who came from the school of unfree competition, the school of party palace intrigues. In his letter, Sakharov presents not only his personal opinion but that of his colleagues, experts in strategic weapons. In particular, he mentions "the official documents [on ABM] sent to the Central Committee of the CPSU by Comrades Yu. Khariton and E. Zababakhin [the scientific heads of both nuclear Installations]." Apparently there had been no reaction to those official documents, which prompted Sakharov to add his voice.

He must have remembered his contribution to the 1963 treaty banning atmospheric testing. It was a similar situation: One problem, whose resolution is doable and important, is taken out of an overwhelming complex of problems. It is important in and of itself and also as a step toward the stability of peaceful coexistence.

Sakharov addressed his letter to Mikhail Suslov, member of the Politburo and Chairman of the Commission of Foreign Affairs of the Supreme Soviet USSR, and asked him to show it to Kosygin and Brezhnev. Sakharov had met

Suslov in 1958. In their first conversation, he gave an entire "lecture" in response to Suslov's question about genetics. The Politburo member listened attentively, "asking questions and taking notes." Sakharov thought that his lecture might have had an effect when Khrushchev's fall was followed by an end to Lysenko's biology as state policy. Therefore, he could have expected that Suslov would take this letter seriously, especially because Sakharov was expressing more than his own opinion.

> Johnson and McNamara are apparently bringing up this issue [ABM moratorium] out of election considerations, but objectively, in my opinion and the opinion of many of the major workers at our institute, it corresponds to the fundamental interests of Soviet policy, because of a number of technical, economic, and political considerations.

Sakharov then laid out those considerations. His point of departure was that the USSR had a "significantly smaller technological, economic and scientific potential than the US": in GDP (gross domestic product), two and a half times; in computer production, fifteen to thirty times; in financing of science, three to five times; and in efficiency of expenditures, several times. "The gap in these and other important indices, except GDP, is growing," he stressed. "This difference forces the USSR and the US to evaluate differently the possibility of creating offensive and defensive weapons." In offensive weapons the concept of sufficiency exists, but not in ABM, where the result of competition "is determined by the technological and economic potentials."

Because an ABM system is much more expensive than an equally powerful offensive system (three to ten times more, as Sakharov writes), the Soviet Union, as the side that is weaker technologically and economically, will have to develop offensive means only:

> Work on ABM systems in the US is quite advanced, and there is a solid technological backup of work done on several systems. While it is not enough for an effective solution to the ABM problem, it is a symptom that the period of approximate and unstable parity of retaliation means that began in 1957 is not eternal, the violation of that equilibrium is possible, or the illusion of its violation. Would we really give up the chance for general regulation while it exists?

That is why it is necessary "'to take the Americans at their word,' both in the sense of real limitations of the weapons race, in which we are more interested than the US, and in the propaganda sense, to strengthen the idea of peaceful coexistence."

Sakharov feels that an open discussion of the issues of a moratorium in the Soviet press would give support to groups of "foreign scientific and engineering intelligentsia which under beneficial conditions can be a force to restrain the 'ultras' and 'hawks.' These groups have played an important role in the preparation of the Moscow Treaty on Test Ban."

He added a "discussion article on the topic"—the article for *Literaturnaya Gazeta*—and asked for permission to publish it.

The Failed Dialogue in *Literaturnaya Gazeta*

In the article "World Science and World Politics," Sakharov leaves out the harsh words and figures about the USSR's technological and economic weakness, but his main point does not really depend on it: The ABM race will significantly increase the threat of nuclear war because of the illusion of impunity. The role of scientists is to reveal and prevent that threat. He recalls events of the recent past—the 1963 Moscow Treaty that

> stopped the poisoning of the atmosphere that doomed annually over a hundred thousand of our contemporaries and descendants, weakened the weapons race, and led to a clarification of the political position of numerous countries. It can rightfully be considered the step in the escalation of peace in the 1960s. I want to remind you that the conclusion of the Moscow Treaty had been prepared by a broad worldwide campaign by the intelligentsia.

When Sakharov moves to the ABM issue in response to the journalist's questions, his reasons seem too specific for a newspaper journalist, which suggests that the initiator of the dialogue was not Henri. It is more likely that the physicist wanted to use the prominent journalist's help in making public the opinion of experts on a current issue that threatened world war. Probably it was the military technical details in the manuscript that worried the editors at *Literaturnaya Gazeta* and required a sanction from above, rather than the radicalness of Sakharov's conclusions, like this one: "The credo of progressive scientists and progressive intelligentsia throughout the world is the open and unprejudiced discussion of all problems, including the most acute ones." And in response to the question: "And if the American political leaders continue to play with fire?" the progressive Soviet scientist replied, "Then, I think, it's up to the American working class, the American people, and not least of all the intelligentsia and scientists."

He ended the article with an appeal: "Shoulder to shoulder with the working class, resisting imperialist reactionaries, nationalism, adventurism, and dogmatism, scientists and the intelligentsia must realize their power as one of the main supports of the idea of peaceful coexistence."

It is clear that here Sakharov still considered himself part of the system and a defender of socialism. The leaders of Soviet socialism, however, did not need his advice. Suslov's secretary informed Sakharov that it was not appropriate to publish the article because "it has several statements that might be misunderstood."

Sakharov was soon to see a confirmation of the timeliness of his advice. The Soviet government had a little more than a month to "take the Americans at their word." Under pressure from U.S. hawks and the icy silence from the USSR, in September 1967 McNamara announced the U.S. decision to build the first ABM system.

Sakharov also saw that the "foreign scientific and technological intelligentsia" were ready to restrain their hawks. In March 1968, Hans Bethe and Richard Garwin, prominent physicists who were close to the military-industrial complex, analyzed the problem of ABMs in *Scientific American*, proving the danger and pointlessness of the new form of arms race.[12]

By the spring of 1968, Sakharov could not trust the country's leaders the way he had Khrushchev. The new "collective" leadership had made no contributions equal to Khrushchev's dismantling of Stalinism and his attempts—albeit awkward ones—to renew public life. On the contrary, after Khrushchev was removed from office, signs of creeping rehabilitation of Stalin began to appear. This Soviet leadership tossed aside the nineteen-page work—the fruit of serious reflection by Sakharov and his colleagues on the vitally important question of war and peace. As Marshal Nedelin had twelve years earlier, they recommended that Sakharov stick to [ahem!] fortifying weapons and let the Politburo decide where to guide them.

What did it matter to Sakharov that a Politburo member said that it was inappropriate to publish the article? Sakharov cared more about the authority of the argument than the argument of an authority. That approach was responsible for all his achievements in science. It was natural to stick to his habits outside science as well. No one was countering his military-political analysis with any substantive reasoning. A few months later Sakharov returned to the idea of a discussion article, to the need for open discussion of an explosive set of problems, the fuse of which he considered to be the ABM system.

But before turning to his new article, his famous "Reflections on Progress, Peaceful Coexistence, and Intellectual Freedom," we have to answer a simple question: Why did Sakharov never tell the history of his unexpected public statement that would have explained the seriousness of his reasons and the responsibility of his intentions at the turning point in his life?

He did not tell the story because he was an honest man in his ordinary life and in its most extraordinary parts.

"I write with certain omissions about my life and work between 1948 and 1968, required by a commitment to maintain secrecy. I consider myself bound for life by this commitment to keep state and military secrets, which I undertook voluntarily in 1948, no matter what life may bring," he wrote in the years of his exile in Gorky. He treated his pledge with absolute honesty.

Elena Bonner, the dearest person to him, shared his exile. They had been prepared to die together in hunger strikes, but he never did tell her that the

Installation where he spent two decades was not far from Gorky, even though he told her many things about his life at the Installation and wrote about it in his memoirs, which she retyped. After the manuscript was stolen by the KGB, he compiled a chronology of the main events of his life as an aid in restoring his lost work. The chronology mentions his letters to the Central Committee, but not the 1967 letter on antiballistic missile defense.[13]

He kept the state secret even from the KGB. Once, when he was already a dissident, a colleague from the Installation visited him, and the conversation touched on their former work. Sakharov, accustomed to KGB eavesdropping, stopped his visitor. "You and I have access to secret information. But those who are listening in do not. Let's talk about something else."[14] He wasn't making a joke. Sakharov knew that different departments of the KGB were in charge of protecting state secrets and the surveillance of dissidents. The KGB never complained about Sakharov as the bearer of secrets.[15]

This is all to say that his 1967 letter to the Central Committee was secret, and Sakharov could not reveal it. Therefore, in his *Memoirs*, the subject of the secret letter is separated from the article for *Literaturnaya Gazeta*. There is only one place, three chapters earlier, where we could guess at a connection: "My professional work as a weapons designer ended only with my dismissal in 1968. The discussions of that period, in particular about antiballistic missile defense (ABM), I describe in other parts of the book."[16]

At Pushkin Monument
on Constitution Day

Sakharov's break with weapons designing was heard throughout the world in July 1968 when his article "Reflections on Progress, Peaceful Coexistence, and Intellectual Freedom" was smuggled to the West and published in the *New York Times*. He was brought to this step not only by his "internal" letter to the Central Committee but also by some external events in preceding years. The first was his speech against Lysenkoism in June 1964. The three-minute statement at the Academy of Sciences broke through the veil of secrecy. He was recognized by the intelligentsia close to academic science.

The secret physicist, academician, and three-time Hero of Socialist Labor was a desirable ally for reform-minded intellectuals in their noble quest. Their goal most often was socialism—the type varied: "Leninist," "true," "democratic," and "with a human face." The common denominator was "non-Stalinist."

The people who came to Sakharov after his three-minute speech were very different in the degree of public activity and freedom of thought they permitted themselves. At one extreme were those who were quite happy with the Khrushchev framework of exposing Stalinism. The journalist Ernst Henri (1904–1990) was one such "conservative reformer."[17] A physicist friend brought him to meet Sakharov in January 1966.[18] At that time, on the eve of the first post-Khrushchev party congress, there were signs the country's leaders were planning to somehow rehabilitate Stalin. Henri had prepared a letter to Brezhnev, warning him against it. It was felt that this letter, signed by prominent figures in Soviet science and culture, would support the progressive forces in the government. Besides Sakharov, the other twenty-five signatories included the physicists Artsimovich, Kapitsa, Leontovich, and Tamm, and famous writers, directors, actors, and artists.[19]

"Henri was in no way a dissident," Sakharov stressed, even allowing that the initiative for the letter had come from somewhere higher up. He did not consider the phrase "progressive circles of the KGB" an oxymoron.

At that time he also met real dissidents and began reading *samizdat*. Both the word and the phenomena became an important part of the sixties. *Samizdat* (literally, "self-publish") was a play on Politizdat, the main Soviet publishers of ideological literature. Books were "published" without printing presses and without censors. It was varied literature—ranging from poetry from inaccessible editions to out-and-out anti-Soviet works. The role of the printing press was taken by typewriters. Each typewriter made five copies, and each subsequent "generation" multiplied the print run by five. The Moscow intelligentsia made great use of the self-publishing free press. In its first decade *samizdat* released around two thousand works.

Uncensored facts about Soviet history and the Soviet present were revealed to Sakharov, who had been isolated from this reality. "Even if things were not always objective in these accounts, at first what was more important was my release from the closed circle in which I had lived."

He was so isolated that he had not even noticed the very publicized trial of two writers arrested in September 1965 and convicted in February 1966 for publishing their works abroad and not repenting. "During the trial of Sinyavsky and Daniel I was still pretty much 'on the sidelines,' and knew practically nothing about it." Yet this was the case that began the dissident movement in the USSR.[20] Nor did he know about the first organized action of the movement, a demonstration in one Moscow's central squares, by the Pushkin Monument, on December 5, 1965, Constitution Day, a state holiday. It was a silent demonstration, the participants unfurling their banners at a signal: "Respect the Constitution" and "We Demand Openness for the Sinyavsky and Daniel Trial." The initiator was the mathematician Aleksandr Esenin-Volpin.[21]

A year later Sakharov received an anonymous invitation to take part in the second such demonstration. The idea was to come on Constitution Day 1966 to the Pushkin Monument "five-ten minutes before six p.m. and exactly at six, remove out hats with the others, as a sign of respect for the Constitution and to stand in silence, bare-headed, for one minute."

Sakharov liked that idea. He placed it among "the many other very original and fruitful ideas" of Esenin-Volpin.[22] It is almost as if he were discussing science or technology. A markedly constitutional, nonaggressive demonstration blessed by the proximity of the great poet was something Sakharov could easily consider a manifestation of "political creativity of democracy," which he had mentioned a few months earlier in his futurological article.[23] He decided to take part in that creativity. For the first time in his life.

> Near the monument was a group of several dozen people, all of whom were unfamiliar to me. Some were speaking quietly. At 6 about half of them took of their hats, as did I, and as agreed, we were silent (as I later realized, the other half worked for the KGB). Once they put their hats back on, the people did not leave. I walked over to the monument and loudly read the inscription on one of the sides of the base:

> > Long will I be by my people beloved,
> > For awakening kind feelings with my lyre,
> > For celebrating freedom in this cruel age
> > and calling for mercy for those who have fallen.

> Then I left with the majority.

The "majority" was composed of a small minority of the capital's intelligentsia. They were united by a desire for openness in public life, or to put it simply, for freedom. Twenty years later "glasnost," which in Russian means publicity and openness, became a political term and entered foreign languages; it also spelled the end of the Soviet regime. But for the silent demonstrators glasnost was a means of improving Soviet socialism, not overthrowing it.

Sakharov was prepared to improve socialism in the permitted areas of public life. In early 1967 he got involved in protecting Lake Baikal from industrial pollution. Prominent scientists, writers, and engineers made up the committee in defense of Baikal under the auspices of the Central Committee of the Komsomol [Communist Youth League]. He considered his work for that cause without results but very important for his understanding of the relationship between ecology and the social structure of society. He personally saw that independent ecological expertise was impossible when science and industry have the same boss—the government.

The years 1966 and 1967 saw many collective appeals to the authorities in defense of individuals and with proposals for improvements in Soviet life. These petitions were usually signed by major figures in science and the arts, and for a time signing these did not lead to repercussions. People signed for various reasons, from sincere belief to interest in prestige. A very few moved on to being dissidents. The rest stopped being visible when it became dangerous—a few months before the occupation of Czechoslovakia in 1968.

A good social barometer is the quantity and "quality" of the signatures on a petition for freedom of the press in October 1967.[24] This is a weighty document (nine typed pages) with references to the first decrees of the Soviet regime, to the Constitution, and to the party's program. The petition proposed passing a law on the press and doing away with censorship. This attack on the powers of the totalitarian state did not frighten the 125 people who signed the letter. Its initiators broke down the fifty most celebrated signers this way: academicians (7), corresponding members (1), members of the Writers' Union (16), directors (3), members of the Artists' Union (7), members of the Composers' Union (8), and Doctors of Sciences (8).

Zeldovich's signature seems the most free thinking, placed as it is above the rest and with a comment, "With the criticism of the present situation and the need for the elaboration of a new law on the press, I agree" (the petition included a "working draft of the law on dissemination, finding and obtaining information"). Sakharov's signature is distinguished only by a more detailed description of his social status: "Academician, three times Hero of Socialist Labor." Only the people at whom the petition was directed knew just how much he was set apart from the others by his professional responsibilities.

In February 1967 the authorities had received a personal closed letter from Sakharov in defense of four people whose crime had been a documentary report on the Sinyavsky-Daniel trial. As a result the Minister of MedMash Slavsky said, "Sakharov is a good scientist, he has done much and we have rewarded him well. But he is an unstable politician, and we will take measures." Sakharov's salary was cut by 50 percent.

It is unlikely that Sakharov's July 1967 letter on antiballistic missile defense was in the same file at the Central Committee with his February appeal and October petition. But in his own reflections in early 1968, all three letters to the government were in the same place. All three were without result. But the most eloquent failure was the secret appeal in July. The subject was within Sakharov's professional competence and it led him to general conclusions about his country, its regime, and its position in the world.

Sakharov knew what a heavy burden the military-industrial complex was for the country. When the Tsar Bomb needed a Tsar Parachute, the production of nylon stockings was stopped. And when mercury was needed for some bomb aspect, they ceased production of thermometers for a long time. There was a popular Soviet cliché, "The country spared nothing for . . ." But was it in fact that the government did not spare the country? Sakharov cared about the country. The gap between the economy of the USSR and of the United States was explainable—the destructive war with the Nazis and Stalin's tyranny had taken their toll. Accepting the need for nuclear balance in order to prevent war, Sakharov was proud that he had found an inexpensive solution for his country, the H-bomb. "Assertions about the high cost of thermonuclear explosives is a myth. In any weapons systems, they are the cheapest component," he wrote in his July article for *Literaturnaya Gazeta*.

But he discovered that the government was prepare to allocate a huge part of the budget for a new weapons system—ABM—without checking, without showing any interest in why the experts thought it not only a waste of the nation's money but also a dangerous enterprise.

Sakharov could allow that he did not know everything about the behavior of the four people he defended in February; they might have violated some letter of the law. He could allow that creating real freedom of the press was not as simple as it seemed to the signers of the October petition. But when it came to ABMs he was among the most informed.

When he wrote in the July article for *Literaturnaya Gazeta* that scientists must become one of the main supports of peaceful coexistence, "countering imperialist reactionaries, nationalism, adventurism, and dogmatism," he was addressing those words to the West. Now he was seeing that there were things to resist inside the country. It seemed that the Soviet hawks were as much in need of restraint, but the Soviet press was not available for an open discussion of that vital issue.

16

REFLECTIONS ON INTELLECTUAL
FREEDOM IN 1968

In February 1968 Sakharov began work on a new article, "Reflections on Progress, Peaceful Coexistence, and Intellectual Freedom." Now he was free of his coauthor Henri, the in-no-way-a-dissident. He was going to set down his understanding of the world on the basis of his competence and sense of personal responsibility.

"Moving Away from the Brink
Means Overcoming Our Divisions"

Sakharov begins his article with a warning of its character as a discussion document and concludes the same way: his goal is "open, sincere discussion." The person inviting the discussion did not consider himself a "specialist in social issues." However, his starting point was an area where he could consider himself a leading expert—the threat of thermonuclear war as a result of the anti-ballistic missile (ABM) problem. Even in his letter to the Central Committee a half year earlier he had expanded an examination of military technology to state economy and policy because that expansion was required by the nature of the issue. In "Reflections" he went even farther beyond the framework of the military strategic problem, but with the same aim: to find a path to stable peaceful coexistence and to understand where that coexistence should lead.

One of the variants of his article began with an epigraph from Schiller: *Nur die Fulle fuhrt zur Klarheit* [Only fullness leads to clarity].[1] It was suitable for the theoretical goal he set for himself. The prospect of the destruction of humanity in a nuclear war—a global catastrophe—demanded a comprehensive approach.

The practical—moral—goal is suited to the Goethe epigraph on which he settled and with which "Reflections" reached the *samizdat* readers: "He alone is worthy of life and freedom / Who each day does battle for them anew!"

The solution of the double goal took several months. He wrote "primarily at the Installation" in the evenings. He redid it several times, and yet later considered it "imperfect ('raw') in form." And as for the content: "Basically it is a compilation of liberal—humanistic and 'science-cratic'—ideas, based on available information and personal experience."[2]

The imperfect form is in fact a hindrance, and science-centricity of the author is apparent from the very first paragraph, where the author expresses concern that "the scientific method of managing politics, economics, art, education, and military affairs has not yet become reality." Right in the next sentence he explains that he considers a method scientific if it is "based on a profound study of facts, theories, and views presuming a nonbiased and open discussion, impartial in its conclusions."[3] So, here "scientific" is just a synonym for "best" and it is simple to guess the author's feelings for science.

Another characteristic is its "socialist" approach. The author calls his views "essentially socialist," without specifying what he means by that. In any case, it is no longer a synonym for the very best, since socialism did not stop the establishment of Stalinism and Maoism, which Sakharov put in the same category as fascism, nor did it keep a new ruling class of the "bureaucratic nomenklatura elite" from forming in the USSR.

Sakharov seems to be relying on certain tenets of the "theory of scientific socialism" as it was preached in the Soviet Union. For instance, he writes that "the capitalist world had to give rise to the socialist," obviously does not sympathize with "the egotistical principle of private property," clearly makes his allegiance known: "we, the socialist camp," and uses the names of Lenin and Marx unequivocally positively. However, the evil name of Stalin appears much more frequently.

He is speaking from socialist feelings rather than from ossified socialist commandments. In Russia, then, the feelings of human solidarity, social justice, and respect for labor were tied for many people with the word "socialism."

Feelings aside, for Sakharov the key point was the fact that none of the founders of socialism had foreseen: humanity was facing the real possibility of total suicide within a half hour, the travel time for a nuclear missile. That possibility had been created by scientific and technological progress and it could not be undone. According to those founders, socialism would beat capitalism by higher productivity of labor, just as capitalism had vanquished feudalism. And the Soviet political education system had a ready-made answer for why capitalism in its final, imperialist stage hindered scientific, technological, and—more so—social progress.

As a student Sakharov never did well in political studies:

Of my university subjects I had trouble only with Marxism-Leninism, I got "D"s, which I later improved. The reason was not ideological, it never occurred to me in those years to doubt Marxism as an ideology in the struggle to liberate humanity; materialism also seemed an exhaustive philosophy to me. But I was upset by the naturophilosophical mindsets that were transported without any reworking into the twentieth century of strict science.

Academician Sakharov, regarding materialistically the facts in the area of his professional expertise, strategic weaponry, could not award victory to socialism in labor productivity. The most that he could give socialism was to call it a draw. The draw in the competition of the two systems suggested that they were both stable. That strengthened the key fact, the threat of global suicide, but also dictated the necessary step to be taken: "Every rational being finding itself on the brink first tries to get away from that brink and only then think about satisfying all its other needs. For mankind, moving away from the brink means overcoming its divisions."[4]

Sakharov's main conclusion was that the convergence of the socialist and capitalist systems was the only alternative to the death of mankind. And the entire article, with its historical and political panorama, is devoted to spelling out the possibilities and ways of that convergence.

He does not pretend to be original in using "a term accepted in Western literature." Adding "socialist" to convergence is not that original either. What is original is what he means by "socialist convergence." Western founders of the concept of convergence assumed that it would be brought about by scientific and technological progress. The governments of both systems would be forced to rely on it more and more. They would have to use the hands and heads of professionals with similar training—scientific and technological managers, whose importance would grow in both systems. From here would come the growing similarity of the new industrial capitalism and the new industrial socialism. One of the most eloquent adherents of this hope, John Galbraith, called his 1967 book *The New Industrial State*.[5]

For Sakharov the need for convergence came not from the similarity of new scientific and technological establishments but from mortal danger. The danger born of scientific and technological progress hung over both opposed systems on a single planet.

As to "growing importance" of sci-tech managers, Sakharov was a highly placed manager and he found that even he was not important enough to be heard by the government on a vital issue. That is why he went public and became the first to start building a bridge of convergence from the socialist shore. He had never been to the other, capitalist shore, and therefore had to rely on secondhand information from the press, with all its filters and distorting mirrors. In 1968, the capitalist shore did not look fit for a convergence bridge—the Vietnam war, the assassination of Martin Luther King, who won the Nobel Prize for Peace in 1964. And in the spring of 1968, the Prague Spring, when Czechoslovakia began building "socialism with a human face," an inhabitant of the socialist camp could convince himself that socialism had creative potential and that the future belonged to it.

Sakharov knew what rebuke to expect from his circle of free-thinking but prosocialist acquaintances. Expressing high regard for Roy Medvedev's unpublished book on Stalinism, written "from socialist and Marxist positions," he added, "Probably the author will not receive the same compliments from Comrade Medvedev, who will find elements of 'Westernism' in his views. Well, nothing like a good controversy."[6]

At that point Sakharov had spent two years interacting with the dissident milieu. He read *samizdat* and took part in the inevitable discussions arising among free-thinking intellectuals. He had a free, or in his definition, "scientific," approach to "facts, theories, and views." His "Reflections" are the result of his thinking. However, Sakharov knew things that his new acquaintances did not and that led him to think about convergence with the West, rather than trying to perfect socialism. He had professional knowledge in a limited but vital sphere of state and international life: the strategic opposition of the two systems and the unstable balance in nuclear-missile arms. That knowledge brought a wider perspective. Sakharov had a clear perception of the scientific and technological competition of the two systems and of intellectual freedom and government dealing in the sphere of strategic balance. But the sphere itself was so significant for the life of the society that it permitted him to judge the society as a whole. Sakharov knew that socialism might not have time to be perfected if the strategic opposition turned into nuclear war.

The Invention of the
Social Theorist

Sakharov formulated his conclusions this way:

> The continuing development of production forces under capitalism is for any nondogmatic Marxist a fact of priority and principle significance, for this fact is the theoretical basis of peaceful coexistence. Since capitalism is not in an economic dead-end, it has not turned to the venture of a desperate war. Both the capitalist and the socialist systems have opportunities to develop for a long time, borrowing positive traits from each other (and in fact moving closer together in essential relations).
>
> I can imagine the howls about revisionism and a blunting of the class approach, mockery of my political naivete and immaturity, but the facts speak of a real development of production forces in the US and other capitalist countries, of a real use of socialism's social principles by capitalists, and of real improvements in the situation of workers. *And most importantly, the facts say that on any other path besides a deepening coexistence and cooperation between the two systems and two spheres, with a reduction of contradictions, and with mutual aid, that on any other path disaster awaits us. There is no other choice.*

Sakharov's last point, which I have highlighted, is the most important indeed. The social theorist established a general strategy for the survival of humanity—convergence.

The social inventor, however, needed a concrete mechanism to implement the strategy. Sakharov found it in individual freedom, in human rights. He subordinated the process of convergence not to some specially invented law of history but to the Universal Declaration of Human Rights adopted by the General Assembly of the United Nations back in 1948. In particular: "Everyone has the right to freedom of opinion and expression; this right includes freedom to hold opinion without interference and to seek, receive and impart information and ideas through any media and regardless of frontiers."

He saw intellectual freedom as an effective instrument of convergence: "I think that gradually the leaders of the capitalist and socialist systems will be forced to accept the point of view of the majority of the world. Intellectual freedom of society will facilitate the evolution of this transformation toward tolerance, flexibility, and freedom from dogmatism, fear, and adventurism."

Intellectual freedom would permit scientific and technical experts to explain "the power of things" to leaders and society as a whole, so that the society could react in a timely and proper way to the swift changes in that scientific and technological force.

Sakharov addressed his analysis to "nondogmatic Marxists," that is, intellectuals who were "politically enlightened" within the system of Soviet education. Another way of putting it is that he addressed people whose brains had been well washed. They studied mandatory classes in Marxism-Leninism in school and college and then went through various forms of required adult education at work. So naturally a few classical tenets and expressions of the founders remained in their memory, on the level of folk sayings and proverbs. Everyone knew Lenin's maxims that "the electron is as inexhaustible as the atom" and that building socialism must "combine Russian revolutionary sweep and American efficiency." Recalling his college troubles with Marxism-Leninism, Sakharov felt that the main problem was his "inability to read and memorize words rather than ideas."

In Marxist ideas, "everyone knew" that the determining force in the development of society was its material life ("basis"), that is, the combination of the society's production forces and those relations that tie people in the process of production. A society's economic life divides people into classes and economic interests determine the political and cultural life ("superstructure"). Capitalism gives rise to the working class, which will grow, strengthen, and bury its "parent," since the workers, who do not own any means of production, have the greatest interest in freeing humanity from exploitation and in establishing social justice. And the true interests of the working class are best known to its avant-garde, the Communist Party, which is leading the country for the well-being of all working people.

But what did the nondogmatic Marxist Sakharov get from all of this? Almost nothing, except a materialistic view of history. The production forces of the steam engine of Marx's day had developed into thermonuclear megaton devices. And as a direct result, a brand-new economic interest of the working class had come about, in addition to the demand for an eight-hour day and so on— not to perish in the nuclear fire with their families and all the other classes of society.

In Vernadsky's language this change in production forces could be characterized as scientific and technological progress giving people power on a geological scale, equivalent to an earthquake, volcano eruption, or glacier movement. And that changed the meaning of many old concepts. Can the term "war" apply to something that can take place in a few minutes at the will of a few people and will kill every living thing on the planet? There won't be time to organize an antiwar rally or to dodge the draft.

To deal with this brand-new situation, Sakharov proposed using intellectual freedom, which in Marxist terms belongs to the "superstructure" and therefore could only come with an appropriate change in the "basis." Any Marxist prepared to accept this would have to be very nondogmatic. Probably not even a Marxist at all, unless you take seriously Lenin's words, which everyone who had gone through political education courses knew, that "Marxism is not dogma but a guide for action."

In saying that capitalism and socialism had come to a draw, Sakharov was in fact admitting socialism's economic defeat. He compared the situation with two skiers traveling in deep snow: the "stars-and-stripes" skier is laying a track of scientific and technological progress, while the "red" one is following a prepared path.

Sakharov grants socialism priority only in "moral ideals." Since he does not clarify his understanding of the term, he is referring to the well-known ideals of social justice, freedom, equality, and brotherhood. But these ideals, according to Sakharov, still had to be moved from slogans to real life so that "in a moral comparison of capitalism and socialism" people do not "first recall the limitations on intellectual freedom under socialism or, even worse, the fascist-like regimes."[7]

Tying socialism's future only to the moral attractiveness of its ideas was a total rejection of Marxism as far as official "scientific socialism" was concerned. Socialist morality, as a superstructure phenomenon, had to come as a result of changes in the economic basis. The law of history would do its work. It had to be helped, or at least not hindered by homegrown utopian theories.

All of Sakharov's arguments could easily be dismissed as a utopian path to a utopian idea by official socialists. Not only official ones. Roy Medvedev thought so, too, if we believe KGB Chairman Yuri Andropov's report to the Central Committee in the summer of 1968.[8]

Physics and Politics of
the Nuclear Age

On this utopian path Sakharov's true predecessors were not the founders of "scientific socialism" but the founders of twentieth-century physics—Albert Einstein and Niels Bohr. After the emergence of nuclear weapons, they came to far-reaching political conclusions. Both saw that in the nuclear age the concept of state sovereignty lost its customary historical meaning. Both saw the alternatives as "peace or total annihilation."

Einstein saw the solution in the creation of a world government, and Bohr in the establishment of an "open world." They can be considered different sides of the same solution: one addresses the organization of a world community; the other, the quality of a society permitting such an organization. Both recipes were more ideal than real politics, but both offered a fundamental solution to a complex of political problems of the nuclear age. How much time would be required—five, fifty, or five hundred years—was a question of "applied" rather than fundamental politics. These two political ideas of Sakharov's great colleagues in physics played a part in his political thought.

The radioactive fallout from testing blurred the concept of sovereignty— radiocarbon paid no attention to state borders. This physics-politics problem in 1958 was the start of Sakharov's immersion in world politics. His familiarity with the social mechanisms of the pollution of Lake Baikal helped him comprehend the global problem of "geohygiene"—or global contamination. It would require that humanity overcome its divisions—"otherwise, the USSR will poison the US with their wastes and the US will poison the USSR with theirs."

In 1986 history presented a lucid example. With the Chernobyl catastrophe, the Soviet Union violated the sovereignty of several European countries, but no one called this noxious violation of state borders aggression. That means that the very concepts of sovereignty and state borders are gradually losing meaning in the nuclear age.

Sakharov mentioned another global problem hanging over mankind, the "threat of hunger." He trusted the prognoses of specialists who felt that if special measures were not taken, by the end of the 1970s the food crises in some countries of the Third World would blend

> into a sea of hunger, unbearable suffering and despair, grief, death, and fury of hundreds of millions of people. This is a tragic threat to all humanity. A catastrophe of this scale cannot but have the most profound consequences throughout the world, for every human being, provoking waves of wars and hatred and a general decline in the standard of living for the entire world.

What did Sakharov propose?

> In the author's opinion, we need a "tax" on developed countries of about 20 percent of their national income for about 15 years. The introduction of this tax will automatically lead to a significant reduction on expenditures on weapons. The influence of such mutual aid will be substantive on the stabilization and improvement of the situation in the most underdeveloped countries restricting the influence of extremists of all types.

Sakharov would first of all tax the United States and the USSR, who spend more than the rest of the world on weapons. Fortunately, the prognosis he trusted did not come to pass. In the late 1960s the Third World underwent a green revolution, the introduction of new high-yield varieties of rice and wheat. This time, scientific and technological progress handled the looming catastrophe.

That fit in with the general interconnection indicated by Sakharov. Only scientific progress is capable of overcoming poverty on a world scale. But the development of science and technology bears two threats on a geological scale—the danger of instant self-destruction and not-so-instant self-poisoning. For humanity to contain these threats, it must be open and have the opportunity for self-consciousness. That is in the purview only of free people whose universal rights are protected by law.

Sakharov says that poverty and the fury of millions of people is fertile soil for extremists and warns that a tragic result of an uncontrollable course of events "will leave a tragic, cynical, and anti-communist mark on the life of future generations."

He is still in the thrall of the communist lexicon, or more accurately, in the thrall of the emotions of a centuries-old tradition of human solidarity that Communism tried to expropriate. He is in the thrall of a tradition that in the mind of many remarkable people found its expression in Marxism. It was the passionate hope of Sakharov's teacher, Igor Tamm. The twentieth century gave us the aphorism, "Socialism is the favorite folly of clever men." There should be an addition of "clever and good men—so good that they consider everyone else to be as good."

Marxism is right about many things, including that the criterion of truth is practice. And communist practice showed Sakharov the falseness, the unscientific nature, of Soviet "scientific socialism." Remaining in the thrall of socialist feelings, by 1968 he had freed himself of communist revolutionary ideology.

That ideology is contradicted by Sakharov's pro-communist-sounding warning. According to the canons of communism, disasters and the wrath of the masses form the main components of a revolutionary situation. The worse, the better, the sooner the revolution will come. And once it comes, the masses will be freed of their chains and quickly build a free society. According to Marx, "Revolution is the locomotive of history."

Sakharov corrected that formula to read: "Evolution, not revolution, is the best 'locomotive of history.'" He confessed himself to be a "convinced evolutionary, reformer, and principled foe of violent revolutionary changes of the social structure, which have always lead to the destruction of the economic and legal system, to mass suffering, lawlessness, and horror."[9]

The convinced evolutionary based his belief on fully materialistic, even Marxist, considerations. In the United States, the

> total consumption of the "rich" is less than 20 percent, that is, less than the total growth of national consumption over 5 years. Since a revolution will stop economic development for more than 5 years it cannot be considered economically advantageous for the working people. I am not even talking about the people's bloodshed that is inevitable in a revolution.[10]

Sakharov's proposal—to place individual rights at the base of international policy—looked unscientific in the times of Marx and Lenin. How could the behavior of a huge object, society, be determined by the characteristics of its microparticle, man? It would be as if the behavior of a star depended on the characteristics of its component particles.

But by Brezhnev-Kosygin times physicists knew that a star's behavior did depend on the characteristics of its microparticles, otherwise there would be no stars of enormous density known as white dwarfs. The ideas in physics that occupied Sakharov on the eve of 1968 were exactly about this: how the characteristics of the "individuals," the elementary particles, determined the "social" characteristics of the universe. His social postulate was in the spirit of his physics. Sakharov's idea to place human rights at the foundation of international politics, for all its apparent naiveté in the face of Soviet reality, spoke of the "microscopic" premise of the "open world" that could survive in the nuclear age.

By tying the microscopic conditions of social life with the characteristics of human civilization on the mega scale, Sakharov combined peaceful coexistence, progress, and the rights of the individual to intellectual freedom. "In essence these are the same themes that seven years later are in the title of my Nobel lecture, 'Peace, Progress, and Human Rights,'" he wrote in 1981, after a year in exile in Gorky. He began his Nobel lecture by stating that those three goals "are indissolubly linked: it is impossible to achieve one of them if the others are ignored."[11]

Did he realize in the spring of 1968 what lay ahead? The Nobel Prize for Peace seven years later, and a few years after that, seven years of exile? He had imagined something quite different. At the end of "Reflections" he described "the most optimistic" variant of events—by 1980 the realists vanquish the Stalinists in the ideological struggle. Instead, in 1980 the Stalinists stripped him of all his state prizes and sent him into exile without a trial.

"He Looked Perfectly Happy"

Sakharov wrote his "Reflections" with the same seriousness that he gave to everything else in his life, but also with a merry intellectual freedom that is essential to the creation of any truly serious work. His intellectual freedom can be seen, for instance, in the way he found a place in his political piece for asteroids, "moved with the help of nuclear blasts onto new orbits." He had written about asteroids first in his 1966 futurological article; perhaps he warmed to this idea not so much because of the prospects for space flight as because it would be a peaceful application of the blasts to which he had devoted so much of his inventiveness. Another stroke of the physicist's free manner is visible in the acronyms RRS and AME that he used for "Russian revolutionary sweep and American efficiency." He used it only once, but apparently he had plans to play with it some more.

But we need not limit ourselves to stylistic crumbs to judge the mood in which Sakharov prepared his first public appearance. We have the testimony of an eyewitness who saw Sakharov in 1968, during the months and weeks in question.

The eyewitness is Vladimir Kartsev, then a young engineering physicist and budding author, who had written his first popular science book.[12] It was on the history of magnetism from ancient times to the latest achievements in science. One such achievement was a record magnetic field obtained in 1964 by realizing an idea Sakharov proposed in 1951—an inch-long magnet with that field could hold the *Titanic* suspended.[13] Besides scholarly publications, there was an article called "Records of Magnetic Fields" in a leading national newspaper.[14]

The young author asked the newly public academician to write a preface to his book. He sent the request to him at the Academy of Sciences. Sakharov soon called him for the manuscript.

Vladimir Kartsev remembers: "About a week had passed. I had decided that there would be no preface. I called him and asked, 'Would an outline or rough draft help you?' He was terribly offended and said, 'I always write my own work.' And he wrote it."[15]

It was his first preface, and it must have elicited special feelings, as suggested by this sentence: "The author of the preface began his work as an inventor with the construction of an instrument for magnetic control of tempering, cracks, and thickness of nonmagnetic coverings and is convinced that even now, a quarter century later, in almost any area of technology there is an inexhaustible field of work for . . ."[16] He crossed out this unfinished phrase, finding it perhaps too personal.

In this brief preface the most interesting part is the author. He is not concerned with the reader's level of knowledge, mentioning the "Schroedinger-Pauli equation" and the "zero-mass interaction" as generally known things. Even without understanding these things, the reader gets an impression of the real people who discovered nature's mysteries of magnetism. They did not discover everything: in his two-page preface, Sakharov listed six unsolved problems in the physics of magnetism. He also did not shirk from pointing out Kartsev's error: "Sakharov caught me out with my Lenin. In trying to justify a seeming inexplicability of magnetic phenomena, I mentioned in my book Lenin's well-known thesis about the electron's 'inexhaustibility.' That was a mistake in physics, it was a simple misunderstanding of physics, and he corrected me as a physicist."[17] Sakharov had noted that in magnetic materials "there are no fundamental unknowns, and perhaps the author needlessly mentions the inexhaustibility of the electron."

But he showed an understanding of the genre, noting that "the book has the humor and flair required for popularization of science." They met a few times. Kartsev remembered him as being "energetic, with a radiant smile" and without any hauteur. "I was astonished that here I was, just some Ph.D., and he treated me very seriously and with a certain envy and respect. He said that he envied me for being able to write and that he also wanted to start popularizing science. I was at his house not far from the Kurchatov Institute. His children and wife were there . . . The house created a sunny, light impression. A very radiant, up mood, optimistic. Not a shadow of sadness or disillusionment. He looked perfectly happy."[18]

At one of their meetings, Sakharov showed him his 1968 political work. "So in my book I inscribed: 'To dear Andrei Sakharov with wishes that he realizes all his projects,' implying several things, including his political seekings and his desire to write popular science. I thought that he had written those reflections for himself. I did not know that it would be so widely disseminated."

Kartsev saved a copy of Sakharov's popular article, "Symmetry of the Universe," inscribed: "To V. Kartsev as a sign of the author's respect and friendship. 30/IV 68 A. Sakharov." The date is important. Only a few days earlier, Sakharov had completed his "Reflections." "On the last Friday of April [April 26, 1968] I flew back to Moscow from the Installation for the May holidays, the typed manuscript in my briefcase."

How could a man about to take a most serious political step be so involved in popular physics, giving a reprint of his article on cosmology, listing unsolved problems in the physics of magnetism? He also accepted a dissertation work on gravitation theory for review in May.[19] The answer is that Sakharov's work on cosmology, the letter to Suslov on ABMs, popular articles in physics,[20] and "Reflections" are various components of a strong creative surge he experienced in 1966–1968. His successful return to theoretical physics strengthened his self-confidence and became a prerequisite for his debut in political commentary. But there is no reason to think that he had planned to move into that area totally. He had other, no less interesting tasks in mind.

But in April 1968 his main task was to complete the article, his greatest literary effort since his dissertation in 1947.

science in progress

 novosti press agency monthly
PUBLISHED MONTHLY IN RUSSIAN, ENGLISH, FRENCH, GERMAN AND SPANISH

EDITED BY V. GOLDANSKI, B. KUZNETSOV,
N. MELNIKOV, F. OVCHARENKO, V. PARIN,
B. PONTEKORVO AND N. SEMENOV

THE SYMMETRY OF THE UNIVERSE

By Academician Andrei Sakharov (Physics)

> This article is published by
> courtesy of the Editorial Board of
> the "Budushcheye Nauki" (Future of
> Science) annual, to be brought out
> at the end of 1967 by Znaniye Pub-
> lishers).

Ten thousand million years ago the density of matter in the Universe was vastly greater, not only than the present-day average density of $10^{-29} g/cm^3$ but also the density of matter in atomic nuclei $(10^{14} g/cm^3)$.

We cannot explain the picture of the Universe as it is today, without possessing definite notions (or at least hypotheses) of the initial conditions of its evolution. The following are questions to which no single-term answers have yet been forthcoming:

1. What was the state of affairs prior to the moment of maximum density?

2. What was the degree of non-homogeneity of primary super-dense matter?

3. Did this matter contain particles and anti-particles in equal quantities or was it non-symmetrical in this respect?

4. Was this matter absolutely cold or "infinitely" hot?

Figure 16.1 Sakharov's inscription on the English translation of his popular science article "The Symmetry of the Universe." The date is only a few days after the completion of "Reflections" and just before their release in *samizdat*. It is clear that Sakharov did not live by politics alone, even at the moment he was entering the public affairs arena, but first and foremost by science.

Sakharov's Dismissal

He wrote "Reflections on Progress, Peaceful Coexistence, and Intellectual Free-dom" at the Installation at night. And at the Installation, he had it typed. He did everything openly. After all, he was planning to send it to the Central Com-mittee as well (which he did in June). In his position, secret activity was just impossible. He respected the professionalism of the security service that pro-tected the scientists' secret work. And his personality was not made for the underground.

The fate of the manuscript he had sent to Suslov a year earlier did not instill much hope for open publication for "Reflections," but he was prepared to en-sure that himself through *samizdat*. The first readers saw it in May, made a few comments, and pronounced it a historic event. The author continued the final touches on the article. On May 18, a Saturday, he dropped in on Khariton, the scientific chief of the Installation, at his dacha to talk about a few things.

> I told him, among the other subjects of conversation, that I was writing an article on the problems of war and peace, ecology, and freedom of expres-sion. He asked me what I was going to do with it when it was done. I replied: "I'll give it to *samizdat*." He got very upset and said, "For God's sake, don't do that." I replied, "I'm afraid it's too late to change anything."

Two weeks later they were traveling to the Installation in Khariton's personal railway car when Khariton "began a conversation that was obviously difficult for him." He said that he had been called in by Andropov (who took over the KGB in 1967) and told that Sakharov's manuscript was being disseminated il-legally and that if it got abroad, it would be a terrible blow to the country. Sakharov immediately suggested that Khariton read the article himself. They met the next morning and Sakharov inquired, "'Well, what do you think?' 'It's terrible.' 'The form is terrible?' Khariton laughed grimly. 'I'm not talking about the form. The content is terrible.' I said, 'The content corresponds to my con-victions and I fully accept responsibility for the dissemination of this work.'"

On July 10, at the Installation, during the evening broadcast of a Western radio station, Sakharov learned that his essay had been published in a Dutch newspaper. The announcer reported the main positions of the article and that the author had worked on the development of the Soviet hydrogen bomb. His call for democratization, freedom of expression, and convergence as an alter-native to total destruction had reached the public.

In the morning he flew back to Moscow and never again returned to his office. A few weeks later, Khariton passed on to him the orders from the minis-ter to remain in Moscow. This was the second time in the history of the Soviet nuclear project that a person of such high scientific rank was removed from work. In December 1945, at his own request, Academician Kapitsa stopped working on the project. In Sakharov's understanding,

Kapitsa then argued not from ideological positions, but out of disagreement on organizational problems and an unwillingness to subordinate himself to people he considered less qualified scientifically. Therefore he was not accused of antipatriotism or sabotage but of being undisciplined or, as they said in Beria's apparat, of hooliganism. I believe, however, that this wasn't merely a loophole, but an actual mixture of heterogeneous reasons, but in what combination it is hard to say.

There is no doubt that in the 1968 version of "Beria's apparat" Sakharov was also considered a hooligan. Sakharov had not asked to be removed from weapons work, and that is but one of the profound differences between him and Kapitsa. But the two hooligan academicians shared a sense of personal dignity and an inability to subordinate themselves to a power they did not trust.

Peace and War in 1968

We have been dealing with events that forced Sakharov to turn his life around sharply. But had he managed to influence the course of world events as early as 1968? The documents that would provide a conclusive answer to this question are still hidden in the archives of the Politburo, but the chronology of events linked directly to Sakharov's secret letter to the Central Committee in 1967 and to the main threat to the world that he wrestled with in "Reflections on Progress, Peaceful Coexistence, and Intellectual Freedom" is quite revealing.

The decision to create an ABM system around Moscow was taken in 1960, and a special Construction Design Office was created. The project was approved in 1962. In the spring of 1967 the Soviet Army introduced a new branch of the armed forces—antimissile and space defense. The ABM designers promised "the Party and the government" to have their system completed by the fiftieth anniversary of the October Revolution, November 7, 1967, but managed to have it done only a decade later. To help them, a resolution on increasing work on ABM systems was promulgated in May 1968.

And in this flurry of antimissile activity in the USSR, the American leaders proposed an ABM moratorium early in 1967. In June, during a summit meeting in Glassboro, New Jersey, they tried unsuccessfully to persuade Kosygin of the need for a moratorium. On May 27, 1968, KGB head Yuri Andropov sent a copy of "Reflections" obtained by his people to the Central Committee, and on Brezhnev's orders, the members of the Politburo read the text.[21]

The first sign that the Soviet government had rethought its position appeared in July 1968—that is, after Sakharov "rapped on the table." The negotiations began in November 1969 and led to the signing in May 1972 of the SALT I (Strategic Arms Limitation Talks) treaty. The most important part of the treaty was the limit on the testing and deployment of ABM systems.

To what degree did Sakharov succeed in persuading the leadership to change Soviet policy on the ABM issue? Did the Politburo review Sakharov's secret letter of the previous year and did they call in other experts to explain how they could possibly be against strengthening defense capability?

Sakharov's 1967 letter is marked: "Comrade Brezhnev L. I.—seen." But had Comrade Brezhnev understood the arguments by Academician Sakharov? Or did he rely on unidentified "aides of the General Secretary" who in turn took advice from their own advisors who had their fingers in the ABM pie?

In any case, had the collective Soviet leadership realized that the ABM moratorium was not a trick of the American imperialists but something that was in the best interests of the USSR, Sakharov could well have earned a fourth star— for courage and valor in performing his work and civic duty—or the Lenin Prize "for strengthening peace among nations." If Sakharov's unprecedented appeal to the world prompted the Politburo to review its decision, he had already earned the Nobel Prize for Peace.

When he wrote in 1968 in "Reflections" that humanity was on the brink, it was more than a metaphor for him. Like his American colleague Hans Bethe, he saw the abyss into which humanity would plunge if even one of the two superpowers gave in to the illusion that national security could be guaranteed by a strategic antiballistic missile defense. What the strategic physicist Sakharov wrote about in his 1967 letter to the Central Party Committee and what his American colleagues wrote about in a 1968 article in *Scientific American* was the inexorable logic that turns the illusion of strategic defense into real steps headed for that abyss.

If the reader does not feel competent to judge the logic and illusions held at the highest government levels of the time, it is important to remember two characteristics of Bethe's scientific profile: he constructed the theory of the thermonuclear energy of stars (for which he received the Nobel Prize) and is known for his sober-mindedness and sense of social responsibility. And we must remember that two experts living on opposite sides of the Iron Curtain at this perilous moment came to the same conclusion independently of each other. In addition, these two outstanding theoretical physicists were also professional developers of strategic weapons and in that capacity had a professional mastery of the appropriate information. Even after he ceased working in weapons development and research, Sakharov remained a professional in strategic weaponry, profoundly concerned by the global problems created by this apocalyptic technology. In the thick of his human rights activity, he still followed developments in the field closely. This is evident in his article "The Danger of Thermonuclear War," written in exile in Gorky in 1983.[22] This piece is noteworthy for the diligence with which the equation for strategic balance is presented and the enormous rigor of the discussion of "solutions" for the equation. Sakharov's professionalism was manifested also in his speeches at the Moscow Forum on a Non-nuclear World and the Survival of Humanity in February 1987, just a few weeks after his return from exile.[23]

A popular film can help us look at the strategic military situation of 1967–1968. Let us imagine how the crew of the *Titanic* would have treated the warning of some highbrow theorist specializing in icebergs just fifteen minutes before the historic collision. A collision that could result in the sinking of this mighty ship would have been unthinkable, both for the passengers and the crew. Theoretical physicists gave analogous warnings at the critical moment in the Cold War. Why did they see more clearly than the rest? Perhaps it is because theoretical physicists are used to thinking about the unthinkable. About movement at the speed of light, about the initial explosion of the universe. As Lev Landau used to say, they sometimes manage to understand things that they can't imagine.

Thus, if Andrei Sakharov and Hans Bethe were right in their analysis of the world military situation, then in 1968 the world, unbeknownst to itself, turned away from the iceberg of nuclear war.

The difference between capitalism and socialism gave Bethe the opportunity—without particular personal risk—to bring his analysis to the attention of his government and society. Sakharov lived in a country where often the only way to fill the breach was with one's body. But without the steps he took at great personal risk, the Soviet leaders would have continued to listen to the opinion of their aides, who listened to their aides, who were antimissile experts. And the ocean liner of humanity would have continued on its steady course toward the iceberg in the night.

The Soviet government did not have to award the academician a fourth star—it would have been enough to tell him that he was right. In that case, the biography of Soviet dissident No. 1 might have been different. But of course, if the Soviet leadership had considered making such an admission, the history of the entire country would have been different.

IV

A Humanitarian Physicist

The year 1968 was a turning point for Sakharov. On his own, he stepped into the open out of his secret military technology life. Fate added major changes to his personal life. In the course of two years almost everything changed: family, work place, the shape of his life. His circle of acquaintances widened extraordinarily, and his scientific colleagues were now only a small part of it.

The ivory tower surrounded by the barbed wire of MedMash was left behind. The top-secret physicist was now in the thick of public life, which waxed and waned along with changes in the attention from the secret police. Western journalists and politicians wanted the physicist's opinion on politics.

The last twenty years of Sakharov's life cannot be compared to the previous decades in number of events and human contacts. He described it as "a life in which every year had to be counted as three." A historically consistent account of the events of that tripled twenty-year period would require a separate book. It would have to recount how Sakharov became the foremost figure in the human rights movement in the Soviet Union. How he stood up to the state police and propaganda machine. How he acquired new friends, many of whom disappeared, forced into prisons or exile aboard. How he parted with old acquaintances who were frightened by his new life. How in January 1980, after the Soviet invasion of Afghanistan, he was exiled to Gorky, a city closed to foreigners, for seven years of isolation, hunger strikes, and force feeding. How throughout that time he said and wrote what he thought about peace, progress, and human rights. Finally, it would have to detail the last three years of greater freedom, the first semifree elections in the country, and the final seven months of his life as a deputy of the parliament and one of the leaders of the democratic opposition. National recognition. And national mourning in December 1989.

The inner content of Sakharov's life, however, did not change that much over those years. Under the influence of new experiences and reflections, he clarified and elaborated his understanding of the world and lived in accordance with it. It is the inner side of Andrei Sakharov's life that will be the main subject henceforth. Only a few events out of a huge number will be chosen to follow the course of his life and to understand what was happening to him.

17

SAKHAROV AND SOLZHENITSYN

The Physics and Geometry of Russian History

"Reflections" in the
New York Times

Publication of "Reflections on Progress, Peaceful Coexistence, and Intellectual Freedom" was an event not only in the author's life but also worldwide. It seemed so unlikely that the editors of the *New York Times* had doubts about the text sent by their Moscow correspondent.[1] At first, on July 11, 1968, they published only an account of the text. They explained cautiously that the author "helped develop the Soviet hydrogen bomb."[2] On July 22 they dared to print Sakharov's text in full (allocating three full pages to it), but the introduction did not in any way make clear that Sakharov was one of the main creators of Soviet nuclear weapons. They mentioned only that he was a nuclear physicist.

"Reflections" was reprinted many times in many languages; there were about thirty publications in 1968.[3] That same year it came out in the United States as a hardcover book. The introduction and copious notes (longer than the text itself) were written by Harrison E. Salisbury, a preeminent journalist and Russia expert. He was the first to state that Sakharov more than anyone else could be called the creator of the Soviet hydrogen bomb.[4]

In his commentary Salisbury relates many interesting things about Russia, but his remarks about the text show a superficiality or, rather, too much common (Western) sense. He wrote that the scientific-technological intelligentsia "is in many respects one of the most influential within Soviet society," since this intelligentsia was responsible for making Russia a nuclear power, placing the first man into orbit, giving Russia its intercontinental ballistic missiles, and creating the enormous Soviet educational-scientific and industrial establishment.[5] Sakharov used to think so, too, until he learned that the influence of the intelligentsia ends where they begin to talk about how "to guide" the country rather than how "to fortify" it.

Salisbury felt that Sakharov's thoughts

> are now known to most members of the Academy of Sciences, to most par-
> ticipants in the leading institutes of physics, mathematics, and physical sci-
> ences, to a broad cross section of the University intelligentsia not only in
> cities like Moscow, Kiev, and Leningrad but in other areas of scientific re-
> search, such as the Siberian centers of Novosibirsk and Irkutsk.

That was far from true.

Salisbury also failed to hear Sakharov's concrete ideas through his pro-socialist tone: "Sakharov is critical not of Marxism, per se, but of the failure of Marxists to evolve a scientific method of directing society. His 'thoughts' contain no criticism of the classic hypotheses of Marxism as such; merely devastating criticisms of what so-called Marxists such as Stalin, Mao, and their like have wrought." What classic Marxist hypotheses were supposed to be compatible with Sakharov's convergence and his declaration of socialism's lack of victory in the economic competition with capitalism?

Salisbury comes to one more erroneous conclusion, but this is one for which he cannot be blamed. He notes that Sakharov's analysis of the danger of world nuclear suicide

> coincides precisely with the published documentation of American nuclear
> scientists . . . His familiarity with American literature in this field is empha-
> sized by his use of the critical article in the March 1968, *Scientific American*,
> by the American physicists Richard L. Garwin and Hans A. Bethe, on anti-
> missile systems . . . Summarizing the Garwin-Bethe arguments, Sakharov
> arrives at the same view as the Americans.

This makes the reader think that having read the March 1968 issue of *Scientific American*, the Soviet physicist joined his American colleagues in their thinking. But what if Suslov had allowed *Literaturnaya Gazeta* to print Sakharov's article in June 1967? Then would we have to suspect Bethe of not being original? It is not so rare in science for two scientists to analyze a situation and independently arrive at like conclusions. This happens much less frequently in journalism.

Salisbury gathered a lot of information on Sakharov's scientific and public work for his introduction, including the incomprehensible titles of his articles from Soviet physics journals and a notation "short work on U-meson reactions." This typo for μ-mesons went unnoticed. More important, the article missed Sakharov's true political debut—the 1958 article on the radioactive dangers of testing. It could have shed light on the origins of "Reflections." Instead, an insignificant work on physics and mathematics schools for gifted children,[6] written with Zeldovich on the latter's initiative, is presented as Sakharov's political debut.

The *New York Times* publication of the "Reflections" was closer to guessing its origins, using a photo of "Soviet Premier Aleksei N. Kosygin and President Johnson at Glassboro, N.J., in June, 1967." At that meeting the Americans were unable to persuade Kosygin that an antiballistic missile moratorium was vitally important for both countries. That was what prompted Sakharov to write his July 1967 letter to the Central Committee.

The author of the newspaper's introduction noted: "Dr. Sakharov and others who share in his views may have been instrumental in persuading the Soviet leadership to discuss limitations of offensive and defensive missile systems with the United States. Agreement to hold such talks was announced July 1 by President Johnson."[7] But the author might not have suspected that the article he introduced had played a role in achieving the agreement.

Thus, Sakharov's real and personal reasons remained invisible, while his gentle "convergence" tone kept the Western journalist from noticing the radical difference between his ideas and Soviet Marxism-Leninism.

A popular encyclopedia yearbook for 1969 mentions Sakharov's essay (without his name) in the article "Communism" as a sign that

> many Leninist-Stalinist ideas seemed to be changing. Both Lenin and Stalin had argued that a war between capitalism and communism was inevitable; Stalin believed that the two social systems could not live in peace. In 1969 [*sic*], however, a pamphlet entitled "Thoughts on Progress, Peaceful Coexistence, and Intellectual Freedom" circulated in Soviet intellectual circles. It maintained that the two social systems could coexist, that war between capitalism and communism was not inevitable, and that the two systems were gradually beginning to approximate each other.[8]

Sakharov's name meant nothing at the time to the average American reader. His main point—that the two social systems *have to converge to prevent world war, or world suicide*—was turned into a formula of "peaceful coexistence" that was often repeated by Soviet leaders, and thus cheapened.

In the same yearbook, the "Physics" article mentions among the most important events of the year:

> Early in 1969, Soviet scientists at the Kurchatov Institute in Moscow succeeded in controlling thermonuclear fusion reactions for as long as 50 thousandths of a second within a giant plasma generator called Tokamak 3. Their success, which was later confirmed by a team of British scientists, promised mankind an unbelievably powerful source of energy . . . The Soviets were working on a larger plasma generator named Tokamak 10. The U.S. Atomic Energy Commission announced plans in late 1969 to build a similar plasma-making machine called Ormak.[9]

The Russian acronym "Tokamak" entered the English language in the 1960s and is still a living English word, while the American "Ormak" was not similarly honored. But the average American did not know that the foundation for the success of Soviet physics had been laid by the same man who wrote "Reflections on Progress, Peaceful Coexistence, and Intellectual Freedom."

Another defender of intellectual freedom in the Soviet Union was much better known in the West. A major event in Russian literature was described in a 1969 yearbook as "the expulsion of Russia's greatest living novelist, Aleksandr I. Solzhenitsyn, from the Writers' Union." The photo caption explained that he was expelled "because of his repeated criticism of Stalinism and allegedly pro-Western sympathies" and that "Solzhenitsyn, still protesting against the lack of intellectual freedom in the Soviet Union, reportedly termed Russia a 'sick society.'"[10]

Solzhenitsyn's voice was first heard throughout the land in 1962, when the literary journal *Novy Mir* published "One Day in the Life of Ivan Denisovich." In the West the author's wrath against the crimes of Stalinism came through most loudly. But for Russians, the voice spoke for those who had vanished in the camps. Solzhenitsyn had gone through the Gulag, and his artistic talent had survived, and perhaps even matured, in the hard-labor camp. Sakharov spoke for many when, in "Reflections," he chastised the censors for keeping Solzhenitsyn's books, "imbued with great artistic and moral power," from readers. Sakharov's main subject in "Reflections" is the present and future. The subject of Solzhenitsyn's books of those days was the truth about the past.

The physicist learned what the famous writer thought of his essay at their first meeting in the late summer of 1968. Both men and their host left accounts of this meeting. All three had a friend in common, who worked at FIAN. Through this friend, Solzhenitsyn suggested to Sakharov that they meet. Sakharov chose the apartment of his colleague and close friend, Yevgeny Feinberg, as the place.

Feinberg had already met Solzhenitsyn by then:

I had been a witness and participant in three of Solzhenitsyn's attempts to find a real ally among the academy physicists with an attractive public reputation. He was met with sincere awe, readiness to help (say, by typing his unpublished works), but that wasn't what he wanted. Now he had come to meet with a man from the same milieu, but who had already performed a great exploit, who had crossed the line . . . My wife and I decided to set the table in our living room. When Solzhenitsyn saw that, he was displeased and said, "What is this, a reception?" We realized we had chosen the wrong style.[11]

The tactful hosts withdrew:

Of course, I realized that Solzhenitsyn was there only to see Sakharov and he didn't need anyone else. But still, our former meetings and my sense of duty as host (after all, Sakharov had wanted to meet at my house for some reason) made me look in a few times, once with tea. I'd stand around for a minute or so then sense that Solzhenitsyn wanted me out of there, so I would leave. They chatted sitting next to each other and half-turned to face each other. Solzhenitsyn, resting his elbow on the table, persistently lectured Sakharov about something. He responded with slow-spoken brief responses and as was his habit, listened more than he spoke.

But Feinberg bore no grudge:

It would have been naive and incorrect to perceive Solzhenitsyn's behavior as being simply impolite or hostile. We have to remember that at that time he was completely obsessed by his Work, and that was combined with the all-destroying focus of his precise actions (truly a combination of "American efficiency" and "Russian counter-revolutionary sweep"). Everything else was cast aside.[12]

Through the Eyes of Solzhenitsyn

Seven years after the meeting, Solzhenitsyn recollected:

I first met Sakharov at the end of August 1968, immediately after our occupation of Czechoslovakia and shortly after the appearance of his memorandum ["Reflections"]. . . . Merely to see him, to hear his first words, is to be charmed: his tall figure, his look of absolute candor, his warm gentle smile, his bright glance, his pleasantly throaty voice, the thick blurring of the r's to which you soon grow accustomed. In spite of the stuffy heat, he was dressed with old-fashioned meticulousness—a carefully knotted tie, a tight collar, a jacket which he unbuttoned only in the course of our conversation—a style apparently inherited from his family, old-fashioned Moscow intelligentsia. We sat together for four hours that evening. . . . I was, perhaps, insufficiently polite, and too insistently critical, although I was not aware of it at the time. I didn't thank him, didn't congratulate him, did nothing but criticize and dispute and pull to pieces what he had said in his memorandum, and all this with no plan or system. I had somehow not realized that I would need one. It was then, in the two hours during which I made such a bad job of criticizing him, that he conquered me! He was not in the least offended, although I gave him reason enough; he answered mildly, tried to explain himself with an embarrassed little smile, but refused to be the least bit offended—the mark of a large and generous nature. . . . I did not make allowances for this at the time, driven as I was by my labor-camp past, and kept pointing out flaws in his reasoning and in the ordering of his facts.[13]

Solzhenitsyn's labor-camp past began in 1945 and lasted almost ten years. He was arrested for writing a friend that he considered Stalin the principal cause of the country's woes. During his camp years he extended that view to Lenin and all socialist values that reigned over so many of his countrymen. He found a very different foundation for his worldview in filial devotion to Orthodox Russia. But at the time of his meeting with Sakharov, Solzhenitsyn kept that to himself.

Subsequently, Solzhenitsyn turned his critical remarks to Sakharov into an essay, which he gave to him. It was called "The Torments of Free Speech."[14] The title reflects not only the torment of the budding writer Sakharov but also the torment of the reader Solzhenitsyn. For him, a man of letters, the word—precise, living, full-blooded—was much more than a vessel for an idea. The style of "Reflections"—in Sakharov's own assessment "eclectic and in places pretentious"—had to irritate Solzhenitsyn by its literary immaturity and its Soviet clichés.

But Solzhenitsyn was perceptive and no aesthete, and he could see that it was merely a Soviet shell sticking to the wings of a just-hatched chick rather than a rotten egg that had cracked. With all his heart, he wanted to help the chick spread its wings, for he had sensed their potential power immediately. So that the free flight of thought would not be hindered. So that the thought could fly high to see the answers to the eternal Russian questions: Who is to blame? What is to be done?

Through the Eyes of Sakharov

Sakharov wrote about meeting Solzhenitsyn a decade later:

> With lively blue eyes and a reddish beard, temperamental speech (almost patter) with an unusually high-pitched voice that contrasted with his measured, precise movements, he seemed a living mass of concentrated and focused energy . . .
>
> I basically listened attentively and he spoke—as usual, passionately and without any vacillations in opinions and conclusions. He began by complimenting my step and its historical significance—breaking the conspiracy of silence of the people near the top of the pyramid. Then he sharply formulated where he disagrees with me. There can be no talk of convergence . . . The West is not interested in our democratization, it is itself confused in its purely material progress and permissiveness, but socialism could destroy it completely. Our leaders are soulless automatons who have their teeth clamped on their power and benefits, and they won't loosen their hold without the fist. I understated Stalin's crimes and mistakenly separate him from Lenin—it is a single process of destruction and perversion that began with the first days of the revolution and continues now; changes in scope and form are not fundamental. Sixty million people died from the terror, starvation, and disease (as their result) . . . The number I gave (over 10 million) dead in the camps is understated. It is wrong to dream of a multiparty system—we need a partiless system, for every party is violence over the convictions of its members for the sake of the interests of its leaders. It is wrong

to dream of scientifically regulated progress. Scientists, engineers form an enormous force, but the basis should be a spiritual goal, and without that any scientific regulation is delusional, a path to suffocating in the smoke and pollution of cities . . .

I told him that there was, of course, a lot of truth in his remarks. But my essay reflected my convictions. It is meant to be constructive, and it seems to me, that caused some of the simplification . . . If I wrote something that was wrong, I hope to fix it in the future. But first I have many things to ponder.

By 1968, Sakharov had pondered many things, elaborating his understanding of the country and the world. He had begun from a naive, "popular" Stalinism. If he hadn't told the story himself, no one would believe that in 1953 he "got carried away" by the national mourning for Stalin and wrote to his wife something like: "I am under the impression of the death of a great man. I think of his humanity." He soon was ashamed of his reaction and that shame is the most accurate part in his recollections of that period. Memory of that shame might have caused him to exaggerate a bit.

It's obvious that he did not have the old letter to his wife before him and that the cited phrase is not an exact quotation. We know that Sakharov is sometimes too strict toward himself (for instance, his story about the "atrocious" supertorpedo) in his recollections and self-analysis, and therefore, inaccurate. The account of a coworker who remembers a completely different reaction from Sakharov also suggests that he is chastising himself too much in the *Memoirs*. Sakharov gave a calm and sober opinion about the situation after Stalin's death, according to his colleague, that nothing terrible would happen, that "society is such a complex system and the flywheels are set to keep things turning as before."[15]

What is hard to doubt is that Sakharov, like most of his colleagues, was certain that he was guaranteeing a peaceful life for his country, which had lived through a terrible war and was trying to embody luminous ideals.

"I created an illusory world to justify myself." That was his later harsh self-diagnosis.

I very quickly banished Stalin from that world (perhaps I had let him in for a very short time and not fully, more to show off in those rather emotionally distorted days after his death). But state, country, and Communist ideals remained. It took years for me to understand and feel how much substitution, speculation, deceit, and lack of correspondence with reality there was in those concepts. At first I thought, despite everything that I saw with my own eyes, that the Soviet state was a breakthrough into the future, a kind of protoimage (albeit a still imperfect one) for all countries (such is the power of mass ideology). Then I came to view our state on equal terms with the rest: that is to say, they all have flaws—bureaucracy, social inequality, secret police, crime and reciprocally harsh courts, police and jailers, armies and military strategies, espionage and counterespionage, the desire to expand their sphere of influence under the guise of guaranteeing security, and a distrust of the actions and intentions of other states. That could be called the theory of symmetry: all governments and regimes to a first approximation are bad, all peoples are oppressed, and all are threatened by common dangers.

Apparently Sakharov was at that stage of his political evolution when he met with Solzhenitsyn. It took several more years for him to compare his country to a "gigantic concentration camp" and to find the appropriate description for its system: "totalitarian socialism."[16] His political "theory of symmetry" required amendment:

We cannot speak about symmetry between a cancer cell and a normal one. Yet our state is similar to a cancer cell—with its messianism and expansionism, its totalitarian suppression of dissent, the authoritarian structure of power, with a total absence of public control in the most important decisions in domestic and foreign policy, a closed society that does not inform its citizens of anything substantial, closed to the outside world, without freedom of travel or the exchange of information.

Sakharov's political evolution was helped by his country, which showed him how it treated its citizens. In 1973 the Soviet state unleashed the full force of its propaganda against dissenters. State journalists, mobilized "representatives of the scientific community," and organized "ordinary Soviet people" expressed everything they were ordered to think about Sakharov and Solzhenitsyn in the newspapers. No one was bothered by the fact that these people, so offended by the writers' improper thoughts, had no legal way to become familiar with those thoughts.

Since Sakharov knew what had prompted his own free thinking, the lesson was a good one. He learned another lesson from Solzhenitsyn's book *The Gulag Archipelago*. It was published in the West in December 1973, and Sakharov read it in early January 1974:

Solzhenitsyn's book stunned me. From the first pages the angry, bitter, and sarcastic narrative created the grim world of gray camps surrounded by barbed wire, the offices of the investigators and the torture chambers bathed in merciless light, the Stolypin railroad cars, the icy mines of Kolyma and Norilsk—this was the plight of many millions of our citizens, the reverse side of the enthusiastic unity and labor achievements celebrated in songs and presented in newspapers.

Solzhenitsyn's book was the first to show the cancerous growth of totalitarianism in its full historical scale and with the terrifying specificity of daily life. It was shown by a witness and a great artist. The authorities could not tolerate this untimely truth. On February 12, 1974, Solzhenitsyn was arrested and forced to leave the country. Sakharov and a few other brave souls defended the writer.

Only after Solzhenitsyn had been expelled from the USSR, Sakharov read Solzhenitsyn's "Letter to the Leaders of the Soviet Union," an elaborate explanation of the writer's views on the future of the country and world. The leaders of the Soviet Union read it six months earlier. Solzhenitsyn sent his forty-page letter to Brezhnev on September 5, 1973. At a Politburo meeting on September 17 the leader characterized that "statement to the Central Committee of the CPSU" with one word: "delirium."[17]

Sakharov took the document much more seriously, and after the letter was published in the West on March 3, 1974, he decided to respond to it openly. His article, "On Aleksandr Solzhenitsyn's Letter to the Leaders of the Soviet Union," is dated April 3. Sakharov begins by noting the writer's exceptional role in the country's spiritual life. "Solzhenitsyn's role was very vividly manifested in his novella 'One Day in the Life of Ivan Denisovich' and now in the great book 'Gulag Archipelago,' to which I bow down."[18] He concludes: "The publication of Solzhenitsyn's letter is an important public event, yet one more example of free discussion on fundamental issues. Solzhenitsyn, despite the fact that some traits of his worldview seem erroneous to me, is a giant in the struggle for human dignity in today's tragic world."

Sakharov's objections were to the most basic elements of Solzhenitsyn's program. Sakharov rejected the national egotism in the form of Russian isolationism, convinced that in the modern world there wasn't a single "key issue that can be solved on a national scale." He meant first of all the danger of nuclear and ecological suicide.

He did not share Solzhenitsyn's certainty that the greatest external threat to Russia then was war with China, and felt that he exaggerated the role of ideology in the conflict of the two Communist superpowers. Nor did he feel any sympathy for Solzhenitsyn's vision of an impending dead end for Western civilization, among whose noxious products the writer considered the ideas of unlimited scientific progress and Marxism. Sakharov did not agree that Marxism alone distorted "the healthy Russian line of development."

> For me in general the very distinction of ideas into Western and Russian is incomprehensible. For me, with a scientific, rationalist approach to social and natural phenomena there is only the division of ideas and concepts into true and erroneous. And where is that healthy Russian line of development? Was there ever a single moment in the history of Russia, or any other country, when it was capable of developing without contradictions and cataclysms?[19]

Summing up in the *Memoirs* his differences with Solzhenitsyn in their views on the West and on scientific progress, Sakharov agrees with the bitter truth "about the alienation in the West, about dangerous illusions, about the cheap populism, short-sightedness, egoism and cowardice of some politicians, and about its vulnerability to all kinds of subversive actions." But he also feels that

Western society is basically healthy and dynamic, capable of overcoming the difficulties that life continually brings. The alienation for me is the reverse side of pluralism, freedom and respect for the individual, those most important sources of a society's strength and flexibility. On the whole, and especially in the hour of tribulation and trial, I believe it is much more important to retain fidelity to these principles than to have an imposed and mechanical unity, which is useful of course for expansion but is historically fruitless. In the final analysis, the living and vital is what conquers and survives.

Fully aware of the grave danger posed by scientific-technological progress, he nevertheless saw it as the only instrument for bettering the living conditions on the planet: "And if humanity as a whole is a healthy organism, and I believe that it is, then progress, science, and wise and serious attention to arising problems will help us deal with the dangers. Having set out on the path of progress a few millennia ago, humanity cannot now stop nor should it, I believe."

"Always Alone"

Their ideological differences saddened Solzhenitsyn: "Russia finally got her long-awaited miracle—Sakharov, and nothing sickened that miracle more than the awakening the consciousness of Russia!"[20]

The two remarkable Russian men who first met in late summer 1968 belonged to the same generation. In almost everything else we could call them opposites, if we remember Niels Bohr's philosophical motto: opposites are not contradictory but complementary. The way odd and even and right and left are complementary.

Solzhenitsyn and Sakharov came to their meeting from very different directions. One came from the very bottom of Soviet life, where he was known as "zek [prisoner] Shch 262" and where he had to use all his strength just to survive in body and soul. To keep from being squashed by the heavy hand of fate, he turned himself into a fist.

The other was from the peak of comfort and respect that the Soviet system could offer and he seemed to feel free to say what he thought. He lived in an intellectually rarefied atmosphere and his job brought him in contact with the people who ran the country. The danger threatening humankind became his personal problem.

One lived in the most national part of culture—the language, collecting words that came down through the ages from folk lexicon. To bring them back to life, Solzhenitsyn compiled *The Russian Dictionary of Linguistic Expansion*. The government's foreign policy was too foreign for him, deflecting resources and attention from what was important—domestic policy.

The other lived in the most international part of the culture. The physicist senses the supranationality of his science as directly as the writer feels within the ethnic boundaries of his linguistic element. Opening a scientific journal in English, the physicist sees that an unknown person in a completely unknown life is thinking about the same problem he is, and that his colleague had come up with something better for the solution but had also overlooked a few things. Therefore, a physicist does not need to be convinced of the unity of humankind and can easily say: "Humankind can develop painlessly only if it regards itself demographically as a single whole, as one family, without divisions into nations in any sense but historical and traditional."[21]

In the fall of 1968 Sakharov for the first time participated in an international conference on physics and spoke with foreign scholars "in a horrible mix of English and German." However, they had a common tongue—the language of physics. Could Solzhenitsyn have talked about his life's work—Russian literature—in a horrible mix of languages?

Besides the differences in life experiences there were also psychological differences. We know of one from Sakharov, who described his manner in the 1950s: "outwardly modest, but in fact quite the opposite." Solzhenitsyn, who for all his fiery intractability, had to have a Christian humility.

Their social views seem to fall into the nineteenth century Russian categories of Westernizers and Slavophiles. However, Solzhenitsyn's Slavophilism reveals the focus of the Western missionary conqueror. He rose from the prison camp dust, took on an enormously difficult mission, and did not bow beneath his burden. Complementarily, the Westernizer Sakharov resembles a Russian type familiar from classical Russian literature, something of a Pierre Bezukhov in *War and Peace*. He did not undertake a cause, although he agreed, almost unwillingly, to do what others could not.

Given these differences, it is amazing that Sakharov and Solzhenitsyn felt deep mutual respect and admiration and that they were protective of each other. Solzhenitsyn, who had received the Nobel Prize for Literature in 1970, proposed Sakharov for the Peace Prize in 1973, at the height of the anti-Sakharov campaign in the Soviet Union.[22]

On February 12, 1974, the day Solzhenitsyn was arrested, Sakharov spoke to Canadian radio and television:

> I am speaking from Solzhenitsyn's apartment. I am stunned by his arrest. Solzhenitsyn's friends have gathered here. I am certain that the arrest of Aleksandr Solzhenitsyn is revenge for his book, which exposes the viciousness of the prisons and camps. If the authorities had treated this book as a description of past woes and thereby separated themselves from that shameful past, we could hope that it would not be resurrected. We perceive Solzhenitsyn's arrest not only as an insult to Russian literature, but as insult to the memory of the millions who died in the camps and in whose name he speaks.[23]

For all their mutual support, their differences divided the free-thinking intelligentsia into factions. There were very few people who knew, respected, and admired both. One was Lydia Chukovskaya. "I don't know how she can love you and Solzhenitsyn at the same time," Sakharov quoted a woman who was a close friend of his wife.

For Chukovskaya their personalities were much more important than their philosophical, historical, and political differences. She was neither philosopher, historian, nor politician. She was a writer. She could judge Solzhenitsyn's talent for herself. She considered his best work to be *Gulag Archipelago*, which she saw not as "information" about a horrible side of Soviet life but an artistic puzzle. How could that inartistic material—the torment of the insulted and injured—become a poem of such lyric power? She could only guess at Sakharov's scientific talent, but she was helped by her relationship with another physicist, her husband, who was executed in the cellar of a Leningrad prison in 1938.

In her two great contemporaries she saw, besides their talents, freedom of thought and uncompromising conscience. That was enough for her to love them and defend the right of her countrymen to hear them out.[24]

This is how she saw Sakharov:

> He spoke a bit dryly, rather academic, yet at the same time there was something antiquated, folkloric, and old-Moscow in his speech. He spoke slowly, as if seeking the most precise word. It was easy to interrupt him, everyone managed to express himself faster, and he readily allowed others to take over the thread of conversation . . .
>
> He was always alone, inside himself. Yes, he had a wife, a loving family, friends, students, followers, common work on human rights, the clatter of the typewriter, meetings with correspondents, telephone calls from all over—calls that woke him at 6 in the morning. In what sense do I mean his solitude? Akhmatova used to say that sometimes when she carried on a conversation she continued writing poetry. Sometimes I would catch her mumbling undertone in a general conversation. I couldn't hear Sakharov's thoughts through his solitude. But I am certain that as I looked at him in the middle of a noisy general conversation that profound and solitary spiritual work was going on inside him, even amidst the general hubbub. Surrounded by people, he was alone with himself, solving a mathematical, philosophical, moral, or global problem—and as he thought, he considered most profoundly the plight of each and every individual human being.[25]

Akhmatova described that mumbling inner life at work in a poem:

> It's like this: a kind of lassitude;
> An insistent chiming clock, the ebbing roll
> Of distant thunder echo in my ears.
> The grievous moans and complaints
> Of unfamiliar, stifled voices appear,
> And space narrows to a kind of magic circle.
> A single sound rises triumphant

From this abyss of whispering and ringing.
Such irrevocable quiet surrounds it
That you can hear it boldly walking o'er the earth
And the grasses growing in the woods.
But now I hear some words and tiny bells—
Signaling faint rhythms
And I begin to fathom.
And then simply dictated lines,
Come to rest in the snow-white pages of my notebook.[26]

Lydia Chukovskaya, a writer, sought comparison for Sakharov in a beloved poet. But what could there be in common between a poet's whispering and the thinking of a physicist, especially since each did personal solitary work? What was there in common between looking into the spiritual life of a human being and the aspiration to understand the workings of nature?

The Book of Nature is written in the language of mathematics, said Galileo, the first real physicist. The lucky ones who read the language find that the genre of this writing is poetry, not bookkeeping. Unlike the poet, the physicist alone cannot hear the voices trapped in experiments or the signal peal of the start of a mathematical rhyme, nor can he alone write down the lines of a finished physics theory. Collective work is required from both experimenters and theorists.

And yet without hearing Andrei Sakharov's thoughts through his solitude, Lydia Chukovskaya correctly placed his solitary spiritual work next to the work of a poet. The creation of something new happens that way, be it in physics, poetry, or the philosophy of history.

The Views of Physicist and Mathematician

The fact that two of Russia's greatest free thinkers had such divergent views on the philosophy of Russian history, consistent and profoundly held, refers to the eternal question: "What is truth?"

It is a basic question in history of science. Even in physics, where everything seems cut and dried and quantifiable, there are seemingly irreconcilable differences between outstanding scientists. One of the most famous such discussions involved Einstein and Bohr. They never did reach agreement on the future of quantum theory, to which both had made fundamental contributions.

The basis of such disagreements in science lies in personal differences, outside of science, in the realm of foresight, predisposition, and personal worldview. Two free-thinking, honest men who respect each other and have highly developed aesthetic sensibilities will hardly agree on a ballet performance if one is congenitally blind and the other congenitally deaf. The "defects" can be replaced by congenital supersenses—for example, the ability to see ultraviolet light or hear ultrasound, and the result will be the same. They would have to come to mutual understanding in a limited area and accept differences in others.

A more suitable example comes from personal differences in science that are rooted in the differences between a physics and mathematics worldviews. The two are profoundly different, albeit mutually beneficial: mathematicians borrow from observations of the physical world and give physicists a language in which they can read the Book of Nature. The difference lies in the fact that there is only one Book of Nature but many books of mathematics. The truth of physics always relates to natural phenomena; it is *obliged* to correspond to them, and therefore it is always one truth, as long as it is correct in terms of physics. There are many mathematical truths, the only requirement being that they are self-consistent—have no inner contradictions.

A physicist, however high his thoughts may fly, always feels beneath his feet the planet to which he must return. The mathematician can forget about where he started and find himself on another planet or even a different universe. The mathematician is in fact better off forgetting his earthy launching pad, for it is easier then to do his solitary spiritual work. In this way, Nikolai Lobachevsky discovered the "imaginary"—non-Euclidean—geometry. But such a forgetfulness more than once let physicists go astray.

Does this psychology of physicists versus mathematicians have any bearing on the historical and political differences between Sakharov and Solzhenitsyn? As strange as this parallel may seem at first glance, it is actually rather easy to call the writer a mathematician. In his self-descriptions, Solzhenitsyn referred to his first legal profession alongside his illegal profession of *zek* [prison inmate].[27] He recalled the gusto with which he taught mathematics in school. But not every classmate he had in his student years in the math department of Rostov University could be considered a mathematician by manner of thinking. And that is the essential part. For the two types of world perception described above existed long before the sciences of physics and mathematics. Two types of research thought, two types of attitude to the world, the truth, and the laws of nature are manifested in physics and mathematics. But not only in them. A writer, too, can be a researcher—Solzhenitsyn subtitled *The Gulag Archipelago* "an attempt at literary research."

One of the most general disagreements between Sakharov and Solzhenitsyn was about the concept of socialism. Sakharov was rebuked by Solzhenitsyn's adherents for not ridding himself of an uncritical attitude toward the main vocabulary word of Soviet ideology. In fact, there were two other key words in Soviet history and two equations that Solzhenitsyn reached in his thinking about that history:

$$\text{Stalinism} = \text{Leninism} = \text{socialism}$$

This double equation could be called the basic theorem of scientific antisocialism, or simply Solzhenitsyn's theorem. Sakharov was considered wrong for not accepting this theorem or even trying to find the fundamentally false axiom that caused the woes of Russia in the twentieth century.

Dora Shturman expressed this puzzlement with Sakharov with great clarity and passion. She was convinced that Solzhenitsyn had proved his theorem with his artistic-historical methods (*The Gulag Archipelago* was meant to be merely the prelude of the proof that was embodied in the multivolume *Red Circle*):

> Strange: a free artist approached this issue in a scrupulously scientific manner, while the scientist was profoundly emotional, not appealing either to scientific logic or the law in its fundamental bases, or statistics, or the history of the issue. All these "abstractions" simply did not interest him for a long time. He was more interested in physics. And after physics—helping concrete individuals, saving them.[28]

Why did the theorist, professionally trained to analyze the essence of phenomena, have no interest in scientific proof or disproof regarding the fatal theorem? Who better could refute the "scientific" basis of Soviet totalitarianism? Why did the fearless theorist stop, not even halfway, but in the first third of the path, settling for the plague of Stalinism rather than for original cause, the plague germ?[29] Why didn't he break out of the "human rights daily grind" and take on "a cause his own size—the major system-forming problems"?[30] After all, the formation of the universe was his size.

Perhaps the answer is that the theorist was a physicist. A physicist must be able to distinguish between a real physical object and theoretical discussions, even if many clever books are devoted to the discussions. A physicist knows, for instance, that in the early twentieth century no encyclopedia would omit an expansive article on "Ether." The concept was at the center of physics then. But in 1905 a physics theory emerged that solved several difficult problems without using the concept of "ether." There was no room for that word in the new theory of relativity.

The concepts of "Stalinism," "Leninism," and "socialism" had very different degrees of physical reality for Sakharov. The greatest reality was Stalinism. The state system created by Stalin decades earlier had outlived him and survived the exposure of his cult of personality and the removal from the Mausoleum and burial of his mummified corpse.

Leninism in the real sense took up only a five-year period in Russian history. And there were so many sharp turns in that period that it would be hard to characterize it in just one word: from military communism, the abolition of money, total nationalization, and the expectation of world revolution any day, to the establishment of the New Economic Policy "seriously and for a long time" with some restoration of private property and with exhortations to learn trading. The responsibility for the events of that five-year period is shared with Lenin by other historical figures, including Nicholas II, who gave up his Russian empire to the typhoon of world war.

And finally, "socialism" as used by Sakharov had at least three meanings: simply as part of common names, for example, Union of Soviet Socialist Republics, Hero of Socialist Labor; the ceaseless flow of texts that the authorities of the world's first country with a socialist name used to brainwash their citizens; and the feelings generated by the word "socialism" in Igor Tamm, the feelings that arose back in the time of tsarist Russia and had grown strong in the British homeland of socialism—when he came to study at Edinburgh University in 1913, he joined the student socialist club.[31] Those feeling were about compassion and social justice, and the dream to heal the sick capitalist society by providing all human beings with social freedom and dignity.

Sakharov's peers learned about the socialist dream when they were very young. And in general, they were enthusiastic about "making the dream real," as a popular Soviet song put it. They might have had other feelings about the theory of this dream when it was a required course in the state educational system. As a student, Sakharov had trouble passing "scientific socialism." He recalled the poet Esenin's line, "I didn't read those books in any weather," as he regarded the tomes by the founders. (Solzhenitsyn did read them in very bad weather: "It's like a joke: it was the fall of 1941, the mortal war was raging, and I—for the umpteenth time and still unsuccessful—was trying to understand the wisdom of *Kapital*.")[32]

Sakharov's trust in Tamm with his socialist convictions kept him from thinking about the real meaning of the word "socialism" until 1968:

> In 1968 I thought I understood what socialism was and I thought that socialism was good. But gradually I stopped understanding many things, and I developed doubts in the correctness of our economic foundations, doubts whether our system had anything except empty words and propaganda for domestic and international consumption.[33]

Having said that in 1973, Sakharov could no longer seriously accept *The Great Soviet Encyclopedia*'s statement that socialism "won in the USSR as a result of the Great October Socialist Revolution."[34] The system that won in the USSR brought about neither justice, nor freedom, nor dignity.

The Encyclopedia Britannica, without checking with the Central Committees of any parties, begins the explanation of "socialism" as follows: "There is no precise canon on which the various adherents of contemporary socialist movements agree." You would need a very large table to hold representatives of every form of socialism—Russian revolutionary socialist, Swedish social-democrats, German national socialist, Christian, Arab, market, and other kinds. The only common denominator on which they would all agree is probably the principle that the state should take care of those who cannot take care of themselves, because of age or health. But in our day even the most virulent nonsocialist would agree with that principle. Under other names, this kind of socialism was much more widespread in the capitalist West than in the "first country of socialism."

Sakharov wrote about socialist elements within capitalist society in 1968, "the real use by capitalists of the social principles of socialism, the real improvement in the situation of workers."[35] And this strengthened his idea of convergence. He wasn't trying to grow a hybrid of Stalinist socialism with Dickensian capitalism, after all. He was talking about bringing people living in two existing state systems closer, whatever terms were used.

The reason he expanded his considerations to the area of international politics was very real to him—the threat of global suicide. He estimated the reality of that threat using scientific logic, and statistics, and history. For his scientific analysis of the strategic military situation, as it was for Hans Bethe's, what was important was not the concept of socialism but the idea of two strategic enemies, closed to each other and therefore with mutual distrust and the lowest expectations. The Soviet and American experts added their professional understanding of the technical capabilities of nuclear attack and antimissile defense. It did not matter in the least whether the two enemies were called striped and polka-dotted, or socialist and capitalist. The essential aspect in a society, for this strategic analysis, was its capability for openness, for coming closer and, therefore, for trust.

The theoretical physicist Sakharov saw the key to an open society in human rights. The great flaw in a closed, totalitarian society was the habitual disregard for the rights of the individual, habitual both for the state and its citizens. Therefore Sakharov gave himself up to the "human rights daily grind," to helping and saving individuals. And that is why he was so little interested in the theoretical cleansing of scientific socialism and the historical examination of scientific antisocialism.

In order to help his homeland, and thereby the entire world, concrete practical actions had to be taken, a practical inculcation of human rights into the life of the society, beginning with rights to intellectual freedom. Such a practical act was *The Gulag Archipelago*, which Sakharov admired greatly.

Sakharov did not minimize the difficulty of the task he had set for himself; he felt it was more than he could handle. In 1973 when asked by a journalist what could be done, he replied, "I believe almost nothing can be done. Because the system is very stable internally. The less free a system is, the better it is preserved internally."[36]

Solzhenitsyn recalls Sakharov's words of that period: "All our activity has meaning only as an expression of moral need" and comments, "I could not argue with the content, it's just that my entire life, contrary to reason, I never experienced that hopelessness but on the contrary some kind of dumb faith in victory."[37] It was not a question of reason or dumb faith. Sakharov was much better informed, by virtue of his work, about the way the upper echelon of power operated in the Stalinist state.

In *Memoirs*, Sakharov recounted how Stalin's death briefly swept him up into the ranks of populist Stalinists and he wrote to his wife about the deceased's humanity. But a young colleague of his remembered a very different reaction:

When Stalin died, everyone was stunned, and now from Sakharov's *Memoirs* we know that he sent his wife some letter in which he expressed his sorrow. But I remembered something quite different he said. We all had the impression that something was about to happen, but Sakharov said: "Nothing at all will happen, society is such a complex system, all the flywheels are set to keep things turning as usual. It's all connected." And in fact, a year passed, and things kept spinning. And moreover, the attempts to make changes now are coming up against a brick wall . . . It is all wrapped in barbed wire, as if the people don't even want anything else.[38]

Despite all this, Sakharov's concrete actions in human rights were what he felt as a moral need and what he saw as the only force capable of transforming Stalinist socialism into an open and vital society.

More specifically and most acutely Solzhenitsyn and Sakharov differed in their attitudes toward the human right to leave one's native country. A cartoonishly scientific way to present their difference is to imagine two bordering countries—Visorland and Topland—whose distinct historical backgrounds are distilled into diverse geometric natures. They are parts of a Baseball Cap planet, bordering each other like a plane visor and a spherical top.

From a geometric point of view, the two countries have nothing in common except their common border. On the plane visor, the sum of the angles of any triangle equals 180 degrees. On the spherical top, it is always more than 180 degrees. In Visorland, simple Euclidean geometry holds, while here in Topland, non-Euclidean geometry prevails. There is a clear border between them. The foreign, flat geometrical laws do not apply to us living on the sphere. Let the flatlanders do what they want in their country, but they shouldn't try to bring their laws into our spherical life. They just don't work here, for the two different ways of life are too different. A mathematician could see things that way.

But a physicist, looking at the nearness of two large natural, albeit social, systems, would have doubts that a clear border between the two lands could make any physical long-term sense, even if that border is protected by both states. He would try to find qualities in common in both countries and take into account their inevitable interaction. And he would think about how to make that interaction not a threat to the peaceful coexistence of people in those different geometries.

A physicist could offer freedom to choose the geometry of residence—or the right to leave one's country as well as to return to it—as a principle for creating good neighbors. Even if he loves his native geometry, does not want to leave it, and only wants to make it a better place to live. He could hope that the freedom of choice would prompt the legislators of the two states to take into account not only the basic geometry, but also the simple physical feelings of those for whom the laws are made.

Let us move from geometry to life. Sakharov considered the right to leave one's own country (recognized by the Universal Declaration of Human Rights) an essential priority. This aroused the greatest objections from Solzhenitsyn. He felt that the primary issue was a free life in your own country.[39]

In Sakharov's view, a decision to leave the country is a person's personal affair, and the reasons for moving—family matter, economic, religious, and so on—are no business of the state. It is the combination of free emigration and free return together that creates the right, proclaimed in the United Nations Universal Declaration of Human Rights:

> This right—equally with the freedoms of conviction and dissemination of information, religious freedom, free speech and press, to form associations, and to strike—is of profound significance and forms the basis of spiritual and material freedom of the individual and simultaneously makes society open and democratic and promotes international trust and security . . .
>
> Those who have a monopoly on the bodies and souls of people in the country cannot allow those bodies and souls to slip out of their control through free emigration. This would truly demand democratic and socio-economic changes inside the country.

The last sentence was written by the inventor, thinking about how to make lofty moral ideas a reality, rather than by the pure theorist.

Who was right? The theoretical physicist in the area of political practice or the writer with his imaginary political geometry? Or was it people like Lydia Chukovskaya, who wanted to hear both? To listen to them and to follow how the argument of two great citizens of Russia and the world gave birth to truths.

Solzhenitsyn helped Sakharov clarify certain inherent qualities in Soviet socialism that had been hidden from him. Apparently Sakharov and Russia's post-Soviet history taught Solzhenitsyn something, for on December 10, 1998, at a reception at the Swedish Embassy he said that respect for the living variety of opinion is necessary for the unity of humanity. Sakharov would have signed that statement with both hands.

18

ON THE OTHER SIDE

At the time of Sakharov's meeting with Solzhenitsyn in August 1968, the freedom of choice of which country to live in did not even have theoretical meaning for him. And in practice, he did not have the freedom of choice of where he worked. For all his scientific degrees and state honors, he remained a subject of the empire of MedMash. Once the minister decreed the Installation off-limits to Sakharov, the forty-seven-year-old academician spent almost a year in Moscow, unemployed but still receiving his salary. The minister did not know what to do with his "mad politician," for he was part of the Central Committee's nomenklatura, and any decision regarding Sakharov had to be confirmed at the highest government level.

But Sakharov had no particular reasons to be unhappy with his strange position. He had done his moral duty. And in reward the burden of being an administrator in the bomb business was removed without any effort on his part. He could devote himself to pure science.

In September he attended his first international conference (in Tbilisi, Georgia). He gave a paper on gravitation as elasticity of the vacuum. He met many theorists personally, among them the American John Wheeler. The two former (anti)colleagues in top secret H-bomb physics spoke of the secrets of the universe. Sakharov's hypothesis on the vacuum nature of gravitation made such a powerful impression on Wheeler that he retold it in his articles and books.[1]

That same autumn Sakharov studied the work of a young theorist, Boris Altshuler, preparing to appear for the first time as an opponent in a dissertation defense in nonsecret pure physics. The defense of the dissertation "General Covariant Boundary Conditions for the Einstein Equations, Quantization of Gravity and Cosmology" took place at FIAN on January 6, 1969. Sakharov's review made clear that the opponent had worked seriously, posing difficult questions, expressing his doubts, and setting the limits of his own competence: "I hope that people with a better understanding of axiomatic field theory will add to this issue."[2]

One of the members of the Scientific Council would remember that defense for thirty years. He had never heard an academician opponent say that a candidate's dissertation had been difficult for him, that he had not understood it fully, but that what he did understand was more than enough for granting the degree.[3]

In the fall of 1968, life kept Sakharov from working on science peacefully. No one who had any hopes for a democratic evolution of Soviet socialism felt peaceful then. Soviet tanks invaded Czechoslovakia, crushing the young shoots of freedom born that spring, the Prague Spring. Socialism with a human face vanished from Eastern Europe, chased out of the socialist camp. A dark shadow fell over the intelligentsia.

The Hard Winter of 1969

Things were very grim for Sakharov that winter. His wife Klava was gravely ill. In January they diagnosed neglected stomach cancer. She had only a few months to live, and she was forty-nine.

They had celebrated their silver anniversary in 1967. They began counting on November 10, 1942, when they first saw each other at the laboratory of the Ulyanovsk factory. Their romantic relationship grew over potatoes, which they planted and harvested together, a typical activity in wartime. What was atypical was Sakharov's proposal, which he made by letter.[4] Before they were married at the civil registry office in the summer of 1943, her father blessed the bride and groom with an icon, according to Orthodox Christian tradition.

Andrei was grateful to his wife for "periods of happiness, sometimes entire years," although their family life was not without its clouds. After their first child was born, Klava stopped working, first to take care of the baby and then to be able to follow her husband as he alternated between Moscow and the Installation. This limited her life, already hampered by the conditions of the closed Installation. She was not consoled by the fact that in academic circles a wife who stayed home to take care of family and house was much more common than in other milieus. But it was her decision. She had the strength of personality to make decisions on her own.[5] Yet her husband felt guilty: He had not managed to insist that she work, "because I did not fully understand its importance and was not certain that she could manage, I could not overcome her complexes in this and other regards, could not create the psychological atmosphere in the family that would generate more happiness and for Klava, the will to live."

Talking about Tamm's wife, Sakharov mentioned a conversation she had with Klava: "Trying to ease the doubts that tormented Klava (completely without basis), she said: men often love unevenly, sometimes their love weakens, almost vanishes, but it returns." Sakharov learned of this conversation from his wife, which speaks for their closeness even on such a difficult issue.

But there is no evidence of their shared interests outside family life. Klava's attitude toward her husband's new, public activity was to be expected of a woman concerned for the well-being of her husband and family. He brought home drafts of "Reflections" to work on at home. "Klava understood the significance of that work and its possible consequences for the family—her attitude was ambivalent. But she gave me total freedom of action." It is understandable that she considered it playing at eccentricity when her husband, a man of high state rank with whom she attended a reception at the Kremlin, suddenly announced that he was joining a demonstration on Pushkin Square in defense of the Constitution.

Sakharov judged himself harshly, albeit soberly:

Unfortunately, in my personal life (in my relations with Klava and later with the children, after her death) I often avoided difficult and sensitive subjects, which I felt psychologically unable to resolve, protecting myself from that, choosing the path of least resistance (however I did not spare my physical strength or time). I later tormented myself, feeling guilty and therefore making new mistakes. A guilt complex is a bad advisor. But on the other hand, I probably would have been able to do little in these seemingly insoluble personal affairs, but by avoiding them, I managed to be active in life as a whole.

These thoughts apparently visited him in his most difficult weeks, when he also lost his sense of balance:

In a state of despair and grief in the face of Klava's inexorable death, I grabbed at straws—someone told me that there was a woman in Kaluga with a miracle cure for cancer . . . The inventor of the vaccine was a fanatical woman, a doctor by education, who for several years (she was retired) had been making her preparation at home. She gave me a box of ampoules, categorically refusing to take money.

The miracle did not come to pass. On March 8, 1969, after horrible suffering, Klava passed away. The youngest of their three children, Dmitri, was eleven. Tatiana, the elder daughter, was married. Nineteen-year-old Luba took on the role of mistress of the Sakharov household.

For several months Sakharov "lived in a daze, doing nothing either in science or public affairs." He dedicated his articles "Antiquarks in the Universe" and "A Multisheeted Cosmological Model" to his wife.[6]

He never played chess again—it must have been too painful a reminder of his traditional evening match with his late wife. (There is a photograph of the couple playing chess, taken not long before the tragic turn. Judging from the photo, Klava is quite pleased with the board. The academician is deep in thought.)

Sakharov's state kept him from remembering (or even perceiving) an important fact—that he had found an ally at the Academy of Sciences, a very celebrated ally and one with a hooligan's reputation as far as the Ministry of MedMash was concerned. On May 7, 1969, Pyotr Kapitsa wrote a letter (his first and only) to Sakharov: "I am sending you a transcription (edited) of my speech at the meeting of the Presidium of the Academy of Sciences on 28 February of this year, about which I had told you and in which I raise the question of discussing your article at the meeting of the Presidium. I hope this transcript will be of interest to you."[7]

It looks as if Sakharov remembered the conversation mentioned here, but put it a year later. "Kapitsa said that he was astonished and pleased reading my 'Reflections.' According to him, he was amazed that I, a man of a completely different generation and experience, think about many things and understand many things in the same way he does."

The transcript that Kapitsa had reworked for publication in the journal *Voprosy filosofii* contained the following:

As is well known, the struggle of views is the basis for the development of all creativity, and an example of the fear of it which has developed among our workers in the social sciences is their attitude to the well-known article by Academician A. Sakharov. One of the questions raised in that article deals with the principles on which the interrelationship of capitalism and socialism must be based in order to prevent nuclear war, which will undoubtedly end in global catastrophe. In contemporary conditions this question is exceptionally important, since its proper solution will determine the existence of all of humanity. We know that abroad this article has been analyzed thoroughly in the most varied strata of society and that proponents and opponents of Sakharov's proposals on the question of relations between the two systems have appeared. Obviously it is only in the process of discussing these questions that a vital solution to these questions can be found. Therefore it is completely incomprehensible why our ideologists are still ignoring an examination of the issues raised by Sakharov.[8]

And Kapitsa proposed that the Presidium of the Academy of Sciences pay attention to the ideological basis of a socialist society and "begin with an examination of the basic questions posed in the article by Academician A. Sakharov." The editors of *Voprosy filosofii* edited all references to Sakharov out of Kapitsa's article.[9]

Apparently Sakharov was still deep in his daze, brought on by his wife's death. Otherwise he would not have missed this important event, for more than anything Sakharov sought open discussion, which was exactly what Kapitsa proposed. Nor did Sakharov remember accurately the circumstances around his return to FIAN in that difficult spring of 1969.

From MedMash to FIAN

A month after his wife's death, Evgeny Feinberg came to see Sakharov and in the name of their teacher, Tamm (who was bedridden by then), invited him to return to FIAN.[10] On April 15, 1969, Sakharov wrote to the director of FIAN with a request to transfer back as a senior scientist in the department of theoretical physics. He explained:

> At the present time I am not working for the Ministry of Medium Machine-Building on what has been my main project in the years 1948–1968. In FIAN I would like to work in the theory of elementary particles. I will need some time to liquidate the blanks in my knowledge in this field. I ask that you make an official request of the Ministry of Medium Machine Building for my transfer to FIAN.[11]

It took the ministry six weeks to release him. He was the deputy scientific chief of the Installation, and any decision regarding him had to be endorsed by the government.

MedMash tried to figure out what to do with its personnel problem. One person approached Lev Feoktistov, then deputy director of Installation No. 2, who knew Sakharov from when they both were at Installation No. 1. The question was, "How do we neutralize Sakharov?"[12] He proposed two variants, one that took into account Sakharov's turn to issues of war and peace:

> There is much that is not understood in how to deal with nuclear weapons, should they be developed, should there be testing, the struggle for peace, what can counter all that, what ideological approaches. But it should be done in a more concrete way: without leaving him all alone, so to speak. My proposal: let's establish an institute on Problems of Nuclear War or something like that. Make him the head, and the institute will elaborate the ideology of nuclear war, weapons, and clearer views will arise. But it will all be controlled!

The other variant was more traditional:

> Sakharov, after all, is with Tamm the founder of the controlled thermonuclear reactor. And it is wrong to have the coryphaeus on the sidelines of noble work. I know his personality—he's not very communicative. Let's create exceptional conditions for him—perhaps not even in Moscow, to cut off his influence, but somewhere else, say, near Gorky. We create an institute for him and say: "You're sick of military bombs—but here is peaceful nuclear energy in all its inexhaustible glory. Go ahead, compete with the Kurchatov Institute, with other institutes. Bring in anyone you want, we'll add a few graduating classes from the University." And here I didn't quite

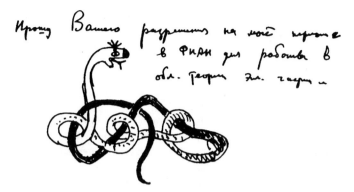

Figure 18.1 Draft of Sakharov's letter to the Ministry of MedMash, 1969: "I request your permission to be transferred to FIAN to work in the field of elementary particle theory." It is not clear what the unemployed physicist was depicting by the drawing of the serpent—a specific person or the entire military-industrial complex.

lie, but I wrote enthusiastically, hoping that maybe, if the plan went ahead, I would end up there myself, because I was also sick of bombs.[13]

But nothing came of it. Sakharov had left the problems of the controlled thermonuclear reactor, and the study of war and peace issues headed by the uncontrollable physicist was too dangerous for the Central Committee.

Finally, on June 3, 1969, Sakharov was released from the world of secret technology into open pure science, even though he remained a bearer of secrets forever. It wasn't an easy decision for FIAN, either. The administration was not pleased that he was asking for such a low position. They were planning to create a new department for him to head.[14] But Sakharov wanted to return to the department of theoretical physics at a level where he could quietly work to "liquidate the blanks in [his] knowledge."

In July 1969 he was back at FIAN, where he had begun his path in science twenty years earlier. He went back to the Installation one last time in August— he was allowed to pick up his things and empty the house where his family had lived for many years.

Sakharov donated all the money he had saved. The family spent much less than his income, plus he also had the money from his Stalin and Lenin Prizes. Their modest lifestyle did not change with his promotions. The only nonessentials in which they indulged were a few high-tech innovations: tape recorder, camera, and a children's telescope and microscope. That enormous sum, equal to thirty times his salary, Sakharov donated for the construction of a cancer hospital and to the Red Cross—in fact to the state. A few years later he wrote that he had done it "under the influence of impulses that do not seem substantive to me now."[15]

He never did explain his impulses, letting people draw the conclusion that he did not want MedMash money.[16] But he clearly understood that his thermonuclear inventions had already given the government "something immeasurably greater" than this money. He regretted not knowing then what his teacher had done with his "excess" funds:

> Tamm was in need most of his life. The Stalin Prize brought him sufficiency. But he set aside of portion to help talented people in need; he asked for recommendations for recipients—but they never knew where the money came from. I feel great shame that I didn't have the same idea or something like it (I learned about his deed, actually, several of them, only after his death).

Sakharov regretted his decision:

> My outwardly "generous" and "noble" deed seems wrong to me now. I lost control over how a great part of my money was spent in turning it over to a "faceless" state. A few months later (still in 1969) I learned of the existence of private aid to families of political prisoners and began making regular donations, but my abilities were more limited. I lost the opportunity to give financial aid to some of my relatives, who could have used it, and actually to anyone else, except my brother and my children. It was a kind of indolence of feeling on my part. And lastly, I lost a lot in my position in my future fight against the state. But in 1969, while I could understand that struggle in my mind, my perception of myself was still in that state—not agreeing with everything about it, and sharply criticizing something in its past and present, and giving advice regarding its future—but from within and with the sense that it was my state.

The leaders of the country had much greater reason to consider the state theirs, in the manner of Louis XIV of France: *L'état, c'est moi.*

"The Group May Be Small, but It's Harmful"

"The group may be small, but it's harmful." This was the expression of one of the Politburo members at a meeting in 1972 regarding the people Sakharov "groups around himself."[17] They devoted the entire top secret meeting, in Brezhnev's words, to "the murky activity taking place behind the back of the working class, the peasants, and our intelligentsia, taking place against their interests, against the interests of our socialist state and our party."[18]

By then, the Soviet leaders of the working class unanimously agreed that things had "to be ended" with Solzhenitsyn, although they had not yet decided whether he should be just "kicked out of Moscow" or "sent beyond the borders of our country." Opinions were divided on Sakharov. Suslov was certain "appealing to Sakharov or asking him was too late. It would yield absolutely nothing." Podgorny still thought that "we have to fight for this man. He is a different kind of person. He's no Solzhenitsyn. By the way, this is what [president of the Academy of Sciences] Comrade Keldysh asks. After all, Sakharov is a three-time Hero of Socialist Labor. He is the creator of the hydrogen bomb."[19]

Sakharov had become a headache for the Politburo in 1968. In June 1969 the chief censor reported to the Central Committee that in the Western press in the first half of 1969 Sakharov remained a central figure "in the so-called resistance movement to the regime," that "the bourgeois press is hailing 'Reflections on Progress, Peaceful Coexistence, and Intellectual Freedom' and its author, calling him the 'ideological inspiration' of the Soviet intelligentsia" and that according to the West German newspaper Die Zeit, Sakharov "is a priceless gift for the Soviet people and the entire world."

The censor gave no suggestion on how to handle that gift, but he did his job protecting state secrets. "The number of anti-Soviet and propaganda publications from Western countries addressed to A. Sakharov himself has increased, and the Censor's Office is confiscating them."[20] The picture of Sakharov as a central figure grouping people around himself corresponded to political thinking—be it Soviet or anti-Soviet—but it did not correspond to reality, historical and psychological.

The "harmful group" consisted of a few individuals, each with personality, views, and a separate path to this activity. Each path presumed willpower and developed individuality. These people were called dissidents. Sakharov preferred the term free thinkers, for that was the trait that attracted him to them. But they defined themselves as people trying to behave like free people in an unfree land.[21]

The majority of them had made the journey from free thinking to free action before Sakharov appeared among them. They called him "The Academician," without fear of confusion—there were no others among them. Another factor was their respect for his education and for the fact that he was a physicist, a profession highly esteemed in the country. They appreciated him as a priceless gift and tired to protect him from the riskier aspect of his activity. The infamous Article 190(1), introduced in the Criminal Code in 1966, considered any samizdat as "dissemination of intentionally false thoughts discrediting the Soviet system" and prescribed "up to three years" for it. The Academician was not made for conspiracies, which is probably why Sakharov did not know about the samizdat publication of his article (rejected by Literaturnaya Gazeta) in 1967. Roy Medvedev, who placed the article in his "Political Diary" (a samizdat periodical for a select group), permitted himself to edit it without the author's permission.[22]

The Academician was not hiding in an ivory tower, but he certainly wasn't grouping people around himself, either. If would even be incorrect to say that people were grouping around him. He was open for communication and sought it—after the break with the Installation and the death of his wife, he might have felt emptiness. But his innate shyness kept him from reaching out. The initiative came from more enterprising people.

In early 1970, he replied to an invitation from one of his new acquaintances, the physicist Valentin Turchin, to appeal to the rulers with an explanation that the democratization of the country was necessary for its scientific, technical, and economic progress. The idea was that this "pragmatic" approach could open up a wider conversation with the authorities. The letter was composed by people devoted to evolution rather than revolution: "Democratization must be gradual in order to avoid any complications or problems. At the same time it must be profound, consistently implemented in accordance with a thoroughly prepared program." The letter's mission, according to its authors, was "to promote a wide and open discussion of the most important problems."[23]

The authors' other idea was to have prominent people, first of all academicians, sign the letter. Nothing came of it—no one was willing. Sakharov then asked Roy Medvedev to be the third signatory, and in that version it was sent to the country's leaders on March 19, 1970. There was no conspiracy here.

Soon after, Sakharov was called in to the Central Committee to see the head of the science department, to chat about democratization. He was not yet blacklisted, and he was supposed to be propagandized and fought for. The official expounded the party line and demonstrated the party's paternal concern. But to no avail. Sakharov did not hide his thought from the Central Committee, and he certainly did not wish to hide them from his fellow countrymen. So the text of the letter went into *samizdat*.

This kind of openness did not fit Soviet frameworks. The highest guardian of those frameworks, the chief of the KGB, reported to the Central Committee on April 20, 1970, that "in the course of the last two years foreign ideological centers with anti-Soviet aims have distributed Academician Sakharov's 'Reflections on Progress, Peaceful Coexistence, and Intellectual Freedom,'" and that in March of the current year Sakharov participated in the composition of a letter of "politically harmful content" and that "he was making contact with politically dubious people." Therefore, "in order to get timely information on Sakharov's intentions, and discover the ties that provoke him to take hostile actions," the KGB chief requested permission "to install secret listening devices in Sakharov's apartment."[24]

Opening a Closed World

Thus, beginning in the spring of 1970, the KGB watched Sakharov from two sides, as the bearer of old military secrets and the bearer of new politically harmful views. Sakharov had long grown accustomed to the first form of surveillance and considered it proper. When a close colleague asked him as he left MedMash, "Do you think you're being tailed?" Sakharov replied, "There should definitely be one. They must be sure that I haven't gone to the American Embassy."[25] He had no intention of breaking the vow he had made in 1948, and he recognized the state's right to protect such explosive secrets.

Nor was he concerned by the surveillance over his political reliability: "We have nothing to hide, we are not engaged in secret activity, and we do not want to waste time on thinking about an army of those high-paid 'watchers'"—especially because there were subjects on which to expend his spiritual strength.

In May 1970, Sakharov encountered one of the most vile forms of taming dissidence—forced hospitalization in a mental institution cum prison. He became involved in the cases of Pyotr Grigorenko and Zhores Medvedev. The first was a war veteran, a general fired for free thinking, "a man of astonishing life, courage, and kindness." Sakharov did not know him personally at that point but had read his *samizdat* work on Stalin's responsibility for the losses in the early months of the war with Hitler. The Soviet general, Ukrainian by birth, took up the cause of the Crimean Tatars, a people repressed under Stalin and not allowed to return to their homeland. Their native lands were reserved for the elite—there were too many health spas for the nomenklatura there. Grigorenko was arrested during a trial of Crimean Tatar activists. In the middle of May, Sakharov joined the appeals to free Grigorenko.

The biologist Zhores Medvedev came to Sakharov's house right after his sensational speech against Lysenko at the Academy of Sciences in 1964. Zhores Medvedev was studying the history of Lysenkoism, and he introduced Sakharov to his brother, the historian Roy Medvedev. In late May 1970, Roy turned to Sakharov for help—Zhores had been put into a psychiatric hospital with a diagnosis based "on an analysis of his works that allegedly proved a split personality (both biology and politics)." On that basis, any person interested in something beyond his or her direct work requirements could be declared mentally ill.

In his own words, Sakharov "threw himself into battle." He composed a petition in the defense of Zhores Medvedev and collected signatures at an international biological conference taking place then in Moscow. The affair was not complicated by side issues, like the plight of the Crimean Tatars, and Zhores Medvedev was known in the academic world. Sakharov was joined by biologist Academician Astaurov and physicist Academician Kapitsa. As a result, Zhores Medvedev was released a few weeks later.

It was in the course of this campaign that Sakharov found himself in the milieu of the human rights activists and found friends, including the geophysicist Grigori Podyapolsky and the biologist Sergei Kovalev.

That fall, in 1970, Sakharov saw the court machine used to suppress intellectual freedom. He attended a trial of people charged with distributing *samizdat*. He had met one defendant, the mathematical physicist Revolt Pimenov, at the gravitational conference in Tbilisi in 1968. Now he was to meet the technique developed to smoothly squash dissent.

The trial took place in Kaluga, a few hours' ride from Moscow. The courtroom was filled with suspiciously similar "citizens" who left no space for anyone else except the defendants' families. They were still making exceptions then for Sakharov, and they let him in—the only one of all the free thinkers who had come from the capital. The trial took three days, and at the end the prosecutor came up to Sakharov and asked, "How do you like the trial? I think that the court was very thorough and objective in examining all the circumstances of the case."

> I think he sincerely expected me to express my pleasure with the court and his summation. Even in the eyes of the prosecutor, who knew that I had come as a supporter of the defendants, I still remained "one of us" for him, and praise from a Moscow academician would have flattered him. However, I replied, "I think that the entire trial was absolutely illegal."

A few weeks later, Sakharov and a few dissidents formed the Committee on Human Rights to study and make public the human rights situation in the USSR. Working in the committee gave Sakharov the opportunity through numerous personal contacts to familiarize himself with the basic unfreedoms of the Soviet man, from the inability to choose where to live to how to worship. The committee's regular meetings were a form of socializing, too: "For me, not spoiled by a surfeit of friendship, perhaps this side was the most important."

He developed a very close friendship with Grigory Podyapolsky (1926–1976). Sakharov saw in him a paradoxical combination (that he himself displayed): "Grisha, a gentle and kind man, was firm and impervious to pressure when defending his convictions . . . He is characterized by an intolerance of all violations of human rights and at the same time an exceptional tolerance of people, their convictions, and even their weaknesses."[26]

Working within the law, openly, Sakharov and his fellow activist were trying to create an open society, a society that was aware of its problems, studied them, and tried to solve them. Their goal was not opposition to the authorities, but dialogue on a foundation of principles. No doors were slammed shut. When in November 1970 Khariton told Sakharov that the chief of the KGB wanted to talk to him, Sakharov waited a few weeks and then tried to reach him.

Had he only known how narrow the crack was through which the authorities observed his open human rights activity and how narrow the framework was within which the country's leaders approached the "Sakharov question." General Secretary Brezhnev, in a conversation with his Politburo comrades used these words to define his activity: "This is not only anti-state and anti-Soviet, it is simply some kind of Trotskyite deed."[27]

Had he known how hopelessly cut off the leaders' world was, he would probably have felt even worse. But it was unlikely that he would have changed his intention to live openly. For this corresponded not only to his convictions of a better way to the radiant future but to his human nature, which Solzhenitsyn described as "complete openness."

The closed military science world remained in the past; he had vowed to keep its secrets. But now he took on another responsibility, to open up his country's public life in accordance with human rights that were formally upheld by Soviet laws and were proclaimed by the United Nations.

As he set out to Kaluga in the fall of 1970 to attend the trial of the *samizdat* publishers, his primarily goal was to make the trial open. Zeldovich visited him not long before the trial:

> I have something serious to discuss with you. I like your tract ["Reflections"] and its constructive spirit. You should go to Kirillin and create a group of experts under the aegis of the Council of Ministers to help the country restructure technology and science in a progressive spirit. This is where you can be useful, this would be constructive. I know that you are planning to attend Pimenov's trial. That will put you "on the other side" instantly. You will never be able to do anything useful again. I recommend that you refuse this trip.
>
> I replied that I was already "on the other side." Lots of people, the entire Academy, can give Kirillin advice. I do not know whether what I am planning to do is useful or not. But I have entered on that path irrevocably.

19

ANDREI AND LUSYA

On the eve of the trial in Kaluga in the fall of 1970, Sakharov dropped in on one of his new, free-thinking acquaintances. "There was a beautiful and very businesslike woman there, serious and energetic . . . He did not introduce us, and she paid no attention to me."

Elena Bonner paid no attention to him because she had no idea who he was. She knew the name very well. She had read an émigré publication of "Reflections" back in August 1968, on her first trip to Paris to visit relatives who ended up there during the Russian revolution. Surrounded by the stormy student demonstrations and just before the tragic Prague Autumn, she noted the epigraph Sakharov had used: "He alone is worthy of life and freedom / Who each day does battle for them anew!" She remembered seeing that Goethe quote in the diaries of Zoya Kosmodemyanskaya, the famous teenage partisan executed by the Nazis when she was only eighteen, and thought that kind of romanticism was unusual in an academician and did not really fit the text, which was bold but not very enthralling. Nevertheless, she brought back several copies of the brochure to the Soviet Union.

Bonner and Sakharov were introduced in Kaluga. It took several months for her to notice his attentions. The next fall, in 1971, "our life paths merged," as he put it. In January 1972 they were married. Their friends observed that "merge" was the right word for the ardor of the fiftyish newlyweds.[1]

Beautiful, Serious, and Energetic

Elena Bonner was two years younger than Andrei Sakharov. Her nickname Lusya, used by family and friends, came from her Armenian birth name, Lyusik. She spent her early childhood with her Armenian grandmother, who would offer the traditional Armenian wish whenever the little girl hurt herself playing: "*Kez matakh, dzhanik*"—"may all your pains come to me instead."[2]

She moved in with her Jewish grandmother at fourteen, too old to hear similar wishes in Yiddish. In fact, the Jewish side of her family had spent several generations in Siberia and had become almost totally assimilated in Russian culture. However, the Armenian-Jewish issues were of little interest to her, for she lived in a world of truly international people whose life goal was world socialist revolution. Her beloved stepfather, Gevork Alikhanov, worked in the highest levels of the Comintern, the Communist International.

The only other influence on Lusya was her nanny, Nyura, a Russian village beauty who lived with them as a family member. She taught Lusya folk wisdom and listened to her prayers every evening, taking out a banned icon from beneath her pillow.[3] Many years later, the daughter of international Communist parents could explain to her Russian Orthodox husband that he had misheard some words in a prayer memorized as a child. With such a varied background, she replied in some confusion to a friendly question about whether she was Jewish: "Yes, I mean, no, my mother is Jewish, I'm a Muscovite, but I'm from Leningrad."[4]

The first thing Andrei Sakharov learned about the beautiful, serious, and energetic woman was that "she had been dealing almost all her life with *zeks* [prisoners] and has helped many." That life began for her in 1937 when she was fourteen, when her parents were arrested. The girl from the privileged class of world revolutionaries turned into the daughter of "enemies of the people." She did not believe it for a second and traveled monthly from Leningrad to Moscow to stand in line to hand parcels of food and other necessities to her parents. They had been too busy with world revolution to register the birth of their children. Thus, when she turned sixteen and could get an internal passport, Lusya chose another name for herself—Elena, in honor of a fictional heroine she liked then, Turgenev's Elena Insarova, and the surname of her mother, Ruf Bonner.

They stopped accepting parcels for her father in 1938. That meant that he was dead. But packages to the camps, for her mother and Ruf's comrades, became an integral part of Elena's life for many years. She was a nurse in a hospital train at the front, and after the war, when she was demobilized because of her vision, which was damaged by a serious concussion, she graduated from a Leningrad medical school and worked as a pediatrician.

After Stalin's death in 1953, her mother was rehabilitated and even given a two-room apartment in Moscow. Elena Bonner came to that apartment in the mid-1960s after divorcing her first husband. By the time she met Sakharov, her children were twenty and fourteen. By the time Sakharov moved into the apartment in 1971, Bonner was using her abilities to put together prison parcels to the hilt. The Gulag had opened its gates to take in violators of Articles 70 and 190(1), which replaced the old Stalin Article 58. Academician Sakharov began getting food for the parcels, too.

Sakharov openly adored his wife, his unabashed gazes sometimes embarrassing outsiders. No matter where they were, they sought and found each other's eyes across the room. He loved her beauty, her "unsentimental readiness to help," strength of character, and literary and culinary talents. In a brief characterization of Elena Bonner's role in his life, Sakharov chose the word "humanizing." What he meant by that, he explained at length in his *Memoirs*, which he dedicated to her, using her birthday as the date he completed them. The epilogue, which he finished the day before his death, ends with "We are together."

Memoirs is the most visible manifestation of their relationship. His wife was the greatest influence on his writing. "In the summer of 1978 at Lusya's insistence, after some resistance on my part, which she overcame, I started writing the first drafts of my memoirs." This is the opening sentence of Sakharov's book.

The world that Lusya brought as part of her dowry was very literary. Her friendship with her classmate Seva Bagritsky, son of the famous poet, brought her into literary circles. Poetry was a vital part of her life and she knew numerous poems by heart, including the poetry of Seva, her first love. He was killed in the war in early 1942 at the age of twenty. And twenty years later Lusya and his mother put together a book of his poetry, letters, and diaries.[5] Sakharov called that slim volume "one of the most important things Lusya did in her life."

She herself wrote; she published articles on public medical issues and reviews of new books. After a year in Iraq giving smallpox vaccinations, she wrote a long travel essay for the magazine *Neva* in 1961.[6]

In their first weeks together, Lusya introduced Andrei to living literature—to the poets Bulat Okudzhava, Alexander Galich, and David Samoilov. Their lyrical poems and songs were known to scientists and the rest of Russia from *samizdat* and *magnitizdat* (tapes), not from published books. "On 21 May 1971, when we were still on formal terms, Lusya gave me a royal gift—a typed manuscript of Okudzhava's songs in a home-made green binding." That fall, the author performed for Sakharov. "I was a little anxious going to see Okudzhava, who had a certain romantic aura for me. But it went well. We even found a spiritual contact—naturally, thanks to Lusya . . . I couldn't have imagined anything like this just six months earlier." (Nor could Bulat Okudzhava have imagined that twenty years later at the Central Writers' Club he would be handing out the Sakharov Prize for a Writer's Civil Courage.)

Sakharov and Bonner began their literary collaboration in the first weeks they were together. "I dictated and Lusya typed my letter to the Supreme Soviet on freedom of emigration and that subsequently became a tradition." Two years later in December 1973 Sakharov wrote his first autobiography that was not intended for the personnel department, but as a preface to *Sakharov Speaks*, a collection of his speeches and articles being published in the United States. "Lusya corrected the text and gave good advice."

In late 1975 Elena Bonner was in Italy for eye surgery when it was announced on October 9 that Sakharov had won the Nobel Peace Prize. The
committee noted "Sakharov's fearless personal commitment in upholding the
fundamental principles for peace between men" and the "convincing manner in which Sakharov has emphasized that Man's inviolable rights provide
the only safe foundation for genuine and enduring international cooperation."[7]

Andrei Sakharov was the first Russian recipient of the prize, and Elena
Bonner became the first woman to represent her husband at the award ceremony, for the Soviet authorities did not allow him to travel to Norway to accept it personally.

On the day the prize was awarded, December 10, 1975, Sakharov was in
Vilnius, where human rights activist Sergei Kovalev was being tried.[8] That
evening, in the house of a friend of the defendant, Sakharov listened to a broadcast of the Nobel ceremony from Oslo: "I hear the sound of Lusya's footsteps—
going up the steps. And now she speaks. I get the meaning of the words only
afterward, a few minutes later. At first I respond only to the timbre of her voice,
so near and dear and yet carried up into another solemn and radiant world.
Her low, deep voice, for an instant quivering with emotion!"

Upon her return to Moscow, they discovered that even his excellent memory
could not cope with all the events. This prompted her to start a diary. The one
for 1977 begins with her notations, even for things that involved only him. By
the middle, notes in his handwriting appear, very brief. The 1978 diary is his
alone. Thus, the physicist acquired the writing habit.

In the summer of 1978, he began writing his memoirs. "I consider memoir
literature an important part of the common human memory. This is one of the
reasons I took up this book . . . The other is that with the wide interest in my
person, much of what is written about me, my life, its circumstances, and my
family is often extremely inaccurate, and I endeavor to relate more exactly."

"Extremely inaccurate" is a very academic term for the lies that the state
spread through mass disinformation domestically and abroad. The lies were
limited only by the imagination of their authors and directed primarily against
Elena Bonner. In his book, Sakharov gives specific examples that may seem
overly detailed now, but we must remember that he was writing in his Gorky
exile, defending the person he loved most.

Even though the KGB knew that Sakharov had become "uncontrollable"
two years before he met Elena Bonner, they considered her the greatest obstacle
in their effort to neutralize his harmful activity. There was a serious basis for
their opinion. In any case, it is hard to imagine that *Memoirs* would have appeared without her. She was the initiator, editor, typist, one of the main characters, the main codefendant, and finally, the smuggler who organized the
delivery of the manuscript to the West under the very nose of the KGB.

The book required enormous effort. It is about thousand pages long, and the
KGB stole parts of the manuscript three times, hundreds of pages that had to
be restored.

Helpmate

> Over many years we developed a method of working. Usually I first give her
> an oral account of my latest idea; then she reads the first (handwritten) vari-
> ant and makes her comments and suggestions. The next stage of discussion,
> during the typing of the manuscript, is usually very stormy, I disagree with
> many things, and we argue; finally, I accept some of her changes in the text
> and reject others. Without me she never changes a single word in my docu-
> ments and manuscripts.

Everyone who has ever suffered over the best way to express his thoughts and
feelings dreams of a good editor. A sensitive first reader who points out confusion and ambivalence without imposing his views helps the author express his
own. Seeing Elena Bonner's strength of character in her public appearances
and not knowing the firmness hidden behind Sakharov's mild manner, one
could think that she would not make a suitable editor because she would im-
pose her views, however inadvertently. The ones who think so are usually those
who find things they do not like in Sakharov's *Memoirs*. For example, the KGB
physicist Terletsky did not like the story of how Mikhail Leontovich threw him
down the stairs "and called him a representative of an ancient and dishonorable female profession."[9]

The couple enjoyed the back and forth of stimulating discussion over language, people, and current events: "One night, when over late tea, Bonner
quoted a Russian proverb to make a point, Sakharov said that its meaning was
misconstrued in modern Russian. They argued for a while, jokingly but firmly,
until the unabridged Dal' dictionary was brought out to settle the question. He
was right."[10]

Sakharov's rare combination of traits made the not-very-perceptive think
that he easily succumbed to external influences. This is apparent in the official
character reference for Sakharov at the Installation in 1955. Giving Sakharov
his due for "exceptional depth of thought" and "outstanding works," the administration noted his faults: "He refused without serious reason to be a candidate for deputy of the City Soviet and he gave incorrect, apolitical opinions
(in personnel selection) on the abilities and suitability of certain ethnicities for
theoretical work" [this was most probably a euphemism for Jews]. These faults
were explained by the fact that "Comrade SAKHAROV is easily swayed by others."[11] This leaves unclear how the writer attributes the refusal to take part in
the Soviet governing system to being easily swayed.

In fact, what appeared to be a receptivity to influence was a rare openness to
other people's opinions. This tolerance was the result of self-confidence in his
ability to test ideas without fear of losing his own standards. Solzhenitsyn recalled
with a certain astonishment how Sakharov reacted to his article criticizing "Reflections": "Although (as he admitted) he had read it, and reread it more than
once, with some distress, it left no trace of ill feeling in his attitude to me."[12]

For Sakharov a critical opinion was particularly valuable, for it was the best way to test his own understanding. He gratefully and meticulously mentions people who revealed an idea to him, even if he subsequently parted ways with that person or if that person parted ways with the idea. He was the same in his relations with his editor at home: open to all her ideas, he easily and gratefully accepted the ones that corresponded to his own understanding but was completely immovable if the idea did not suit him.[13]

The best proof of this is that the editor is not at all thrilled by the final version of *Memoirs*. When asked by a reporter if they had argued about it, Elena Bonner replied,

> We crossed words daily, almost as soon as I sat down at the typewriter and he started dictating. To this day I think that I'm a better writer than he is. Andrei manages to make any idea "overserious"! We did not argue, we had fights over his book of memoirs. It's so richly textured, such a cross-section of life—scientific, philosophical, public! Everyone says that Sakharov was kind, but as I reread the book I sense how carefully he avoided emotion in describing people. There are almost no live people in the book. The world of ideas is very interesting and varied, while the people are schematic. Perhaps only Zeldovich is "alive," and that's because Andrei considered him a friend and was stunned when "Zeld" did not respond to his desperate letters from Gorky. But Zeldovich was no worse than the rest. Andrei was indifferent to all those who betrayed him, but here he lost control and as a result created an image.[14]

This editorial eye seems to be too severe. There is another image that is portrayed beautifully in the book—the author himself. Very alive, loving, angry, defensive, the image appears wherever he talks about his closest friend, his wife. She is presented as having only good qualities, which does not appear very realistic. But it shows the author's true feelings.

"Humanization"

Besides love and friendship, his wife also gave him a larger world. For a shy man, this was a great gift, comparable to what he had once gotten from Zeldovich. The latter had tried to open up the world of pure theory for him, while Elena Bonner helped reveal to Sakharov the world of practical humanitarian life: "Lusya with her open and active humanity promotes the humanistic, concrete focus of my public activity, steadfastly and selflessly supports me throughout all these difficult years, often taking the blow upon herself, and helps me in word and deed."

He considered it possible that the authorities

> are planning to depict in the future my public activity as a random delusion brought on by someone's influence, the influence of Lusya, a greedy,

depraved woman, a Jewish criminal, an agent of international Zionism. They need to make me once more a prominent Soviet (Russian—which is important) scientist with invaluable contributions to the Homeland and world science, and to exploit my name to serve the ideological war.

He did not make up this grim possibility. It was being pushed by Soviet propaganda to explain the academician's "hooligan" behavior. A very active explainer was Nikolai Yakovlev, a doctor of history and a prolific writer, author of dozens of books. His *The CIA Against the USSR*, which contains the depiction of his wife that Sakharov mentioned, had three printings (3rd ed., 1983). The government generously financed hundreds of thousands of copies of that portrait. Millions of copies of Soviet magazines added to the image.

One of the most touching episodes in Sakharov's *Memoirs* is his story of how in July 1983, in the fourth year of his exile in Gorky, Yakovlev showed up in his apartment when Elena Bonner was in Moscow. The ostensible reason for the visit was to interview Sakharov on his article "The Danger of Thermonuclear War," just published in the West.[15] Obviously, the visitor was there with the regime's blessings, but Sakharov did not know whether he was merely following instructions or there was a trap intended.

Sakharov was seeing the writer for the first time, and he was amazed by the combination of "arrogance with an almost physical meekness, his unquestioned literary talent and erudition with total lack of principle, falsehood and cynicism."

"In the nineteenth century, I would have to challenge you to a duel," Sakharov stated "quite seriously, without a smile or irony." With the same seriousness he then told the man to his face what he thought of his writing. He ended the meeting in the best traditions of the age of Pushkin:

> Yakovlev: "You can sue me. I have witnesses, data from the prosecutor's office, the court will handle it." I said: "I do not believe in the objectivity of the courts in this—I will simply slap your cheek." As I spoke, I quickly came around the table, he jumped up and managed to defend himself by reaching out his arm and bending down to shield his cheek, thus parrying my first blow, but with a second blow with my left hand (which he had not expected) my fingers touched his flabby cheek. I shouted: "And now get out, instantly!"

Sakharov's right- and left-hand symmetry proved to be unexpectedly useful in this instance.

Sakharov readily agreed that his wife's influence was strong. But in what? In questions of military and technical balance, where he based his thinking on his knowledge and twenty years' experience, her influence extended only to typing his manuscripts without errors. But the manuscript could get to freedom—for publication in *samizdat* and in the West—only with her help. For the first four years of his exile, she was allowed to travel between Gorky and Mos-

Сдано в набор 24.08.83. Подписано к печати 09.11.83.
А 00345. Формат 84×108¹/₃₂. Бумага кн.-журнальная.
Гарнитура «Правда». Печать офсетная.
Усл. печ. л. 24,36. Уч.-изд. л. 22,98. Тираж 750 000 экз.
(1-й завод: 1—250 000). Цена 85 коп.

Набор и фотоформы изготовлены в ордена Ленина
и ордена Октябрьской Революции типографии газеты
«Правда» имени В. И. Ленина. 125865, ГСП, Москва,
А-137, ул. «Правды», 24.

Figure 19.1 The book *TsRU protiv SSSR* [CIA Target, the USSR] was printed in 1983 in 750,000 copies at Pravda Publishing House, in Pravda font, at the Pravda printing house (located, naturally, on Pravda Street). Since *pravda* is Russian for "truth," it is supposed to be the truth, the whole truth, and nothing but the truth. The book purports to expose the views of Sakharov and those "who could objectively set him in the service of the interests of imperialism." How? That will require delving in Sakharov's personal life. "It's as old as the hills—after the death of his wife, a stepmother came into the house and threw out the children. An action considered bad in all times and by all peoples. The oral and written tradition of humanity is filled with horrible tales on the subject." This plausible fairy tale was easier to believe than the true version, in which an academician, two-time State Prize laureate, and three-time Hero of Socialist Labor leaves the apartment that was one of his perks to his children and moves with his wife to his mother-in-law's two-room apartment, where together with his wife's married daughter and son they totaled six residents.

cow, and she used the opportunity to pass his manuscripts to the outside world. The KGB had no doubts that it was her doing—in December 1982 they confiscated Sakharov's manuscripts when they strip-searched her on the Gorky–Moscow train. A mother of two and grandmother of three, she was prepared to do what was necessary and to put up with everything to be Sakharov's connection with the outside world. No other candidates were available.

According to Sakharov, the danger of thermonuclear war was the result of the military-technical standoff between the superpowers and the instability of that equilibrium of fear. But the real strategic problem was the standoff itself, fraught with global suicide. He found the solution in an area seemingly distant from nuclear missiles—in the observance of human rights. It was in this field that he most felt his wife's influence.

Anyone who had not accepted the usual lawlessness in the country—that is, who retained a sense of human dignity—could become victim, defender, and expert in the field of human rights. Anyone who accepted as a self-evident truth that all people are endowed "with certain unalienable Rights, that among these are Life, Liberty and the Pursuit of Happiness"[16] could become an expert. History had proven the truth of this before it became part of the Declaration of Independence of the United States of America on July 4, 1776. For Sakharov, Bonner, and their friends, the pursuit of happiness included establishing that truth in their own country.

Human rights figured in his 1968 "Reflections," but he concentrated on only one right there, to intellectual freedom. The dissident milieu had a greater than average number of people with a sense of personal dignity. In this milieu, Sakharov's understanding of human rights expanded, in particular, in the activity of the Committee on Human Rights, founded in the fall of 1970.

How did Bonner help? Sakharov knew how to grasp complex phenomena in science, high technology, and strategic weaponry, but he was isolated from many simple experiences of daily life in the Soviet Union. Her importance was in familiarizing him with the human—inhuman—side of Soviet life. The union of empirical information and theoretical elaboration led him to a very general yet practical conclusion: "I am convinced that the ideology of defending human rights is the only basis that can unite people independent of their nationality, politics, religion, and social status," wrote Sakharov, and added, "as Lusya put so well in one of her interviews."

Many people who cared about Andrei Sakharov felt that Elena Bonner had overdone the "humanizing" and could not forgive her. They could not forgive her for his hunger strikes. He went on five—in 1974, 1975, 1981, 1984, and 1985. The goal of the first was very general—"to draw attention to the plight of political prisoners"—and he stayed on the hunger strike for six days (Bonner was in the hospital for her eyes then).

The other four hunger strikes were to obtain permission for someone to leave the country. Three times for Bonner (1975, 1984, and 1985) and once for her son's fiancée, Liza Alexeyeva (1981). Bonner and Sakharov were on hunger strikes together in 1975 and 1981, and the last two he did alone, for many months, with forced feeding.

In his *Memoirs*, Sakharov explains why he felt responsibility for the health of his wife and for the plight of Liza Alexeyeva, why he saw them as hostages to his public activity, and what rational logic there was behind his hunger strikes, which seemed illogical to many of his well-wishers. But what the explanations show is that behind his logic there was not a public figure, not a political thinker, not a theoretical physicist, but simply a man defending his own human dignity, his human rights, by every method he could.

Sakharov achieved his goal with every hunger strike but the penultimate one in 1984. At what price? He did not consider it more than the values he was defending. Some people might have wished him a different wife, but there is no evidence at all that he wanted to do any editing of his Lusya or her influence on him.

20

FREEDOM AND RESPONSIBILITY

A Miracle in the Swarms of the Venal Scientific Intelligentsia?

Solzhenitsyn began his portrayal of Sakharov by saying that the creators of the totalitarian state had planned for everything

> to ensure that the voice of freedom would never ring out, or any movement against the current ever set in. They foresaw all eventualities but one—a miracle, an irrational phenomenon, the causes of which could not be divined, detected and cut short beforehand. Just such a miracle occurred when Andrei Sakharov emerged in the Soviet state, among the swarms of corrupt, venal, unprincipled scientific intelligentsia.[1]

Could Sakharov have agreed with that, when his first words in "Reflections" admit that "the author's views were formed in the milieu of the scientific and scientific-technological intelligentsia"? Who was corrupt, venal, or unprincipled? Igor Tamm? Mikhail Leontovich? Nikolai Dmitriev, the defender of ideal socialism, who spoke out against the "real socialist" invasion of Hungary? Or Lev Altshuler, whom Sakharov had to save because he had been too truthful about Lysenko? Or Mates Agrest, who had not made his religious feelings top secret at the godless KB-11?

As a graduate student at FIAN, Sakharov had missed Leonid Mandelstam by only a few weeks, but he learned from his teachers, who had been Mandelstam's students, that "neither in his scientific, or public, or personal life did Leonid Mandelshtam permit himself to take actions if he was not absolutely certain that they were right. Once he had that certainty, he acted in accordance with it, never retreating under any pressure."[2]

Sakharov had missed Pyotr Lebedev at Moscow University by several decades, but he knew how the latter had acted in circumstances presented by Russian history. Weren't these people part of Sakharov's milieu? Just as Lebedev had been born for experimental physics, Sakharov was born for theoretical physics. The circumstances of Russian history and their own lives forced them to make a moral choice that made them public figures.

Of course, Solzhenitsyn did not make up a venal and corrupt scientific intelligentsia. Letters to newspapers condemning Academician Sakharov began with one signed by forty academicians to *Pravda* on August 29, 1973. They were listed in alphabetical order, prompting a former classmate of Sakharov's to write a sarcastic poem about them.[3]

Sakharov was not sarcastic but accurate in his assessment of the academicians, noting that not everyone has "sufficient courage and integrity to resist the temptation and habits of conformity."[4] He knew many of the forty well. Social and intellectual conformity were apportioned variously in them. Social conformity is simple, material. All the signers held administrative posts in the sciences—from president of the Academy of Sciences to laboratory heads. They had something to lose.

The intellectual conformity was deeper. Sakharov's understanding of his country and the world was too far removed from the Soviet standard. It was not that all the academicians considered the situation to be ideal, but that even Igor Tamm, for all his closeness to Sakharov, did not accept some of his social ideas. What can be said of others, who knew Sakharov's views only from how they were presented in the newspapers and had no time to think about the naive foolishness of someone who was not a specialist in social issues?

All radically new ideas are opposed by conformity. Sakharov's idea of the instability of the proton found a similar obstacle—intellectual inertia. It is a healthy defense mechanism in the body of science, part of its immune system that distinguishes between vital ideas and lifeless or even noxious ones. Science has methods for testing new ideas—open discussion and comparison with empirical data.

The Soviet regime had a much stronger immunity against new ideas. And the authorities had no intention of testing Sakharov's social ideas. Pyotr Kapitsa, a member of the Presidium of the Academy of Sciences and a highly respected man, could not even get the Presidium to discuss Sakharov's "Reflections" in their narrow circle. Kapitsa felt that Sakharov's ideas deserved the most serious consideration. This attitude of the experimenter about a theorist a generation younger is more astonishing than the rejection of Sakharov's ideas by many other members of the Academy.[5]

The others did not have the professional inside information that Sakharov did (in the field of strategic weaponry), or they lacked his profound and daring thinking. So it was easier for them to believe that the physicist Sakharov was wrong in his social views and that he had crossed the line of his professional competence. This was exacerbated once he violated the unwritten rules of Soviet academic etiquette. Giving interviews to Western reporters? Holding press conferences? Without official permission? It was unheard of.

It is easy to underestimate the freedom Sakharov took for himself when looking at the Soviet Union in 1973 from the perspective of a society in which interviews and press conferences do not require state permission. But Soviet citizens could see a society like that only through the filter of Soviet mass media.

This is what Sakharov was talking about in his 1973 interviews. "The USSR is a closed, totalitarian society, 'a country behind a mask,' . . . whose actions can be unexpected and extremely dangerous . . . The West must work on reducing the closed nature of soviet society. Only when these conditions are met can detente lead to international security."

He knew that this position would be difficult to understand for people who believed in the peace-loving nature of Soviet policies and the West's treacherous plans. "Truly, if we are for peace, then the more missiles, thermonuclear weapons and shells with nerve gas, the safer it is for our people and, therefore, for everyone. It is not easy to understand that this reasoning applies equally well on the opposite side and thereby becomes absurd." Especially if all you know about the world comes from controlled sources.

Sakharov, who had been in the same situation, recognized this problem of perception—when you look but do not see. But he may have underestimated the difficulty of the problem by using himself as a measure. He was counting on his free speech that would bypass the censors through *samizdat* and Western broadcasts to reach his countrymen.

But for the ordinary radio audience, Sakharov was a mystery, described only in a few lines in an encyclopedia. Other sources of information were rumors—for instance, that Sakharov was in fact a Jew named Tsukerman ("sugar" in Russian is *sakhar* and in Yiddish *tsuker*). In September 1973, when the "organized wrath of the citizens" splashed out on the pages of newspapers, Lydia Chukovskaya captured this rumor in her article "Wrath of the People."[6]

But many of the forty indignant academicians knew Sakharov personally and remembered how he had spoken out against Lysenkoism at a general assembly of the Academy of Sciences. The organizers of Soviet science had added another inducement to the temptations of habitual conformity. Those who did not find the boss's orders enough of an impetus to sign letters were told that a public expression of disagreement with Sakharov would stave off his arrest. The government would see that the rotten apple had not spoiled the barrel, and therefore extreme measures would not have to be taken against the apple. Thus, the signers could even feel that they were saving the rebel academician. That trick worked.

The letter was not signed by Academician Vitaly Ginzburg, who took over the theoretical department at FIAN in 1971 after Tamm's death. But he explains:

I could have mistakenly signed the first letter against Sakharov. I was out
of Moscow. I was lucky. There was nothing in the press yet, [Academy Presi-
dent] Keldysh gathered a group of academicians and said: "You know we
need to defend Andrei Sakharov . . ." The first was a letter from a group of
academicians criticizing him. It didn't have all that vileness yet, and some
people wanted to improve it, but they were not allowed, they were persuaded
to leave things. I think back with horror that I could have signed that first
letter, I could have brought shame upon myself for the rest of my life.[7]

Actually, the first letter is vile enough, even though it ends with the concerned phrase: "We hope that Academician Sakharov will think about his
actions."[8]

The second Academy letter was organized two years later, when Sakharov
was awarded the Nobel Peace Prize. As Ginzburg recalls, acting president
Vladimir Kotelnikov

worked over the academicians one on one, and he suggested that I sign the
second letter against Sakharov. I refused categorically. I decided this: I'll
refuse even if they expel me from the Party. But if they beat me, I'll say
Sakharov is bad. That was my limit. But in fact they did nothing to me. They
did nothing to no one. It was very easy to refuse.[9]

Of those asked to sign, five refused and seventy-two members of the Academy lined up obediently. Reporting to the Central Committee on their work and
requesting permission to publish the statement, the leaders of the Academy of
Sciences tattled on the academicians who refused to sign.

Ginzburg cited "personal considerations," Zeldovich felt that "the letter
should be written in a different spirit and that he is planning to prepare an individual letter," Kantorovich "as a recent laureate of the Nobel Prize [econom-
ics, 1975], felt it untimely to sign a collective letter and is thinking of writing
an individual protest," Kapitsa "feels that it is necessary to call Sakharov in for
an explanation at a meeting of the Presidium of the Academy of Sciences of the
USSR and only after that react in a commensurate manner to his actions," and
Khariton "feels that such a letter should not be sent, since the members of the
Academy of Sciences of the USSR, including him, have already protested the
actions of Academician Sakharov."[10]

Even without looking at the disparity of reasons, the 72 to 5 tally supports
Solzhenitsyn's opinion of the corrupt and venal scientific intelligentsia.[11] They
did not sell out cheaply. Their prosperity and status, which were handed out
by the state, like everything else, put them in the highest elite, on the level of
high party official. The older generation of academicians was aided in selling
out by the fear that remained from Stalinist times. The younger generation
replaced some of the fear with cynicism: "Sakharov and Tamm thought that
they were performing a patriotic deed by establishing nuclear parity in the
world. They were replaced by a generation of scientists who approached this

in a rather cynical way. Yes, we take money to do science, they said, but otherwise that money would be spent on meaningless and even criminal goals."[12] "Maybe subconsciously we simply were not ready to sacrifice the privileges given to us by the system—the chance to do science and enjoy certain well-controlled dosages of foreign trips."[13]

Ginzburg summed up his opinion of such academicians in a survey on "Science and Society" in *Literaturnaya Gazeta*: "There is no basis for maintaining that doing science builds high moral qualities."[14] And Sakharov, apparently, could not accept that fact.

A Nonelitist Individualism

Sakharov belonged to three elites: social, intellectual, and moral. And he accepted the idea of elite payment for work, since "every minute used incorrectly by a major administrator, every lost minute of an arts figure" means a loss for society.[15] But he had no sense of his own eliteness.

A colleague remembers Sakharov's phrase, "If not me, then who?"[16] But in his texts Sakharov used Leo Tolstoy's "I cannot be silent."[17] The sentiments are almost the same, indicating a turning point, but in the former the stress is on understanding with the mind and an (elitist) juxtaposition with others, whereas the latter is an emotional and individual inability not to act.

The banal words, "I have deep respect for all labor: of the worker, the peasant, the teacher, the doctor, the writer, the scientist," for Sakharov mean exactly what they say. In official Soviet ideology, the labor of workers was considered primary. But in the unofficial consciousness among Soviet people in science and the arts, the labor of cultural figures was most important.

In her response to the "organized wrath" of Soviet people in September 1973, Lydia Chukovskaya painted a portrait of the seditious academician:

> A man of a feeling mind and a thinking heart, Andrei Sakharov came to hate bombs and all violence. In appealing to the Soviet government, to the people and governments of the earth, he was the first to think out loud about what is now called "detente of international tension." He wrote several large essays known throughout the world, except to you, comrade Soviet people; essays in which he invited the people of the globe to stop stockpiling bombs and start stockpiling thoughts: how can humanity be saved from the threat of war? Hunger? Disease? Extinction? How to save nature, humanity, and civilization from destruction?
>
> He did something even more significant: he thought about the plight of individuals, of every person, and first of all, the people of our homeland. That is his greatest achievement, because thinking about the fate of the whole world, no matter how important your thoughts are, is easier than helping a single person in trouble. For besides bombs, disease, and hunger all over our planet, and in our homeland in particular, there is a surfeit of prisons,

camps, and—this is our own Soviet contribution—mental institutions where healthy people are forcibly placed.[18]

The KGB chief, reporting to the Central Committee on Chukovskaya's article, quoted it at length:

> CHUKOVSKAYA maintains that between the people and . . . SAKHAROV and SOLZHENITSYN, there is a wall that "is no lower and no less harmful than the Berlin Wall. At the Berlin Wall, which separates one part of the city—and the people—from the other, guards open fire when people try to get over it. Every shot is heard throughout the world and echoes in the heart of every German and non-German. The struggle for the soul of 'the ordinary man,' for the right to bypass the wall of censorship and communicate with him is being waged silently in our country."[19]

Lydia Chukovskaya had begun trying the patience of the KGB in 1966 when she defended persecuted writers.[20] Her 1973 article was the last straw, and on January 9, 1974, she was expelled from the Soviet Writers' Union. That same day, Sakharov responded to the expulsion with an open letter, referring to the "strong and pure voice of Lydia Chukovskaya":

> Her writing is the continuation of the best Russian humanitarian traditions from Herzen to Korolenko. It is never an accusation, always a defense . . . Like her teachers, she knows how and dares to explain what many prefer to keep quiet, protected by titles and honors . . . I am proud of my friendship with Lydia Chukovskaya. I bow before her fearless frankness and kind courage![21]

His feelings for her did not keep him from seeing their differences. Having read in 1978 the manuscript of her book *Process of Elimination*, on the events of 1973–1974, he noted in his diary that he saw an element of mythologizing in her work. "I am no volunteer priest of the idea, but simply a man with an unusual fate. I am against all kinds of self-immolation (for myself and for others, including the people closest to me)."[22]

A day later, after visiting her, he wrote: "The conversation about her book was calm. I told her what I thought." They spoke not only about the book, of course, and the guests had come with a present, since it was the Saturday before Easter. "We brought two eggs, colored in the brilliant way Lusya suggested. (Before boiling them in onion skins, you glue on strips and pieces of adhesive; Lusya did her egg first, for Lydia, then I, for Lyusha; since mine was prettier, Lusya and I exchanged authorship, and agreed to tell Lydia so)."[23]

In his *Memoirs*, he returned to Chukovskaya's 1973 article. "My image in that essay is rather idealized and more focused than I am in fact, and at the same time, a bit more naive and pure." Only a pure person could say something like that.

The other elements of idealization are due to the fact that Chukovskaya did not know many of the real circumstances, many top secret, that led him to the turning point in 1968. But Sakharov saw in that idealization what he called "ideological aberrations." He left behind notes that indicate what he meant by that, and it is clear that he was intentionally vague in the memoirs because he "most extremely would not like to hurt her feelings."

> For Lydia Chukovskaya, as it seems to me, moral and cultural problems are important, and not political ones. This position of hers, active and fearless, is something I understand and respect deeply. But sometimes, it seems to me, Lydia Chukovskaya's opinions reflect unhappy traits of elitism, I suppose, I don't know how else to call it—a loss of a human approach, or breadth and tolerance.[24]

He suggested that "this is the reverse side of the cult of culture" but crossed that out and concluded without any possibility of hurt feelings: "I do not want to go into this, and perhaps I am completely wrong here."

Even if he is wrong, we can see his sense of nonelitism, which set him apart from many in his circle. Kapitsa, who was forcibly detained in the USSR in the 1930s and not allowed to do his work, compared himself to a violin used to hammer nails.[25] There was something similar in the thinking of people who could not understand how Sakharov could expend his energy on day-to-day human rights issues.

During one of his hunger strikes, Sakharov got a telegram: "I share your goals, but I urge you to stop, to sacrifice the personal for the sake of the common." He diagnosed his well-wisher thus: "Totalitarian thinking!" Totalitarian thinking regards a person "as a means of reaching a goal," even if the goals are human rights, and disregards the person's right to make his own decisions. Sakharov and his wife decided to go on a hunger strike "as free people, fully understanding its seriousness, and we both took the responsibility for it, and only we. In some sense, it was our own personal, private affair."

The lack of understanding from people close to him saddened Sakharov both as a private individual and as a social thinker. The public and the private are interrelated in the ideology and psychology of human rights: only by valuing one's own right can a person truly respect the rights of others. In that sense Sakharov was an individualist—a consistent individualist.

In Soviet terms, individualism was a synonym for egoism and the antonym of collectivism. The philosophy of human rights changes that relationship radically, maintaining that the well-being of society as a whole can be achieved only through the observance of the rights of the individual. "The disunity [of Western society] for me is the reverse side of pluralism, freedom, and respect for the individual, those most important sources of a society's strength and flexibility." A consistent individualist has no trouble understanding the social significance of human rights. Sakharov rejected the totalitarian government and the totalitarian unity of his society.

Back in 1958, writing about the danger of radiocarbon, he said, "Only an extreme lack of imagination can let one ignore suffering that occurs out of sight."[26] An individualist with imagination, Sakharov took to heart the suffering of prisoners of conscience and political prisoners in Soviet prisons and camps. For him they were not just numbers, but persons, each with a name and a life story. He related the lives of many people he met in his memoirs. He talked with these people, attended their trials, sometimes as far from Moscow as Omsk (Siberia) and Tashkent (Central Asia). Vladimir Shelkov, Georgi Vins, Mustafa Dzhemilev, Friedrikh Ruppel, Efim Davidovich . . . each embodied a social ill—religious persecution, a ban on the Crimean Tatars returning to their homeland, obstacles to the emigration of Germans and Jews. But for Sakharov each was "simply" a person whose rights were being trampled.

In his 1975 Nobel lecture, Sakharov explained his conviction that peace, progress, and human rights were inextricably intertwined. He listed more than a hundred names in his lecture, adding:

> There is no time to mention all the prisoners I know of, and there are many more whom I do not know, or of whom I have insufficient knowledge. But their names are all implicit in what I have to say, and I should like those whose names I have not announced to forgive me. Every single name, mentioned as well as unmentioned, represents a hard and heroic destiny, years of suffering, years of struggling for human dignity.[27]

Sakharov included similar lists in his other statements. But most frequently he would name one person, talk about only one life. These were not symbolic gestures, but a concrete form of help. In 1975 the Soviet Union signed the Helsinki Accord, which included the recognition of European borders and the recognition of human rights. This gave Western politicians a legal method to ask about human rights inside the USSR. A general interest could be fobbed off with general statements, but the Soviet authorities had to answer a question about a specific person with specific information or otherwise admit that they were violating their responsibility. It gave prisoners of conscience moral support and hope when their names became known around the world.

A Science Underpinning

Sakharov's individualism was rooted in his vocations of researcher and inventor. Both need to act alone and break new ground. That requires an inner voice and the spiritual strength to heed it.

Only a distant observer might think that Sakharov had cooled toward science after 1968. The mass media and disinformation services focused on his nonscientific activities, which were easier to describe than the asymmetry of the vacuum. However, in the twenty years after 1968 Sakharov published approximately as many articles on pure physics as he had in the twenty before. Before 1968 he was distracted from pure science by weapons research, and afterward by human rights activity.

The day he was exiled to Gorky, he was arrested on his way to a seminar at FIAN, and on the day that he returned from exile, he attended a FIAN seminar. Both were Tuesdays, when the theoretical physics seminars are held at FIAN.

Sakharov's feelings about science are evident in his words about his teacher Tamm: "His true passion, which tormented him all his life and gave his life a higher meaning, was fundamental physics. He said a few years before his death, already gravely ill, that his dream was to live long enough to see the New (capitalized) theory of elementary particles that answered 'the damned questions' and to be in a state to understand it."[28]

Theoretical physics is not equally responsive to all ages. The truly innovative ideas usually come to people who are under forty. The more elderly theorists help younger ones to gestate and nurse new ideas. One's individualism helps treat the individualism of other scientists with respect. Science represents a community of individualists, and in the Soviet Union it was a major refuge for individual freedom. This kind of social experience was probably the basis of Sakharov's social views.

People who were close to Sakharov but shared only his views on society and not his passion for science could see how important physics was to him. In a letter from Gorky, writing to someone who could not possibly be suspected of harboring an interest in physics, Sakharov nevertheless mentioned his scientific concerns among others:

> I'm trying to study what has been done by smart people in quantum field theory and so-called superstrings, but this study is extremely difficult, and I often despair of ever reaching the necessary level—I lost too much since 1948, nothing but gaps, and all the years since I managed to do a few things only through luck and "pushiness," often missing completely or working in vain. But now such important events are taking place in physics, the way for which was paved by the work of many hundreds of brilliant scientists in the last decades.[29]

Upon his return from exile in December 1986 in the morning at the Moscow train station, when asked by a reporter what he intended to do, Sakharov replied that first of all he would go to a seminar. And at 3 p.m., he was at FIAN. And when a few weeks later, Herman Feschbach, the American physicist who had done much to garner support for him in the United States, met with him, much of their conversation centered on elementary particles.[30]

The person who was closest to him in his last twenty, political, years, was Elena Bonner. A few years after his death, a journalist started a question to her with "When Sakharov became a dissident, what did he"—and she interrupted him. "He was never a dissident." "What was he then?" asked the interviewer in surprise. "A physicist," she replied

She knew the importance of science for her husband. In 1974 she typed his futurological article, "The World in Fifty Years," and saw what science fiction he considered scientific—flying cities and a "universal information system" (what is now called the Internet).[31] She saw how much passion he put into his scientific articles, correcting and rewriting them. She grew accustomed to the vocabulary. In the years of their exile in Gorky, he could have scientific dialogue only with the rare visit from FIAN colleagues. So the physician began learning theoretical physics—he would read his lectures to her, sometimes over breakfast in the kitchen.

In his article, "The World in Fifty Years," written for the *Saturday Review/ World* in 1974, Andrei Sakharov "gave [himself] the freedom to think about the desired future," and in particular:

> I foresee a universal information system (UIS), which will give everyone access at any given moment to the contents of any book that has ever been published or any magazine or any fact. The UIS will have individual miniature-computer terminals, central-control points for the flood of information, and communication channels incorporating thousands of artificial communications from satellites, cables, and laser lines.
>
> Even the partial realization of the UIS will profoundly affect every person, his leisure activities, and his intellectual and artistic development. Unlike television, the major source of information for many of our contemporaries, the UIS will give each person maximum freedom of choice and will require individual activity.
>
> But the true historic role of the UIS will be to break down the barriers to the exchange of information among countries and people. The complete accessibility of information, particularly in the creation of art, carries the danger of reducing its value. But I am certain that this contradiction will somehow be resolved. Art and its perception are always so individual that the value of personal contact with the work and the artist will always remain. Books will also retain their value. The private library will always exist, because it represents personal, individual choice and beauty and tradition, in the good sense of the word. Personal contact with art and books will always remain a joy.

The academician who gave Sakharov the greatest support was Kapitsa, an individualist with a strong social sense. His support began when Sakharov spoke out against Lysenko at the Academy of Sciences in 1964. Kapitsa quieted down a high party official seated next to him in the Presidium by explaining that the speaker was "the father of our hydrogen bomb." After the publication of "Reflections," Kapitsa tried to make Sakharov's ideas the subject of open discussion. And he tried to get him out of the exile in Gorky and save him from dying in this first hunger strike.[32]

The two men were separated by a generation and a generational approach to public activity. They had entered public life in very different stages of Soviet totalitarianism. Kapitsa acted covertly, which was the only possible way under Stalin—he wrote letters to the country's leader, the "senior comrades," as he called them.[33] Therefore, Sakharov did not know about many of the steps Kapitsa had taken on his behalf. When Kapitsa was asked whether Sakharov should act the way he did, that is, trying to influence higher-ups confidentially, the seventy-six-year-old Kapitsa replied, "Let everyone do what he can. I am too old to act the way Sakharov does."[34]

One of the most powerful personal interests of the individualist Sakharov was public welfare. He forgave a lot in people who wanted the same thing, and he did not particularly care how they treated him: "In 1985, while in Semashko hospital, I watched one of Gorbachev's first television appearances, and I told one of my roommates (KGB agents, I was not allowed to talk to anyone else then): 'It looks as if our country's in luck—we have an intelligent leader.'"

He had been brought to the hospital forcibly in late April, a week after he began his hunger strike, to be force-fed. Yet he regarded the new General Secretary of the Party not from the position of a victim of the state, but as a scientist: "I think that Gorbachev (like Khrushchev) is truly extraordinary in the sense that he could cross the invisible border of 'taboos' that exist in the milieu where most of his career took place."

Sakharov attributed the inconsistency between theory and practice in Gorbachev's perestroika to the inertia of the massive bureaucracy and to the fact that Gorbachev and his allies "were themselves not fully free of the prejudices and dogmas of the system they are trying to rebuild." This is an ordinary situation in the creation of a new theory in physics or a radically new technical device. Sakharov sympathizes with his colleague's difficulties, as if this was not the leader of a superpower and initiator of perestroika who, presiding at the Congress of People's Deputies, interrupted Sakharov's speech on Afghanistan and about whom Sakharov said, to the loud disagreement of the audience, that he supported him, but only conditionally depending on his actions.

Even though he had been returned by Gorbachev to Moscow in December 1986, Sakharov had no illusions about his social position. When an acquaintance congratulated him on finding himself on the top floor of power, Sakharov corrected him: "Next to the top floor, but on the other side of the window."[35]

Figure 20.1 The story of the postcard from the American part of Sakharov's milieu:

"Briefly, the story is this: In February 1981 I noticed an item in the 'Letters' feature of *Physics Today*, signed F. Janouch—a name I hadn't seen before, Research Institute of Physics, Stockholm. Janouch described a 'trick' for sending preprints to Sakharov in Gorky. It stuck in my mind, so some time that summer I made up a packet and sent it off, following Janouch's instructions about requesting a receipt card, assuming of course that this would be the end of it, but what the hell, it cost so little. But a month later the card arrived, postmarked and signed in Cyrillic characters that read, as I found using alphabets from my dictionary, 'Gorky,' 'A. Sakharov' [in fact, it is 'Thank you. A. Sakharov']. I wrote at once telling Janouch the joy that I owed to him. He replied, 'I am very glad that your preprints did reach Andrei Sakharov. According to my information, he is now receiving quite a lot of different preprints, it is essential only to follow the scheme, otherwise it is easy to confiscate them. It would be a good idea, if you could write about it in a letter to *Physics Today*—just to remind our colleagues about the scheme, about the necessity to help Andrei Sakharov and about your experience, to encourage people.' I did as he suggested, and 'the rest is history,' as they say."

(From a Dec. 10, 1997, e-mail message from John Linsley [1925–2002], who studied cosmic rays at the University of New Mexico.)

This did not keep him from openly collaborating with those government efforts that he considered would further well-being and peace, overriding warnings that he could spoil his reputation by supporting the Soviet regime.

Even more telling was Sakharov's attitude toward Khrushchev, who had shouted at him, discounted his thoughtful recommendations, and told writers and artists how to do their work, and who had clung to Lysenko. But it was also Khrushchev who had freed himself of a few dogmas and began liberating the country from the barbed wire of the Gulag and the barbed wire of Stalinist psychology. Even though they were defeated and mocked, Khrushchev's initiatives were enough to elicit Sakharov's gratitude. It was probably this feeling, rather than political considerations, that prompted Sakharov to send condolences to the late leader's family in September 1971.[36]

Sudba in History

The translator of Sakharov's *Memoirs* into English noted: "Russia's greatest Westernizer proved in his writing also to possess a very Russian sense of life. He often used the Russian word for fate, SUDBA, when in English we would simply say 'life.'"[37]

Another English word for *sudba* is "destiny." However, both "fate" and "destiny" sound somewhat pretentious in English. The Russian *sudba* is neither emotionally charged nor pompous.

In his *Memoirs*, Sakharov observes with curiosity and notes diligently apparently insignificant external impulses that preceded major turning points in his life. It is almost as if he wants to lay some of the responsibility for his decisions on his fate: "I did not want to hurry fate, I wanted to leave things to flow naturally, without rushing ahead or trying to be clever in order to stay out of danger," or "fate continued circling me," or "it pushed me to a new understanding and new actions." The year before his death he said, "My destiny turned out to be greater than my person. I merely tried to meet the level of my own destiny."[38]

He did not quite understand why he had interfered in the election of the Lysenkoite Nuzhdin to the Academy of Sciences. "Why did I take such an uncharacteristic step as a public statement at a meeting against the candidacy of a man I did not even know personally?" He attributed it in the end to "impulsiveness; perhaps that was a manifestation of fate."

He gave as the source of an even more important decision, to begin the essay that became his "Reflections," a remark of an acquaintance that it would be good to write a piece "on the role of the intelligentsia in the modern world."

However deep his logical analysis might have been before making a decision on serious questions, he followed his intuition, be it about physics or hunger strikes. It came not only from self-confidence but also from an understanding of the profound complexity of the world that cannot be plumbed by rational logic.

Even if Sakharov's achievements in pure science do not put him on a level of the great physicists, he is similar to them in the power of his intuition and his trust in it. Speaking about his most successful idea in physics (baryon asymmetry), Sakharov noted that the original stimulus came "from intuition and not deduction."

Trusting one's intuition does not guarantee success—Bohr spent many years believing in a hypothesis of nonconservation of energy in microphysics, and Einstein spent many years in vain seeking a geometric unified theory. But when intuition is silent, fundamentally new steps are not taken. Intuition is necessary in all creativity, and the greats are distinguished by their ability to live in spiritual solitude while following their intuition. In the late 1970s, in a conversation with a friend, Sakharov expressed his fear of ending up like Einstein in his late period—wasting a lot of his life on a fruitless theory.[39] Intuition may lead you into a dead end. But what besides intuition can show you the way out?

Being able to explain and persuade others is a different skill. Rarely are outstanding researcher and outstanding popularizer combined in one person. The difference between blazing a path in unknown territory and creating a guidebook for later travelers is too great. Explaining his train of thought and convincing others was never Sakharov's strong suit.

A former classmate who met him again in the mid-1950s was surprised that "Andrei's manner of speaking is nothing like his old, prewar style. Everything was logical, consistent, systematic, without the spontaneous leaps of thought so characteristic of the young Sakharov." Life forced him to be that way, for there was much to explain to his colleagues and, as he himself said, "perhaps, the most difficult, to the generals."[40]

A decade later he had a harder task than explaining the physics of his devices to the generals. Life was forcing him to explain something that was not physics but related to those devices. He could not wait for total academic clarity. The detente created by mortal fear of mutual destruction had to be turned into stable peace. But how? He answered the question for himself first: "Without giving a final answer, we must still constantly think about it and advise others as our minds and conscience prompt. God is your judge, as our grandparents would have said."

Sakharov followed that simple recipe—the guidance of mind and conscience—in the human rights area. It was more complicated here because there were many unknowns in the equation, and the unknown were actual people. His practical philosophy was based on the fact that "life in its causal connections is so complex that pragmatic criteria are often useless, leaving only the moral ones." The moral criteria are not imposed from outside; it is his inner voice, his moral intuition.

People in the human rights movement did not approve of his hunger strike of 1981, and he wrote to Lydia Chukovskaya (after its successful conclusion):

Of course, I was saddened. Apparently I was unable to express clearly and convey even to people close to us our reasons and that sense of unconditional rightness and the knowledge that we had taken the only possible path, which never left us (Lusya and me)—not in the first 13 days, nor in the most difficult days of 4–8 December, when we were separated and knew nothing about each other and they tried to frighten, confuse, and break us—the sense that now gives us happiness and pride. Believe me, of all the things I managed to do in life, little has brought such unconditional, undoubted joy. And another thing, if I feel free, it is in part because I try to base my actions on my specific moral evaluations and do not consider myself bound by anything else. It is all internal, and I understand, of course, that you, standing in an opposite position that I do not fully understand, would not be ready to leave it quickly. But I hope that with time our mutual understanding will be restored.[41]

Sakharov was not very good at explaining his "inner" feelings. But the way he lived his life was proof that he acted on the basis of his own moral assessments and was bound by nothing else. This sometimes was manifested in ways that were unusual. For instance, when he took walks in the woods, he picked up bottles and stuck them into the ground neck down. He explained to his baffled wife that otherwise ants might climb into the bottle and die, unable to find their way out.[42]

Another time, "a wasp got caught in an indoor plant and Andrei went out on the balcony to free it."[43] As a reward for his kindness to the insect, he saw friends waiting for him out on the street, who had come without invitation or permission of the KGB.

Sakharov's openheartedness turned into lack of perception about some secrets of human nature. He overcame that difficulty in the simplest way for himself—the presumption of innocence, or decency, in everyone he met.[44] Even Yakovlev, the writer-for-hire who denounced him, was presumed to be gravely mistaken, since Sakharov tried to explain the falseness of his exposés. And when a vituperative busybody pestered him in the street in Gorky, he allowed that she might simply be misled. She told him she had been in the war and wrathfully demanded he leave his Jewish Elena Bonner and find himself a Russian woman who would not push him toward warmongering. He explained where the true danger of war lay and told her that his wife was wounded in the war and that she was as much Armenian as she was Jewish.

Parallels between Perpendiculars: Sakharov, Oppenheimer, and Teller

The presumption of decency was the basis of Sakharov's attitude toward two American physicists in whom he saw "striking parallels" with his own life, even though the lifelines of Robert Oppenheimer and Edward Teller could be called mutually perpendicular, crossing in the "Oppenheimer Affair" of 1953–1954. Oppenheimer, "the father of the A-bomb," was removed from secret work, and Teller, "the father of the American H-bomb," was supposed to have played a substantive role in that.

Sakharov's lifeline can more easily be considered perpendicular to the lines of both Americans. Sakharov did not believe that he had "known sin," in

Oppenheimer's expression, by creating nuclear weapons. Nor did he try to persuade the government, as did Teller, of the need for a hydrogen bomb.

Sakharov ignored caricatures created during the McCarthy period of the two American physicists. Allegedly, the two, being unhappy with their scientific achievements, switched their ambitions into the public sphere, but one joined the liberal left (who were Stalinists), the other the right-wing hawks (who could turn to neofascism). Although there was an element of truth in each caricature, the element was too small for Sakharov to believe it. He saw both of his American counterparts as tragic figures and felt sympathy for both. Oppenheimer's tragedy was in his personal participation in the decision to use the A-bomb in Japan. Teller's tragedy was that having given his true opinion of Oppenheimer, he helped the government remove the influential expert from the world of military-scientific policy, which brought Teller lifelong condemnation in the scientific milieu he considered his own.

Sakharov amazed many American readers of his memoirs by calling Teller's treatment at the hands of his American colleagues "unfair and even ignoble." And this was not because *Pravda* joined the majority of American physicists, writing that Teller "accused his colleague R. Oppenheimer of treason because the latter was against further development of nuclear weapons."[45]

In Sakharov's view, the situation was quite different. In the late 1940s, Oppenheimer opposed work on the American H-bomb, hoping that the USSR would follow suit and not create their own superweapon (not knowing that Sakharov and his colleagues had already begun research on the hydrogen bomb). Teller did not trust the Soviet regime and was convinced that only American military power could contain the USSR from uncurbed expansion. That is why he called for the speedy creation of an American H-bomb.

Sakharov felt that in this "tragic confrontation of two outstanding people," both deserved equal respect, since "each of them was certain he had right on his side and was morally obligated to go to the end in the name of truth." The positive thing he had to say for Oppenheimer's stand was that "it was not pointless," for the Americans were the first to initiate the quest for the H-bomb. But politically, Sakharov agreed with Teller. His experience with Beria and his successors told Sakharov that the Soviet leaders

> would never give up attempts to create new kinds of weaponry. All steps by the Americans of a temporary or permanent rejection of developing thermonuclear weapons would have been seen either as a clever feint, or as the manifestation of stupidity. In both cases, the reaction would have been the same—avoid the trap and immediately take advantage of the enemy's stupidity.

Sakharov disagreed twice with Teller on issues of principle—on nuclear testing in the atmosphere in the late 1950s (when he considered himself a technical expert totally loyal to the Soviet regime) and on antimissile defense in the

1980s (when he was the chief Soviet dissident). But disagreements on some points did not influence his agreement on others. And Sakharov had good reason to agree with Teller—his personal experience with Soviet leaders and his professional knowledge of strategic weapons.

He assumed that his American colleagues worked on the same combination of professional and moral factors. What he did not have was the sense of the "Promethean" sin of science that brought people nuclear fire. In November 1947 Oppenheimer publicly blamed his profession: "The physicists have known sin; and this is a knowledge which they cannot lose."[46] This statement misled people in the humanities for a long time. In 1988, two years after the catastrophe in Chernobyl, the writer Ales Adamovich questioned Sakharov about the "Oppenheimer complex," the guilt complex of physicists, and could not believe his ears when Sakharov told him he did not have it:

ADAMOVICH: The curse of lies hangs over atomic energy. I suddenly remembered that you are the father of our hydrogen bomb. How do you deal with that?

SAKHAROV: I'm not the only father, of course. It was collective work, but no less terrible for that. Back then we were convinced that the creation of first the atomic bomb (I did not participate in that) and then the thermonuclear was necessary for world balance so that our country could develop in peace and quiet without being under the pressure of overwhelming superiority of the other side. To this day, I cannot rule that out. We—I include the Americans in this—created a weapon that gave humanity a peaceful breather. It is still continuing. But I am convinced that this break is not indefinite. If the nuclear confrontation continues on the monstrous level it has reached, no "word of honor" will help.

ADAMOVICH: We nonscientists have the illusion that physicists must suffer from the Oppenheimer complex, the guilt syndrome. Is that so or not?

SAKHAROV: That is an illusion. We console ourselves with the fact that we are reducing the possibility of war.[47]

Sakharov, like other physicists with humanitarian concerns, had a sense of professional and moral responsibility that prompted him to explain that nuclear weapons were not only a new type of weapon but that nuclear war was the greatest threat for humanity. By the time he had this conversation with Adamovich, Sakharov had been saying for twenty years that science by itself could not avert war or make life better—people had to make the effort. He did not spare his own efforts. And, amazingly, he did not make moral compromises, even though the art of compromise is the usual tool of the practical politician.

An Impractical Politician?

It is all the more astonishing that Sakharov's impractical policy, coordinating only with his own moral evaluation, helped bring about a new social reality just as effectively as the actions of very practical politicians.

The policy of glasnost, proclaimed by Gorbachev in 1985, was accepted wholeheartedly by Sakharov, even though he was incarcerated in a hospital and being force-fed. Eighteen years before that, he finished his first political essay with these words: "With this article the author addresses the leadership of our country and all its citizens as well as all people of goodwill throughout the world. The author is aware of the controversial character of many of his statements. His purpose is open, frank discussion under conditions of publicity."[48]

When he was allowed to take advantage of glasnost (Russian for publicity and openness), his political career was not hampered by his lack of oratorical skills or concerns about his "image." After seven years of exile and isolation, the sixty-eight-year-old Sakharov spent the last seven months of his life as a parliamentarian. Russians who had just been given the vote understood the moral position of his speeches. Many even subscribed to the illusion that the time had come for moral politicians.

Practical politicians, of course, had a practical view of that moral politician and tried to use his international and domestic authority for their own goals—social and personal, noble and careerist.

In January 1987, two weeks after Sakharov's exile was over, Minister of Foreign Affairs Eduard Shevardnadze, a member of the Politburo, sent Gorbachev a long memorandum on Sakharov's political views. The resumé of his works of the 1968–1975 period was accompanied by a laconic remark: "Some aspects of his statements deserve attention."[49] Thus, the leaders of the country obtained an official opportunity to study Sakharov's reflections, twenty years later.

The rest of the country had to wait almost a year to learn Sakharov's views without a filter—his first articles and interview began appearing in late 1987.[50] But his free communication with the country began only in 1989, in the course of the first elections in Soviet history that actually presented a real choice.

The country and the world watched the work of the Congress of People's Deputies that opened on May 25. Its sessions were broadcast live, and for a few weeks the entire country was transfixed by the congress. All work was put on hold. Televisions were on in every office, restaurant, and home. People gathered on the street around parked cars with the radio blaring, so as not to miss anything while away from the TV.

People's Deputy Sakharov addressed the most fundamental problems of perestroika—the restructuring of the state. On the first day of the Congress he spoke out against the manipulations directed at returning power to the old party and state *apparat*. On the final day he proposed passing a Decree on Power,

which began with the repeal of Article 6 of the Constitution, which proclaimed the "leading and directing" role of the Communist Party in society. A crossed-out number 6 was a very popular poster, appearing in every political rally of those heady days. The article was repealed a year later.

The democratically oriented deputies formed the Interregional Group, and one of its five cochairmen was Sakharov. It was a minority of the delegates, and the minority within it consisted of people who fully sensed their responsibility for the country's future.

As a "convinced evolutionary," Sakharov knew that a social explosion was a natural phenomenon that was little dependent on concrete rabble rousers. Water comes to a boil at a certain temperature, even if it is thoroughly distilled to clean it of "rabble rousers." He wanted to avoid a social explosion at all costs, and all his efforts were directed at reconstructing the boiler of the Soviet empire before the contents came to a boil, blowing it up. But the Central Committee boiler room felt that the lid had to be kept on tighter and things would cool off on their own.

One of the stokers, the last Soviet chief of the KGB, Vladimir Kryuchkov, sent Gorbachev a secret report marked "Top Importance" on December 8, 1989 "On the Political Activity of A. D. Sakharov." He knew a lot about the activity of Ascetic and Vixen, the code names used for Sakharov and Bonner in the 583 volumes of material on them. But the material must have lost its import in the perestroika era, and the files were burned, the last seven volumes on September 6, 1989.[51] The factual evidence was accurate, and the KGB chief did not hide his attitude toward the facts:

The election of A. Sakharov as People's Deputy of the USSR was perceived by his allies in our country and aboard as a change of status from "lone human rights activist" to one of the leaders of the legal opposition. This, in particular, has given him the opportunity not only to propagandize personally his ideological schemas but to try to bring them to life through the other members of the Interregional Group of deputies.

In regard to the national question, A. Sakharov starts from "the need to destroy the imperial," in his words, national-constitutional structure of the USSR, which is an "instrument for oppressing other peoples." In their place he proposes an association of independent republics on the basis of a new union agreement . . . He considers the CPSU's platform on the national issue "absolute madness" and that it will inexorably lead to the departure of certain republics from the USSR . . .

Without being personally an organizing link in the activities of the radical deputies and members of non-formal associations, A. Sakharov has become their banner, a moral symbol and author of many political initiatives. Thus, he was one of the first to promote the idea of repealing Article 6 of the Constitution of the USSR, which subsequently became a key demand of the radicals. His "Decree on Power" played a role in the political platform of the Interregional Group and in many voter clubs . . .

Bearing in mind Sakharov's ambitions, and the excessive attention to him in the mass media, we should expect him to continue trying to play the role of "generator of opposition ideas."[52]

The man who signed this report showed his own ambitions eighteen months later by participating in an attempt to overthrow the government, a last bit of the Communist Party's "absolute madness" that blew up the Soviet boiler. One might suspect the KGB chief of exaggerating Sakharov's role, for he certainly did not seem like an orator or hero.

What Sakharov's life looked like to a friend from abroad is described by Nina Bouis, the translator of this book, who was one of the first Westerners to see Sakharov after his return to Moscow from exile in Gorky.[53]

I had come to Moscow to interpret for friends from the Scientists' Institute for Public Information, who were going to interview him. I had already met Elena Bonner in the US after her heart surgery and had translated her book, *Alone Together*. I had also been the translator of Sakharov's 1974 essay on the world fifty years later then for *Saturday Review*. We went to the apartment on Chkalov Street the day that the German amateur pilot Mathias Rust landed his Cessna on Red Square (May 28, 1987), bypassing the border guards and creating a different sort of major disruption of the old Soviet order . . . As we were leaving, Sakharov remembered that he had a message for George Soros, who wanted to open a foundation to further democratization and open society in the Soviet Union. Soros felt that Sakharov's return to Moscow was a bellwether signal that Gorbachev's plans for perestroika were serious.

"Tell him that permission has been granted for him to get a visa and to start a foundation here. But it will be a waste of time—the KGB won't let him do anything worthwhile." It was probably the only inaccurate prediction I ever heard Sakharov make, because the Soros network of foundations changed the lives of thousands of people in the former Soviet Union . . .

Sakharov was unusually prescient, sensitive to insignificant-seeming actions that were clues to behavior to come. He was the first person to criticize Gorbachev—when the world and America in particular were in the grip of Gorbymania. And this was before government troops fired on nationalist demonstrators in the Baltics and Georgia . . .

He devoted a lot of time to his public life. I attended many of the meetings he chaired, for those were heady times in the Soviet Union, you could feel things changing. He was cochairman of Memorial, a group dedicated to restoring the memory of the victims of Stalinism; Moscow Tribune, a club for the capital's intelligentsia who were pro-reform but not dissidents; and the Interregional Group of People's Deputies, the democratic opposition in the Congress. I would sometimes feel frustrated on his behalf that he spent so much time sitting on stages while other people rambled on— but I would remind myself that he was adding gravitas and significance to the proceedings, and in fact, by his very presence, was protecting the speakers.

Even People's Deputies needed protection, because the changes that were occurring in the Soviet Union developed at different rates in different areas. Yuri Afanasyev, one of the most radical movers of perestroika, ended up in court because the local newspaper of his constituency printed a speech he made in Congress, as it was required to do. The paper was then shut down by the local Communist Party administration for publishing "seditious material," which had been broadcast live on national television and radio. Sakharov appeared in court as a character witness for his cochairman of the Interregional Group and Memorial. The trial got a lot of media attention—much more than the dissident trials Sakharov attended in the pre-glasnost years.

Sakharov needed protection, too, from people seeking his help, his signature, his advice, his attention. The only quiet time at the apartment was around 2 A.M. Usually, there were callers all day and evening. Sakharov was perceived as the man who would right every wrong. Since Sakharov was unfailingly polite to everyone who came to him for help, it fell to Bonner to be the gatekeeper and to weed out overzealous journalists, publicity-seekers, KGB plants, and nasty callers. It was an exhausting life, and it gave them very little time alone together. They sometimes joked about missing their life in exile, when they had much more time for work and reading and each other.

Their love was visible—even in a crowded room they sought each other's gaze. When they were with friends at home, they cuddled as they watched the news on TV. Bonner watched his diet, making sure he had his favorite fresh farmer cheese and sour cream, fruit, and vegetables every day. He always did the dishes, claiming to enjoy the simple chore that allowed him to think about other things while he did it. I remember cozy evenings in the kitchen with Bonner and her mother, Ruf, chatting over tea, while they smoked up a storm and he did the washing up. Talk ranged from literature, family, and gardening to politics and human rights. As soon as we got onto gossip, he usually excused himself, pleading work.

Once he was elected a People's Deputy to the Soviet Congress, Sakharov became extremely interested in how things worked in the U.S. Congress. He said that there should be some way to know how deputies voted, so that people could base their own votes on their representatives' behavior. My husband told him about the Congressional Record, and Sakharov immediately went about trying to institute a similar publication in Russia.

The first time he went abroad, to the United States as part of a delegation from the International Foundation for the Survival and Development of Humanity, I took on a bit of the protection duty. Bonner wasn't allowed to accompany him—she had to remain as hostage, even in the more relaxed days of perestroika—and she asked me to help. On the plane, an American television crew kept barging into the first-class cabin and filming him, while he tried to chat with Yevgeny Velikhov, his cochair and traveling companion, and eat his meal. I asked Andrei Dmitrievich if he had given them permission to film him—naturally not, but he had been too polite to complain. The stewards made the press leave the cabin.

When Sakharov spoke at the New York Academy of Sciences, the auditorium was packed. Again, he seemed unfazed by the lights and hubbub, meeting personally with dozens of the people who had agitated in the West for his release from the torture of force feeding and from his exile in Gorky. He addressed the crowd calmly but with passion. When he was in Washington, DC, Sakharov asked to be taken to the little-known Einstein memorial.

The fall and winter of 1989 moved in a whirlwind: there were rallies, meetings, human rights actions, battles in the Congress over the war in Afghanistan and a new Constitution, and Sakharov was involved in all of them. Things got very ugly, because even though Gorbachev had made it possible for Sakharov to return to Moscow, he did not want to hear what Sakharov had to say. Gorbachev chaired most of the sessions of Congress and he began to treat him rudely, cutting off his microphone before he could finish his statements.

Sakharov persevered. He came home late from Congress on December 13, and told his wife that he was going to bed early, because he was tired, and he had to fight again tomorrow.

Early on the morning of December 14, 1989, we were awakened by a telephone call. Sakharov had passed away in the night. We rushed over to the apartment, where we said a prayer and bid him a farewell. It was still very early, and only a few people were there with Elena Bonner. As we left, we ran into a Japanese journalist on the landing, who wanted to know if it was true. It was the beginning of enormous press coverage all over the world . . .

The day of the funeral, 18 December, was terribly cold, with falling sleet, but it did not stop people from coming to say farewell. Official estimates were of a hundred thousand people . . .[54]

The Meaning of Life

The entire question of the meaning of life or the meaning of history is considered meaningless by some. Steven Weinberg, the Nobel laureate theoretical physicist, proclaimed that pointlessness:

> It is almost irresistible for humans to believe that we have some special relation to the universe, that human life is not just a more-or-less farcical outcome of a chain of accidents reaching back to the first three minutes [of the universe], but that we were somehow built in from the beginning . . . The more the universe seems comprehensible, the more it also seems pointless.[55]

But Sakharov had a sense of the meaning of his life. His freedom of thought and trust in his intuition took him far beyond his colleagues in these considerations. His generation of physicists, like Tamm's, managed easily without religious concepts. The son of a physicist, grandson of a lawyer, and great-grandson of a priest, Sakharov described his attitude toward religion in just a few sentences.

My mother was devout. She taught me to say my bedtime prayers ("Our Father," "Hail, Mary") and took me to confession and communion . . . The majority of my other relatives were religious, too. My father, apparently, was not, but I do not remember him ever talking about it. When I was about thirteen, I decided that I did not believe—under the influence of the general atmosphere of life and not without Father's influence, albeit covert. I stopped saying my prayers and went to church very rarely, and as one who no longer believed. Mother was very upset but she did not insist, I don't recall any conversations on the topic.

Now I do not know, deep in my heart, what my actual position is: I do not believe in any dogmas, I do not like official Churches (particularly those that are closely bound with the state or are distinguished primarily by their rituals or fanaticism and intolerance). At the same time, I cannot imagine the Universe and human life without some meaningful element, without a source of spiritual "warmth," lying outside matter and its laws. Probably that feeling could be called religious.

Feeling is a very personal concept. Apparently Sakharov did not need to delve any deeper into it for a long time, remaining a totally nonmilitant atheist with an open heart. When he was thirty, at the start of his life at the Installation, he often visited a profoundly devout mathematician for a chat, and those conversations must have been of interest to him. A few years after that when Sakharov was forced to listen to Marshal Nedelin's joke, which struck him like a whip, he was left with a scar—the story was crude and "semiblasphemous, which was also unpleasant."

In his human rights work he often came across examples of crude suppression of religious freedom, which he perceived as part of the general freedom of convictions: "If I lived in a clerical state, I would probably defend atheism and persecuted followers of other faiths and heretics!"

He considered "religious faith a purely inner, intimate, and free choice, just like atheism." This did not mean that he always kept silent about his inner thoughts. In September 1989, Andrei Sakharov addressed the French Physics Society in Lyons. He titled his lecture "Science and Freedom," two very familiar elements for him.

In science he saw the most important part of civilization. It was in science that he found the real taste of freedom that was unavailable in other areas of Soviet life. By his nature and circumstance, Sakharov had inner freedom. Perhaps that is why he reacted so strongly to any kind of serfdom in others and devoted as much of his life as he did to science in the defense of other people's rights.

He had come to France from a country that was parting with its un-freedom in front of the entire world. Parting with it by having to overcome the resistance of the leadership and the inertia of the masses. Sakharov got it from both sides once he became an "official" politician, a People's Deputy, in the spring of 1989. When he made speeches at the Congress, he was shushed from the presidium and booed by his fellow congressmen.

So at the University of Lyons, he felt particularly free—he was surrounded by colleagues, bound more closely by science than others often are by language or homeland. He did not write a speech. He spoke freely, thinking aloud. It was even a help that his interpreter had to interrupt him and translate a few sentences at a time. He always spoke slowly, coming up with a sentence and then rethinking it, so the enforced pauses for the interpreter to speak gave him time to think through his next statement.

His lecture, which was taped, is probably one of the freest expressions of his thoughts. It is also one of the last. He died three months later. But in the Lyons lecture he was summing up not his own life but the age in which he lived. "In a little more than ten years the twentieth century will end, and we must attempt to evaluate what we will call it, what was its greatest characteristic."[56]

Sakharov related how recently with thousands of his countrymen he stood at a common grave in which the victims of Stalin's terror had been re-interred, and how clergymen of three religions officiated at the funeral. The age of unparalleled terror? The age of world wars? The age of genocide?

Sakharov named a different characteristic as the most important: "The twentieth century is the age of science, of its greatest forward push." He described his vision of the scientific portrait of the world and the three greatest goals of science, which were intertwined: science as the human mind seeking knowledge, as the most powerful productive force, and as the force uniting humanity. He thought aloud about twentieth-century physics and surprised his audience with this worldview prediction:

> Einstein became, and it was no accident, the embodiment of the spirit of the new physics and the new attitude of physics toward society. In Einstein's statements and letters you often find this parallel: God and Nature. This reflects his thinking and the thinking of very many people in science. In the Renaissance, in the eighteenth and nineteenth centuries it seemed that religious thought and scientific thought contradicted each other, mutually exclusive. This juxtaposition was historically justified, it reflected a specific period of social development. But I believe that it has a profound synthetic resolution in the next stage of the development of human consciousness. My profound feeling (not even conviction—the word "conviction" is probably wrong here) is that there is an inner reason in Nature, that Nature as a whole makes sense. I am speaking here of intimate, profound things, but when you are dealing with summing up and with what you want to pass on to people, then it is necessary to speak about this as well.

While feeling the need to share his sense of the world, he did not believe it was necessary to bring it to a logical and consistent form. In some way he combined the idea of the course of events over which man had no control and the sense that a man's actions and his free choice somehow affect that course of events, that a person somehow earns and deserves his fate.

These seemingly incompatible ideas are brought together by the concept of "meaning"—the meaning of existence and of fate. God's only job is to be the guarantor of that meaning, "despite all the apparent pointlessness," in Sakharov's words.[57]

Hope is based on the fact that "fortunately, the future is unpredictable and also—because of quantum effects—uncertain." The consolation he offered to a physicist and human rights activist in a letter written during his exile in Gorky.[58]

Sakharov's conviction seems to contradict the conviction that he wrote down in his quantum-cosmological bet with Frank-Kamenetsky in 1956. The joking bet, entitled "The Problem of Quantum Determinism," was mentioned in the chapter "The Physics of the Universe," (p. 237) with the promise to reveal how much truth there was in the joke. Sakharov had bet that "there is a singular solution of the Schrodinger equation that describes the Universe for all degrees of freedom for all times." For those readers who do not feel at home in quantum physics, the author offers this translation: The Being who can follow everything that happens in the universe can also solve the main equation for it, and as a result learn all the possible variants for the fate of the universe and then choose the most appropriate, that is, the most meaningful one.

Sakharov bears no responsibility for this translation and did not seem to need one. Nor did he seem to need either quantum or classical theology—in any case, a theology that knows precisely how history will end. This was reflected in his attitude toward the duet of intuition and logic.

He liked the words of the Polish philosopher Leszek Kolakowski: "Inconsistency is simply a secret awareness of the contradictions of this world . . . a permanent feeling of possible personal error, or if not that, then of the possibility that one's antagonist is right."[59] The word "inconsistency" did not suit Sakharov, but he could not find another to combine "self-critical dynamism with the presence of certain valuable 'invariants'"—or, in the author's translation from Sakharovese, self-critical reason and moral intuition. He combined them consistently in his own life and felt that it was enough for the happy fate of humanity to strive "to embody the demands of Reason and create a life worthy of ourselves and the Goals we can only indistinctly conjecture."[60] He hoped that then, "having overcome dangers and having achieved the great development in all spheres of life, humanity will manage to keep the human in man."

A Russian proverb holds that "a holy place does not stay empty." The place held by Andrei Sakharov in Russia, however, has been empty since his death in 1989. His countrymen ask themselves the unanswerable question: If he were still alive, would they have averted the major disasters that befell Russia: The putsch of 1991, the dissolution of the Soviet Union, the war in Chechnya, the economic crisis, the way "democrat" turned into a dirty word? Another Russian proverb supplies the answer: "One man in a field is no warrior." But Sakharov proved this saying wrong many times.

A new Russia has to be built in a much more open world without him. Can his experiences in a closed world be of use? When he received the Einstein Peace Prize in November 1988, Sakharov began his speech by stating that the social role of science has grown, but the role remains just as contradictory as public life itself. He said Einstein's lesson was "to resolutely hold, amidst all these contradictions, to moral criteria, perhaps making occasional mistakes, but be ready to subordinate one's actions to these common human criteria of morality."[61]

This could be called Sakharov's lesson, too. He followed that simple principle in very complicated circumstances. Sakharov's greatest lesson for the world was in demonstrating the indissoluble relationship among peace, scientific progress, and human rights.

CHRONOLOGY

1860s	Born: Maria Domukhovskaya, Andrei Sakharov's grandmother, "the soul of the house" where he grew up. Born: Pyotr Lebedev, the first world-class Russian physicist. The word *intelligentsia* is invented in Russia. Barbed wire is invented in America for agricultural use. (A nonagricultural application was found for barbed wire throughout the world, particularly in the Soviet Union.)
1911, February	Pyotr Lebedev, Vladimir Vernadsky, and Kliment Timiryazev, along with a large group of professors, walk out of Moscow University in protest of the government's police tactics.
1916, December	The Physical Institute on Miusskaya Square in Moscow is completed. Its director, until 1931, is Pyotr Lazarev.
1921, May 21	Andrei Sakharov is born.
1925	Leonid Mandelshtam begins work at Moscow University.
1930	Boris Gessen is appointed director of the Physics Institute at Moscow State University.
1931, March	Academician Pyotr Lazarev is arrested and exiled to Sverdlovsk. The mysterious "Physico-Chemical Institute for Special Assignments" is moved into the building of the Physical Institute on Miusskaya Square.
1931, December	Georgi Gamow sends a report suggesting that the Physics-Mathematical Institute of the Academy of Sciences in Leningrad be divided into two indepen-

dent institutions. As a result, a few months later the Physical Institute of the Academy of Sciences (FIAN) is formed, headed by Academician Sergei Vavilov.

1934   FIAN is moved to Moscow, to the building on Miusskaya Square. At the end of the year it is named after Lebedev.

1938   Sakharov enters Moscow University.

1942   Sakharov graduates from Moscow University in the summer (in evacuation in Ashkhabad because of the war). He begins work at the Ulyanovsk Cartridge Factory, where he meets Klava Vikhireva (he marries her in 1943).

1943, spring   The Soviet atomic project begins under the direction of Igor Kurchatov.

1945, February   Sakharov becomes a graduate student under Igor Tamm in the theoretical department of FIAN. His first child, Tatiana, is born.

1945, July 16   The first atomic bomb is tested in the United States.

1945, July 17   Sergei Vavilov becomes president of the Academy of Sciences.

1945, July 24   During the Potsdam Conference, President Harry S. Truman tells Stalin that the United States has "a new weapon of unusual destructive force," without specifying that it is the atomic bomb.

1945, August 6   The atomic bomb is dropped on Hiroshima, and three days later on Nagasaki.

1945, August 20   The Special Committee is formed, headed by Beria. Full-scale work starts on the atomic project in the USSR.

1946, March   Winston Churchill's Iron Curtain speech in Fulton, Missouri, becomes the declaration of the Cold War.

1947, November   Sakharov successfully defends his candidate of science thesis.

1948   Sakharov conducts his pioneering work on muon catalysis.

1948, March   Klaus Fuchs passes information on American hydrogen bomb work.

1948, June   Government resolution on creation within FIAN of an auxiliary group, headed by Tamm, to help Zeldovich's group at the Institute of Chemical Physics work on the hydrogen bomb

1948, August   The session of the All-Union Academy of Agricultural Sciences: Lysenko pogrom of Soviet biology

1948, December	Preparations begin for the All-Union Conference of Physicists.
1949, January	Sakharov's report on Sloyka, the thermonuclear bomb project
1949, March	Cancellation of the All-Union Conference of Physicists
1949, June	Sakharov's first visit to the Installation
1949, July	The Sakharovs' daughter Lubov is born.
1949, August 29	The first Soviet test of the atomic bomb
1950, January 31	Truman's directive to create the H-bomb
1950, February 26	Soviet decision to expand work on the H-bomb
1950, March	Sakharov moves to the Installation.
1950–1952	In parallel with work on thermonuclear weapons, Sakharov proposes the principle of magnetic thermal insulation of plasma for a controlled thermonuclear reaction and the principle of obtaining superstrong magnetic fields in the explosion-magnetic generator.
1952, autumn	Sakharov's first political action. With ten other leading physicists in the atomic project, he signs a letter in support of the publication of Vladimir Fock's article "Against Ignorant Criticism of Contemporary Physics Theories" in response to the June 1952 newspaper article "Against Reactionary Eisensteinism in Physics."
1952, November 1	American thermonuclear test "Mike." On December 2, 1952, Beria mentioned this test in his letter directing that creation of Sloyka "is of the first priority. Judging from certain information that has reached us, there have been tests related to this type of device."
1953, March 5	Death of Stalin
1953, June	Sakharov defends his doctoral dissertation (at the Installation). Beria is arrested.
1953, August 12	The first Soviet test of the thermonuclear bomb Sloyka (Joe-4)
1953, October	Sakharov is elected to the Academy of Sciences.
1954, January	Sakharov is awarded his first star of Hero of Socialist Labor and the Stalin Prize.
1954, January 14	Zeldovich and Sakharov report on "atomic compression of the super-device."
1954, March 1	American thermonuclear test "Bravo." Its 15 megatons made it 2.5 times more powerful than calculated.

1954, spring	Birth of the "Third Idea" (the Soviet version of radiation implosion) and start of work on a full-scale thermonuclear bomb
1955, November 22	The first airdrop test of the thermonuclear bomb. Marshal Nedelin gives Sakharov a lesson in political literacy.
1956, September	Sakharov is awarded his second star of Hero of Socialist Labor and the Lenin Prize.
1957, June	America announces the creation of a "clean" nuclear bomb.
1957, August	The Sakharovs' son Dmitri is born.
1958, May	Sakharov's articles on the radioactive danger of nuclear testing
1961, July	At a meeting with the atomic project scientists, Khrushchev announces his decision to renew nuclear testing. Sakharov openly disagrees, angering the country's leader.
1961, October 30	The most powerful bomb in history (developed under Sakharov) is tested.
1962, March	Sakharov is awarded his third Hero of Socialist Labor star.
1962, September	"The most terrible lesson"—Sakharov could not prevent an unnecessary test.
1962, October	Cuban missile crisis
1963, August	Moscow Test Ban Treaty
1964, June	Sakharov speaks out against Lysenko at the Academy of Sciences elections.
1965	Sakharov's first work on cosmology is published.
1966	Sakharov's first popular science articles
1966, February	Letter to the leaders of the country against the rehabilitation of Stalin. Among the 25 signers are the physicists Artsimovich, Kapitsa, Leontovich, Sakharov, and Tamm.
1966, June	"Prognosis of the Prospects of the Development of Science" is published in the anthology *The Future of Science.*
1966, September	Article on the baryon asymmetry of the universe
1966, December 5	Sakharov joins a demonstration at Pushkin Square in defense of the constitution.
1967, February	Sakharov's first letter in defense of dissidents
1967, July 21	In a secret letter to the Central Committee, Sakharov explains the need to "catch Americans at their word" and accept their proposal "for the rejection of the USA and the USSR of antiballistic

                        missile (ABM) systems defense against a mass attack of a powerful enemy, while retaining those works that are necessary for defense against small-scale aggression."

1967, August	Sakharov's article on the vacuum nature of gravitation is written.
1968, February	Sakharov begins work on "Reflections on Progress, Peaceful Coexistence, and Intellectual Freedom."
1968, May	Sakharov releases "Reflections" to *samizdat* publication. In late May the KGB sends it to the Politburo. In June, the manuscript is sent to Brezhnev.
1968, June 18	At Tamm's request, Sakharov reads his lecture "Evolution of Quantum Theory" at the Academy of Sciences ceremony bestowing a gold medal on Tamm.
1968, July 1	The U.S. president announces the USSR's agreement to start negotiations on arms limitation.
1968, July	"Reflections on Progress, Peaceful Coexistence, and Intellectual Freedom" is published in the West.
1968, July 10	Sakharov's last day in his office at the Installation
1968, August 26	Sakharov's first meeting with Solzhenitsyn
1969, March 8	After a long illness, Sakharov's wife Klava dies.
1969, May	Sakharov is officially fired from the Ministry of Medium Machine Building. In July he becomes senior scientific fellow of the theoretical department at FIAN.
1969, August	Sakharov's last trip to the Installation. He donates his huge savings (equivalent to thirty times his annual salary) "for the construction of a cancer hospital, the fund for children's services at the Installation, and to the International Red Cross to help victims of catastrophes and the starving."
1970, March	The memorandum by Sakharov, Turchin, and Medvedev is sent to the Soviet leaders. It explained that the democratization of the country was necessary for its scientific, technological, and economic progress.
1970, April 20	The KGB chief appeals to the Central Committee for approval to bug Sakharov's apartment.
1970, May–June	Sakharov helps defend Pyotr Grigorenko and Zhores Medvedev and enters the dissident milieu.
1970, autumn	Sakharov attends his first human rights trial. There he meets Elena Bonner (they become a couple a year later).

1971, September	Sakharov appeals to the Supreme Soviet on freedom of choice of where to live.
1972, October	Sakharov's first interview with a Western correspondent
1973, August 21	Sakharov's first press conference
1973, August 23	In an interview with Western journalists, Sakharov says, "How can we speak of trust, if one of the sides is like a giant concentration camp?"
1973, August 29	Letter of 40 academicians to *Pravda*, the start of the newspaper campaign of the "people's wrath"
1973, September	Solzhenitsyn, winner of the Nobel Prize for Literature in 1970, proposes Sakharov for the Nobel Peace Prize. Lydia Chukovskaya's article "Wrath of the People" in defense of Sakharov and Solzhenitsyn is released in *samizdat* (she was expelled from the Writers' Union for this article).
1974	Sakharov gives the prize *Cino del Duca* he wins in France to the Fund for Aid to Children of Political Prisoners.
1974, January	Sakharov reads *Gulag Archipelago*, just published in the West. "We knew the innumerable facts of mass cruelty and illegality in the world of the Gulag, and had an idea of the scale of those crimes. And still, even for us, Solzhenitsyn's book was a shock."
1974, February	Solzhenitsyn is arrested and expelled from the USSR.
1974, April	Sakharov responds to Solzhenitsyn's Letter to the Leaders, published in March. The ideological differences between them, for all their mutual respect, are evident.
1974, June	Sakharov's first (6-day) hunger strike to draw attention to the plight of political prisoners
1975, October 9	The Nobel Peace Prize award to Sakharov is announced.
1975, December 11	Sakharov's Nobel Lecture, "Peace, Progress, and Human Rights," is read in Oslo by Elena Bonner. Andrei Sakharov listens in on the radio in Vilnius, where he is attending the trial of a human rights activist.
1977, January 20	Sakharov sends a letter to President Jimmy Carter concerning the human rights situation in the USSR and Eastern Europe. (Carter responds in an open letter of February 5, promising to seek the release of prisoners of conscience.)

1978, summer	Sakharov begins writing his memoirs.
1980, January	Sakharov publicly criticizes the Soviet invasion of Afghanistan. On January 22, on the way to a seminar at FIAN, Sakharov is arrested and exiled to Gorky.
1983, May 18	At the direction of the Congress, President Ronald Reagan designated May 21, 1983, as "National Andrei Sakharov Day": "Today, we call upon the Soviet leaders to give Andrei Sakharov his freedom. The world needs his learning, his wisdom, his nobility. In observing National Andrei Sakharov Day, May 21st, we urge the American people and all the peoples of the world to speak for him, for in doing so we speak for ourselves, for all mankind, and for all that is good and noble in the human spirit."
1983, June	Sakharov's article "The Danger of Thermonuclear War" is published in the West, and a new campaign is begun against Sakharov and Bonner in the USSR.
1984, May 2	Elena Bonner is arrested. Sakharov begins a hunger strike, demanding permission for his wife to travel to the United States for heart surgery. On May 7 he is forcibly hospitalized and force-fed.
1984, August 10	Elena Bonner is sentenced to 5 years of exile in Gorky.
1984, September 8	Sakharov is released from the hospital, where he was held in isolation for four months.
1985	Franklin Institute Award to Sakharov
1985, April 11	Gorbachev is the new General Secretary of the CPSU.
1985, April 16	Sakharov starts new hunger strike for his wife to travel aboard for medical treatment. On April 21 he is taken to a hospital and force-fed.
1985, October 23	Sakharov leaves the hospital. October 25, Bonner is given permission to travel. In November she flies to the United States, where she has heart surgery. She returns to Gorky in June 1986.
1985, December	The European Parliament establishes the Sakharov Prize for Freedom of Thought, to be given annually for outstanding contributions to human rights. The first prize is given in 1988 to Nelson Mandela and Anatoly Marchenko (posthumously).
1986, December 16	Gorbachev calls Sakharov and tells him that he and his wife may return to Moscow.
1986, December 23	Sakharov returns to Moscow.
1987, February	Sakharov speaks at the Moscow Forum for a Non-Nuclear World and Human Survival.

1987, November	Sakharov's first interview in the Soviet press
1988, October	Sakharov is elected member of the Presidium of the Academy of Sciences.
1988, November	Sakharov's first trip abroad. In the United States meets Edward Teller and receives the Einstein Peace Prize.
1989, January–March	First semi-free elections in Soviet history for Congress of People's Deputies. Sakharov is elected deputy from the Academy of Sciences.
1989, May 25–June 9	Sakharov participates in the First Congress of People's Deputies of the USSR. At the final session he calls on them to pass "Decree on Power," which would, among other things, repeal the one-party political system.
1989, July	Sakharov is elected cochairman of the Interregional Group—the democratic opposition in the Congress.
1989, September 6	The KGB burns the last (?) seven volumes of materials gathered on "Ascetic" and "Vixen" (Sakharov and Bonner). Around 600 volumes had been destroyed previously.
1989, September 27	Sakharov addresses the French Physics Society in Lyons with his lecture "Science and Freedom."
1989, autumn	Sakharov finishes his memoirs, works on his draft of the constitution, and works in the Supreme Soviet.
1989, December 14	Andrei Sakharov dies.
1990	Sakharov's *Memoirs* are published. In Russia the Sakharov Prize for Civic Courage in a Writer is established. The first recipient is Lydia Chukovskaya.

NOTES

Prologue

1. P. N. Lebedev, "Opytnoe issledovanie svetovogo davleniya," *Zhurnal Russkago fiziko-khimicheskago obshchestva*, 1901, t. 33, vyp. 7, otd. 1; *Annalen der Physik*, 1901, Bd. 6, S. 433–458.

2. P. N. Lebedev, *Nauchnaya perepiska*, Moscow: Nauka, 1990.

3. Ibid.

4. K. A. Timiryazev, *Smert' Lebedeva*, Sobr. soch., Moscow: OGIZ, 1939.

5. This sentence is taken from one of the most famous books about the beginning of the nuclear era. R. Jungk, *Brighter than a Thousand Suns*, New York: Harcourt Brace, 1958.

6. V. P. Kartsev, *Vsegda molodaya fizika*, Moscow: Sovetskaya Rossiya, 1983.

7. Lebedev, *Nauchnaya perepiska*.

8. K. Chukovsky, *Sobranie sochinenii*, Vol. 4, Moscow: Terra-Knizny klub, 2001.

9. F. Tyutchev, "Umom Rossiyu ne ponyat . . ." (1866), *Izbrannoe*, Rostov-na-Donu: Feniks, 1996.

10. V. G. Korolenko, *Voina perom*, Moscow: Sovetskaya Rossiya, 1988.

11. Ibid.

12. When Maxim Gorky was elected to the Russian Academy of Sciences in 1902, his election was annulled soon afterward for political reasons, and Korolenko and Chekhov quit the Academy in protest.

13. M. N. Gernet, O. B. Goldovskii, and I. N. Sakharov, eds. *Protiv smertnoi kazni*, Moscow, 1907.

14. Timiryazev, *Smert' Lebedeva*.

Chapter 1

1. These documents are collected in the book: E. G. Bonner, *Vol'nye zametki k rodoslovnoi Andreya Sakharova*, Moscow: Prava cheloveka, 1996. They are further quoted from this book.

2. P. D. Boborykin (1836–1921), Russian writer, Honorary Academician of the St. Petersburg Academy of Sciences (1900), introduced the word *intelligentsia* in the 1860s.

3. A. B. Goldenveizer, *Vblizi Tolstogo*, Moscow-Petrograd, 1923. Tolstoy wrote about listening to Goldenveizer playing the piano in his journal dozens of times.

4. V. I. Lenin, *Poln. sobr. soch.*, Vol. 38.

5. V. I. Vernadsky, *Ocherki i rechi*, Petrograd, 1922

6. V. I. Vernadsky, *Zhizneopisaniye. Izbrannye trudy.Vospominaniya sovremennikov. Suzhdeniya potomkov* (G. P. Aksyonov, comp.), Moscow: Sovremennik, 1993.

7. S. I. Vavilov, *Tridtsat' let sovetskoi nauki*, Moscow: Izd-vo AN SSR, 1947.

8. P. N. Lebedev, *Nauchnaya perepiska*, Moscow: Nauka, 1990.

9. A. K. Timiryazev, *Kinetichkaya teoriya materii*, Moscow: Uchpedgizdat, 1939.

10. Tsentral'nyi istoricheskii arkhiv Moskvy [Central Historical Archives of Moscow], 418-314-827.

11. G. A. Savina, *Napisano v podvalakh OGPU*, Vestnik Rossiiskoi Akademii Nauk, no. 5, 1995.

12. S. I. Vavilov, "Fizicheskii kabinet—Fizicheskaya laboratoriya—Fizicheskii institut AN SSSR," *Uspekhi Fizicheskikh Nauk*, no. 1, 1946.

13. D. Bedny, "Do atomov dobralis," *Pravda*, November 25, 1928.

14. Arkhiv RAN (Archives of the Russian Academy of Sciences) 2-1/1932–6. See G. Gorelik, "Georgi Gamow—zamestitel' direktora FIANa," *Priroda*, no. 8, 1993.

15. G. Gorelik and V. Frenkel, *Matvei Petrovich Bronshtein, 1906–1938*, Moscow: Nauka, 1990, p. 88. P. Kapitsa, Corresponding Member of the Soviet and British Royal Societies, worked in England at that time.

16. Arkhiv GRI (Archives of the State Radium Institute) 315-2-90.

17. Arkhiv RAN 204-1-4.

18. Ibid., 411-3-12-4.

19. S. I. Vavilov, "Fizicheskii institut," *Pravda*, November 5, 1934, p. 6.

Chapter 2

1. G. Landsberg's letter to L. Mandelshtam, July 18, 1924. Arkhiv RAN (Archives of the Russian Academy of Sciences) 1622-1-75.

2. His full title was Consultant of the Radiolaboratory of the Electrotechnical Trust of Weak-Current Plants.

3. *Akademik L. I. Mandelshtam. K 100-tiyu so dnya rozhdeniya*, Moscow: Nauka, 1979.

4. Ibid.

5. S. I. Vavilov, "Istoriya fizicheskogo instituta Moskovskogo Universiteta so vremen revolutsii," 1932 manuscript, Moscow, Library of Moscow University.

6. "Tamm v dnevnikakh i pis'makh," *Kapitsa. Tamm. Semenov. V ocherkakh i pis'makh*, Moscow: Vagrius-Priroda, 1998.

7. *Vospominaniya o I. Ye. Tamme*, Moscow: Izdat, 1995.

8. "Tamm v dnevnikakh i pis'makh."

9. Ibid.

10. Ibid.

11. Ibid.

12. Ibid.

13. Arkhiv RAN 411-3-308.

14. L. Mandelshtam's letters from Odessa to R. von Mises in Strasbourg, October 30, 1918, and in Berlin, March 14, 1922. Richard von Mises Papers, Harvard University Archives, HUG 4574.5.

15. I. I. Sobel'man, "K teorii rasseyaniya sveta v gazakh," *Uspekhi Fizicheskikh Nauk*, 2002, T172

16. L. Mandelshtam, "Ueber die Rauhigkeit frier Fluessigkeitsoberflaechen," *Annalen der Physik*, 1913, B 41, No. 8, S609–624.

17. *Akademik L. I. Mandelshtam.*

18. Ibid.

19. I. E. Tamm, *Sobranie nauchnykh trudov*, Moscow: Nauka, 1975.

20. *Akademik L. I. Mandelshtam.*

21. L. S. and L. I. Mandelshtam's letters to R. von Mises, January 24 and February 9, 1928. Richard von Mises Papers, Harvard University Archives, HUG 4574.5.

22. E. L. Feinberg interview, May 1, 1998; N. A. Belousova interview, December 8, 1998.

23. *Akademik L. I. Mandelshtam.*

24. *Sergei Ivanovich Vavilov. Ocherki i vospominaniya*, Moscow: Nauka, 1991.

25. *Vospominaniya o I. Ye. Tamme*, Moscow: Izdat, 1995.

26. L. I. Mandelshtam, *Polnoe sobranie trudov. T. 1–5*, Moscow: Izd-vo AN SSSR, 1950.

27. *Akademik L. I. Mandelshtam.*

28. L. Graham, "The Socio-Political Roots of Boris Hessen: Soviet Marxism and the History of Science," *Social Studies of Science*, Vol. 15, no. 4, 1985, pp. 705–722. Gessen's lecture "The Social and Economic Roots of Newton's 'Principia'" was published in English several times.

29. G. Gamow, *My World Line*, New York: Viking Press, 1970, p. 94.

30. It was Professor Mlodzievskii (S. M. Shapiro interview, September 30, 1992).

31. Especially because, in addition to articles in scientific journals, Friedmann published the popular book *Mir kak prostranstvo i vremya* [The World as Space and Time].

32. Moscow University Archives 46–1l-52. See G. Gorelik, "Tri marksizma v sovetskoy fizike 30–kh godov," *Priroda*, no. 5, 1993. See also G. Gorelik, "Moskva, fizika, 1937 god.," *VIET*, no. 1, 1992.

33. "Tamm v dnevnikakh i pis'makh."

34. Ibid.

35. B. M. Gessen, "Teoretiko-veroyatnostnoe obosnovaniye ergodicheskoi gipotezy," *Uspekhi Fizicheskikh Nauk*, no. 5, 1929.

36. A. B. Andreev, "Politicheskaya situatsiya na fizmate MGU v kontse 20-kh godov," *VIET*, no. 2, 1993.

37. "Tamm v dnevnikakh i pis'makh."

38. Maksimov exposed Gessen, along with the "hotbed of idealism in physics," in the journal *Uspekhi Fizicheskikh Nauk* [Successes of the Physical Sciences], where Gessen was an editor.

39. B. M. Gessen, *Efir. Bolshaya Sovetskaya Entsiklopedia*, T 65, 1931.

40. Maksimov saved a copy of this letter in his archive (Arkhiv RAN 1515-2-16). Whether the letter ever got to its addressee is unknown, but the letter's existence is confirmed by an entry in V. Vernadsky's journal dated February 14, 1932: "Gamov wrote a letter to Stalin in relation to being persecuted for 'world ether.'"

41. V. O. Yegorshin, "O polozhenii na fronte fiziki i zadachi Obshchestva fizikov-materialistov pri Komakademii," *Za markistsko-leninskoe estestvoznanie*, 1931, p. N1.

42. Arkhiv RAN 2–1a/1937–70.

43. E. L. Feinberg, *Epokha i lichnost. Fiziki. Ocherki i vospominaniya*, Moscow: Fizmatlit, 2003.

Chapter 3

1. A. S. Pushkin, "Ya pamyatnik sebe vozdvig nerukotvornyy . . ." (1836), *Sobranie sochineniy*, Moscow: Pravda, 1969.

2. M. L. Levin, *Zhizn', vospominaniya, tvorchestvo*, Nizhny Novgorod: IPF RAN, 1998.

3. V. Y. Kirpotin, *Nasledie Pushkina i kommunizm* [Pushkin's Legacy and Communism], Moscow: Khudozhestvenaya literatura, 1936.

4. A. S. Pushkin, "Iz Pindemonti" (1836), *Sobranie sochineniy*, Moscow: Pravda, 1969.

5. And government support was not such a terrible thing. Pushkin was published often during Soviet times, too, but at the beginning of the 1980s, when the first unlimited subscription to his three-volume collected works was announced, the edition ended up exceeding 10 million copies.

6. A. S. Pushkin, "O skolko nam otkrytiy chudnyh . . ." (1829), *Sobranie sochineniy*, Moscow: Pravda, 1969.

7. This little masterpiece by Pushkin was an echo of a scene in a three-act play by his contemporary, the Scots poet John Wilson (who is virtually forgotten today).

8. Quoted in Levin, *Zhizn', vospominaniya, tvorchestvo,*

9. Levin, *Zhizn', vospominaniya, tvorchestvo.*

10. Arkhiv RAN (Archives of the Russian Academy of Sciences) 2-2a/1937–70. See G. Gorelik, "Moskva, fizika, 1937 god.," *VIET*, no. 1, 1992.

11. B. Gessen's file in the KGB Archive (no. P-29017) suggests that he was arrested on August 21, 1936, charged with "until recently supporting personal ties with N. Karev and N. Lur'ye, who were arrested in the Trotskyite-Zinovievan Terrorist Center, and that he conducted c/r [counterrevolutionary] Trotskyite work."

12. Leonid Tamm, engineer-chemist, was arrested in the fall of 1936 and became a victim of one of the "open" trials; "Tamm's evidence" appeared in the newspaper.

13. See G. Gorelik, "Moya anti-sovetskaya deyatel'nost (Odin god iz zhizni L. D. Landau)," *Priroda*, no. 11, 1991; G. Gorelik, "The Top Secret Life of Lev Landau," *Scientific American*, August 1997.

14. E. L. Feinberg, *Epokha i lichnost. Fiziki. Ocherki i vospominaniya*, Moscow: Fizmatlit, 2003.

15. L. Chukovskaya, *Zapiski ob Anne Akhmatovoi*, Moscow: Soglasiye, 1997.

16. *Vospominaniya o I. Ye. Tamme*, Moscow: Izdat, 1995.

17. A. M. Yaglom interview, July 21, 1994.

18. L. N. Bell interview, August 30, 1994.

19. A. I. Solzhenitsyn, *Publitsistika*, Yaroslavl: Verkhne-Volzhskoye izdatel'stvo, 1995; D. Samoilov, *Pamyatnye zapiski*, Moscow: Mezhdunarodnye otnosheniya, 1995.

20. S. M. Shapiro interview, September 30, 1992.

21. *On mezhdu nami zhil . . . Vospominaniya o Sakharove* (B. L. Altshuler et al., eds.), Moscow: Praktika, 1996.

22. A. S. Pushkin, "K Chaadayevu" (1818), "Pro sebya" (1820). Both are in A. S. Pushkin, *Sobranie sochineniy*, Moscow: Pravda, 1969.

Chapter 4

1. I. N. Golovin, *I. V. Kurchatov: A Socialist-Realist Biography of the Soviet Nuclear Scientist* (W. H. Dougherty, trans.), Bloomington, IN: Selbstvarlag Press, 1968.

2. Quoted from M. S. Sominskii, *Abram Fedorovich Ioffe*, Moscow: Nauka, 1964.

3. A. F. Ioffe, "Sovetskaya fizika i 15–letie fiziko-tekhnicheskikh insititutov," *Izvestiya*, October 2, 1933.

4. From the open letter published in *Times*, November 27, 1948, and in Russian in *Britansky Soyuznik*, December 12, 1948.

5. Particularly harsh criticism came from British biologist and Nobel laureate Peter Medawar.

6. Journal of V. Vernadsky, March 12, 1938. His journals of 1932–1944 are kept in the Archives of the Russian Academy of Sciences [Arkhiv RAN 518–2, d. 17–24]. The 1938 journal is quoted from the publication by I. I. Mochalov, *Druzhba narodov*, nos. 2 and 3, 1991. See G. Gorelik, "Vernadsky i sovetskiy atomnyy proekt," *Znanie-Sila*, nos. 3 and 4, 1996.

7. Minuvshee, Vyp. 18, Moscow, 1995.

8. V. I. Vernadsky, "Iz pisem raznykh let," *Vestnik Rossiiskoi Akademii Nauk*, no. 5, 1990.

9. Minuvshee, Vyp. 18, Moscow, 1995.

10. In 1939, von Mises found refuge at Harvard University and published his book in the United States: *Von Mises, Richard, Kleines Lehrbuch des Positivismus: Einfeuhrung in die empiristische Wissenschaftsauffassung*, Chicago: The University of Chicago Press, 1939. His *Positivism: A Study in Human Understanding*, Cambridge, MA: Harvard University Press, 1951, deals with issues of law, morality, poetry, and religion, as well as science.

11. A potential auditor, Vitaly Ginzburg, titled his reminiscences of Mandelshtam "One Suggestion from Leonid Isaakovich Mandelshtam" (*O fizike i astrofizike*, Moscow: Kvantum, 1995). The advice was to postpone studying philosophy.

12. I. N. Golovin, *I. V. Kurchatov*, 3rd ed., Moscow: Atomizdat, 1978.

13. M. G. Meshcheryakov, "Vitalii Grigor'evich Khlopin: voskhozhdenie na posledniuiu vershinu," *Priroda*, no. 3, 1993.

14. *Atomnyi proekt SSSR*, Vols. 1 and 2, Moscow: Nauka, 1998–2002; D. Holloway, *Stalin and the Bomb: The Soviet Union and Atomic Energy, 1939–1956*, New Haven, CT: Yale University Press, 1994, pp. 90–95.

15. Arkhiv RAN 2-1/1943–94.

16. M. G. Meshcheryakov interview, March 19, 1993.

Chapter 5

1. Arkhiv RAN (Archives of the Russian Academy of Sciences) 524-9-463.

2. Ibid.

3. L. N. Bell interview, April 28, 1997.

4. *On mezhdu nami zhil . . . Vospominaniya o Sakharove* (B. L. Altshuler et al., eds.), Moscow: Praktika, 1996.

5. The question "came from life"—the life of submarines, where underwater audibility is lost after a storm when the seawater is filled with bubbles.

6. As quoted in L. Brown and H. Rechenberg, *The Origin of the Concept of Nuclear Forces*, Philadelphia, PA: Institute of Physics Publishing, 1996, p. 109.

7. I. R. Gekker, A. N. Starodub, and S. A. Fridman, *Fizicheskii institut Akademii Nauk na Miusskoi ploshchadi (K istorii FIANa)*, preprint no. 78, Moscow, 1989.

8. In addition to the one mentioned: theory of light scattering on crystals and the concept of phonon (1930), theory of light scattering on the relativist electron (1930), the Tamm states in solid-state physics (1932), predicting the magnetic moment of the neutron (1934), and the Tamm-Dankov method for describing the interaction of particles (1945) (D. A. Kirzhnits, "Vekhi nauchnogo tvorchestrva," *Priroda*, no. 7, 1995).

9. E. L. Feinberg, *Epokha i lichnost. Fiziki. Ocherki i vospominaniya*, Moscow: Fizmatlit, 2003.

10. V. G. Kadyshevskii, "Chto by vy khoteli sprosit' u akademika Tamma?" *Priroda*, no. 7, 1995.

11. Feinberg, *Epokha i lichnost.*

12. The laboratories were headed respectively by D. Skobeltsyn, N. Papalexi, G. Landsberg, S. Vavilov, S. Mandelshtam, B. Vul, and N. Andreyev. There were 138 people working at the institute, of whom 63 were scientific associates.

13. Arkhiv RAN 532-1-101; Gekker et al., *Fizicheskii institut Akademii Nauk na Miusskoi ploshchadi.*

14. Arkhiv RAN 532-1-101.

15. A. D. Sakharov, "Generatsiya zhestkoi komponenty kosmicheskikh luchei (ZhETF, 1947)," *Nauchnye trudy*, Moscow: Tsentrkom, 1995.

16. *Mezon* (I. E. Tamm, ed.), Moscow: GITTL, 1947.

17. V. Mayakovsky, "Razgovor s fininspektorom o poezii" (1926), *Lirika*, Moscow: Khudozhestvennaya Literatura, 1967.

18. R. G. Dalitz, "Kandidatskaya dissertatsiya Andreya Sakharova" [Andrei Sakharov's Candidate of Sciences Dissertation], in Sakharov, *Nauchnye trudy.*

19. Arkhiv RAN 532-1-133; Sakharov, *Nauchnye trudy.*

20. Sakharov, *Nauchnye trudy*; Arkhiv RAN 524-9-463.

21. Arkhiv RAN 524-9-463.

22. "Tamm v dnevnikakh i pis'makh," *Kapitsa. Tamm. Semenov. V ocherkakh i pis'makh*, Moscow: Vagrius-Priroda, 1998.

23. A. S. Pushkin, "Iz Pindemonti" (1836), *Sobranie sochineniy*, Moscow: Pravda, 1969.

24. Arkhiv RAN 41-3-533.

Chapter 6

1. V. V. Krylov, "Vybor ili vybory? K istorii izbraniya prezidenta Akademii nauk SSSR. Iyul' 1945," *Istoricheskii Arkhiv*, no. 2, 1996.

2. The other two academicians in the group of eight who were listed without any mention of personal shortcomings were the chemist N. D. Zelinsky and the geologist V. A. Obruchev—both of whom were more than 80 years old.

3. H. G. Dale, "Open Letter to the President of the USSR Academy of Sciences," *Times*, November 26, 1948 [available in Russian in *Britansky Soyuznik*, December 12, 1948].

4. S. I. Vavilov, *Ocherki i vospominaniya*, Moscow: Nauka, 1991.

5. Arkhiv RAN (Archives of the Russian Academy of Sciences) 2-1/1943–94.

6. E. L. Feinberg, *Epokha i lichnost. Fiziki. Ocherki i vospominaniya*, Moscow: Fizmatlit, 2003.

7. As quoted in M. Popovsky, *Delo akademika Vavilova*, Moscow: Kniga, 1991.

8. *Bremya Styda* [The Burden of Shame] is the title Daniil Danin chose for his book about this period (Moscow, 1996).

9. D. Holloway, *Stalin and the Bomb: The Soviet Union and Atomic Energy, 1939–1956*, New Haven, CT: Yale University Press, 1994, ch. 6.

10. Vavilov came to Stalin's office a total of three times: once before on April 15, 1943, and once more afterward on July 13, 1949 ("Posetiteli kremlyovskogo kabineta I.V. Stalina," *Istoricheskii Arkhiv*, no. 4, 1996; no. 4, 1998.).

11. As quoted in Y. N. Smirnov, "Stalin i atomnaya bomba," *VIET*, no. 2, 1994. Kurchatov visited Stalin for the second and last time on January 9, 1947.

12. Holloway, *Stalin and the Bomb*, p. 149.

13. L. N. Bell interview, August 30, 1994.

14. N. A. Dobrotin interview, February 3, 1993. See also *Vospominaniya o V. V. Vekslere*, Moscow: Nauka, 1987.

15. S. M. Rytov, "Academician N. D. Papalexi," *Izvestiya AN SSSR*, ser. fiz., Vol. 12, no. 1, 1948.

16. S. I. Vavilov, "Rech pri otkrytii traurnogo zasedaniya," *Izvestiya AN SSSR*, ser. fiz., Vol. 12, no. 1, 1948.

17. Ibid.

18. Arkhiv RAN 471-1-53; Y. L. Alpert interview, December 31, 1994; V. V. Migulin interview, November 17, 1989.

19. Arkhiv RAN 471-1-53.

20. According to E. L. Feinberg, I. Tamm's candidacy was personally crossed out by A. A. Zhdanov (*Epokha i lichnost*). The specific reasons why Tamm fell into disfavor with higher levels of officialdom in the party, aside from his personal losses during 1937, are not known.

21. Arkhiv RAN 596-3-245.

22. Arkhiv RAN 2-8/1946–74.

23. Arkhiv RAN 2-8/1946–69; 2-8/1946–74.

24. L. N. Bell interview, August 30, 1994. Bell (whose native language was English) was formally dropped from graduate work on March 4, 1947, "in connection with non-completion of the foreign language study plan" [!] (Arkhiv RAN 524-10-28).

Chapter 7

1. Arkhiv RAN (Archives of the Russian Academy of Sciences) 518-2-17.

2. P. L. Kapitsa, *Pis'ma o nauke* (P. E. Rubinin, comp.), Moscow: Moskovskii rabochii, 1989.

3. *Atomnyi proekt SSSR*, Vol. 2, part 1, Moscow: Nauka, 1998–2002.

4. Y. P. Terletsky, "Operatsiya 'Dopros Nilsa Bora,'" *VIET*, no. 2, 1994.

5. As quoted in P. E. Rubinin, "Niels Bohr and Petr Leonidovich Kapitsa," *Physics* [*Uspekhi*], Vol. 40, no. 1, 1997.

6. Rubinin, "Niels Bohr and Petr Leonidovich Kapitsa"; D. Holloway, "Beria, Bohr, and Question of Atomic Intelligence," *Reexamining the Soviet Experience* (D. Holloway and N. Nairmark, eds.), Boulder, CO: Westview Press, 1996.

7. Letter to his wife, A. A. Kapitsa, July 24, 1935. As quoted in Rubinin, "P. E. Kapitsa and Sakharov. Opposition to Totalitarianism at Various Stages of the Soviet Totalitarian State's Development." Manuscript of a speech for the conference Science and Political Authority, Massachusetts Institute of Technology, Cambridge, MA, May 1992.

8. Letter to his wife, A. A. Kapitsa, January 11, 1935. As quoted in P. E. Rubinin, "K istorii odnogo pis'ma P. L. Kapitsy," *Kommunist*, no. 7, 1991.

9. A. A. Kapitsa interview, November 23, 1992.

10. V. V. Krylov, "Vybor ili vybory? K istorii izbraniya presidenta Akademii nauk SSSR. Iyul' 1945," *Istoricheskii Arkhiv*, no. 2, 1996.

11. Vitalii Grigorievich Khlopin, the director of the Radium Institute, did not obey this "command of the order of the day." When he received instructions to purge his institute of undesirable elements of Jewish and German descent, he wrote his own name at the top of the list, and said that this was the only version of the order he could sign (M. G. Meshcheryakov, "Vitalii Grigorievich Khlopin," *Priroda*, no. 3, 1993).

12. B. G. Erozolimskii interview, February 2, 1995.

13. G. D. Latyshev, the name of the author of the ill-fated paper, is mentioned only once in the chapter "Delenie atomnykh yader" [Fission of Atomic Nuclei] in Ioffe's popular book *Osnovnye predstavleniya sovremennoi fizike* [Basic Concepts of Contemporary Physics], Moscow: FTTI, 1949. Accusations against Ioffe included the "Latyshev affair" as well as "lack of realization of the vital importance of studying the foundations of Marxist philosophy." (RGASPI (Rossiiskii gosudarstvennyi Arkhiv sotsial'no-politicheskoi istorii) [Russian State Archive of Sociopolitical History] 17-133-171.)

14. Y. N. Smirnov, "Stalin i atomnaya bomba," *VIET*, no. 2, 1994.

15. Terletsky, "Operatsiya 'Dopros Nilsa Bora.'"

16. This was said by V. A. Davidenko (1914–1983), who knew Kurchatov as far back as the prewar Leningrad PhysTech and who arrived at Laboratory No. 2 in the first few months of its operation in 1945.

17. Beria graduated from the Baku Mechanical and Builders School (*Istochnik*, no. 4, 1996).

18. "Lev Landau: god v tyur'me," *Izvestiya TsK KPSS*, no. 3, 1991.

19. Terletsky, "Operatsiya 'Dopros Nilsa Bora.'"

20. A. I. Kokurin and A. I. Pozharov, "Novyi kurs L.P. Berii. 1953 god," *Istoricheskii Arkhiv*, no. 4, 1996.

21. R. C. Williams, *Klaus Fuchs, Atom Spy*, Cambridge, MA: Harvard University Press, 1987.

22. J. Albright and M. Kunstel, *Bombshell: The Secret Story of America's Unknown Atomic Spy Conspiracy*, New York: Random House, 1997.

Chapter 8

1. Churchill did not doubt the answer to this question and was even prepared to arrest Bohr (see D. Holloway, *Stalin and the Bomb: The Soviet Union and Atomic Energy, 1939–1956*, New Haven, CT: Yale University Press, 1994, p. 238).

2. Who would have believed that Stalin would order Mikhoels killed and that knowing him would become evidence in the "Doctors' Plot"—the Soviet "Beilis Affair"?

3. Kratkii slovar' inostrannykh slov, Moscow, 1951.

4. G. B. Kostyrchenko, *V plenu u krasnogo pharaona*, Moscow: Mezhdunarodnye otnosheniya, 1994, ch. 4.

5. V. I. Vernadsky, "Pis'ma N. N. Petrunkevichu," *Novyi Mir*, no. 12, 1989.

6. *Akademik L. I. Mandelshtam. K 100-tiyu so dnya rozhdeniya*, Moscow: Nauka, 1979.

7. "Brat'ya evrei vsego mira! Vystulpleniya predstavitelei evreiskogo naroda na mitinge, sostoyavshemsya v Moskve 24 avgusta 1941 g," Moscow: Gospolitizdat, 1941.

8. Kostyrchenko, *V plenu u krasnogo pharaona.*

9. Ibid.

10. The 1942 memorandum "on cadres" was prepared under the direction of G. F. Aleksandrov, whom Stalin appointed to the Central Committee Directorate of Agitation and Propoganda straight from graduate school and subsequently made his biographer. Therefore, there is little doubt that the idea itself was approved by Stalin. Aleksandrov's admixture of Jewish blood might account for his anti-Jewish zeal and need to serve the times ("Vospominaniya A. G. Spirkina," *Vestnik*, nos. 12–15, 1997).

11. R. L. Berg, "Palachi i rytsary sovetskoi nauki," *Vremya i my*, no. 65, 1982.

12. A. A. Kapitsa interview, November 23, 1992.

13. *Akademik L. I. Mandelshtam.*

14. E. L. Feinberg interview, November 28, 1992. The impression it made had even more impact because of the "inappropriate" (during atheistic Soviet times) reference to the Old and New Testaments. According to his daughter, A. N. Krylov "wasn't a believer in the least, but had known the Bible and prayers well since childhood, and he always relished quoting from the Bible" (A. A. Kapitsa interview, November 23, 1992).

15. See G. Gorelik, "Fizika universitetskaya i akademicheskaya," *VIET*, no. 1, 1991.

16. I. E. Tamm, "O nekotorykh teoreticheskikh rabotakh A. S. Predvoditeleva," *ZhETF*, no. 4, 1936.

17. A. S. Predvoditelev and N. N. Ukholin, "Pis'mo v redaktsiyu," *Pod Znamenem Marksizma*, no. 4, 1938. Students called Predvoditelev's department "the department of thermal and administrative phenomena" (S. M. Shapiro interview, November 13, 1997).

18. P. L. Kapitsa, *Pis'ma o nauke* (P. E. Rubinin, comp.), Moscow: Moskovskii rabochii, 1989.

19. A. R. Zhebrak, "Soviet Biology," *Science*, Vol. 102, 1945, pp. 357–358.

20. V. Nemchinov, "Protiv nizkopoklonstva!" *Literaturnaya gazeta*, April 10, 1947.

21. V. L. Ginzburg, *O nauke, o sebe i o drugikh*, Moscow: Fizmatlit, 2003.

22. Stalin made Lysenko's text more respectable by replacing "anti-Marxist biology" with the simpler "reactionary biology" and "Soviet biology" with "scientific biology" (K. Rossianov, "Stalin as Lysenko's Editor," *Configurations*, Vol. 1, no. 3, 1993).

23. Y. P. Terletsky interview, April 5, 1993 (conducted by A. V. Andreyev).

24. As quoted in A. S. Sonin, *Fizicheskii idealizm: istoriya ideologicheskoi kampanii*, Moscow: Fizmatlit, 1994.

25. Ibid.

26. Ibid.

27. Ibid.

28. Ibid.

29. See Gorelik, "Fizika universitetskaya i akademicheskaya."

30. V. L. Ginzburg interview, September 25, 1990.

31. G. A. Goncharov, "Termoyaderny proekt SSSR: predystoriya i desyat' let puti k termoyadernoy bombe," *Istoriya sovetskogo Atomnogo proekta. Dokumenty, vospominaniya, issledovaniya* (V. P. Vizgin, ed.), Saint-Petersburg: Russkiy hristianskiy gumanitarnyy institut, 2002.

Chapter 9

1. I. E. Tamm, "Vnutriatomnaya energiya," *Pravda*, April 11, 1946, p. 3.

2. He also authored two popular-science publications on "intra-atomic energy".

3. E. L. Feinberg interview, September 30, 1992; I. Y. Barit interview, October 31, 1992.

4. Tamm's parents remained in occupied Kiev, and his father, of half-German descent, registered as a *Volksdeutsche* [ethnic German] so that he would not be mistaken for a Jew because of his non-Slavic name. FIAN Archives 2–2–320; L. I. Vernskii, "Rodoslovnaya I. E. Tamma," *Priroda*, no. 7, 1995.

5. L. I. Kudinova, *K istorii mirnogo ispol'zovaniya atomnoi energii v SSSR. 1944–1951* (A. V. Shchegel'skii, comp.), Obninsk: Fiziko-energeticheskii institut, 1994.

6. See G. Gorelik, "S chego nachinalas' sovetskaya vodorodnaya bomba," *VIET*, no. 1, 1993.

7. *Atomnyi proekt SSSR*, Vols. 1 and 2, Moscow: Nauka, 1998–2002.

8. G. A. Goncharov, "Termoyaderny proekt SSSR: predystoriya i desyat' let puti k termoyadernoy bombe," *Istoriya sovetskogo Atomnogo proekta. Dokumenty, vospominaniya, issledovaniya* (V. P. Vizgin, ed.), Saint Petersburg: Russkiy hristianskiy gumanitarnyy institut, 2002.

9. Ibid.

10. Ibid.

11. All he could do was discuss his idea in the article "Atomnaya energiya i ee osvobozhdenie" [Atomic Energy and Its Liberation] in the popular science magazine *Priroda* in 1946.

12. H. Bethe and F. Seitz, *How Close Is the Danger? In One World or None* (D. Masters and K. Way, eds.), New York: McGraw-Hill, 1946, pp. 98, 100.

13. There were two other entirely different reasons why Frenkel was a less suitable candidate than Tamm. From the standpoint of the security service, both had serious flaws in their biographies (the fact that both were adherents of socialism and its Soviet implementation had no particular relevance). Tamm's many relatives and close associates who had been subject to repression were comprising for him, whereas Frenkel was famous for his public abuse of dialectical materialism at the All-Union Conference in 1931, when he declared, "Socialism requires the basis historical materialism gives it, but it is unrelated to dialectical materialism, which is an obstacle in the development of a science. Neither Lenin nor Engels are authorities for physicists" (as quoted in A. S. Sonin, *Fizicheskii idealizm: istoriya ideologicheskoi kampanii*, Moscow: Fizmatlit, 1994). The security service viewed Frenkel's type of nonconformism and "unmanageability" as a greater defect than relatives, friends, and students who had been repressed. According to Y. L. Al'pert, a security service officer had asked him to provide a character reference for Tamm. This was a very strange request, if one considers their distance in scientific matters and closeness within the Mandelshtam school. Easier to see this as the security service's need for a undoubtedly positive recommendation in order to balance Tamm's "sins" in its files on him and successfully officially legalize the decision to include Tamm in the atomic project.

14. *Atomnyi proekt SSSR*.

15. R. Rhodes, *The Making of the Atomic Bomb*, New York: Simon and Schuster, 1986, p. 770.

16. In a similar situation Robert Oppenheimer, one of the fathers of the Ameri-

can atomic bomb, said: "When you see something that is technically sweet, you go ahead and do it and you argue about what to do about it only after you have had your technical success" (as quoted in R. Rhodes, *Dark Sun: The Making of the Hydrogen Bomb*, New York: Simon and Schuster, 1995, p. 476).

17. Goncharov, "Termoyaderny proekt SSSR."

18. V. L. Ginzburg interview, March 28, 1992; Y. A. Romanov interview, November 11, 1992.

19. Y. A. Romanov, "Otets sovetskoi vodorodnoi bomby," *Priroda*, no. 8, 1990.

20. A. Einstein, "Reply to the Soviet Scientists," *Bulletin of the Atomic Scientists*, Vol. 4, no. 2, 1948.

21. B. Russell, "Values in the Atomic Age," *The Atomic Age*, London: George Allen and Unwin, 1949, pp. 81–104.

22. Arkhiv RAN (Archives of the Russian Academy of Sciences) 596-2-175.

23. A. P. Znoyko's articles titled "The Periodic Law of Atomic Nuclei. The Specific Charge of a Nucleus and the Periodic System of Isotopes," "Chemical Analogs of the Elements in the Periodic System of Atomic Nuclei," "Isotopes at the End of the Periodic System," and "On Elements 97 and 98" were published in *Reports of the Academy of Sciences of the USSR* in 1949–1950.

24. "Laboratory No. 15 within the physics department of Moscow university was created according to Resolution No. 3736–1565 of the USSR Council of Ministers dated September 7, 1949, to study and develop the periodic regularity of the properties of atomic nuclei discovered by A. P. Znoiko [author's certificate dated July 28, 1947]". (RGANI (Rossiiskii gosudarstvennyi arkhiv noveishei istorii) [Russian State Archive of Contemporary History] 5-17-482 (5718).)

25. H. G. Dale, "Open Letter to the President of the USSR Academy of Sciences," *Times*, November 26, 1948.

26. V. L. Ginzburg, *O fizike i astrofizike*, Moscow: Kvantum, 1995.

27. V. L. Ginzburg interview, April 29, 1998.

28. E. L. Feinberg, *Epokha i lichnost. Fiziki. Ocherki i vospominaniya*, Moscow: Fizmatlit, 2003; B. M. Bolotovsky, Y. N. Vavilov, and A. N. Kirkin, "Sergei Ivanovich Vavilov—uchenyi i chelovek," *Uspekhi Fizicheskikh Nauk*, no. 5, 1998.

29. A. A. Kapitsa interview, February 16, 1990.

Chapter 10

1. A government resolution dated April 9, 1946, ordered allocation of the site of "factory No. 550 to KB-11." In government resolutions the classified 215 sq. km area, contiguous to KB-11 and including Sarov, was called "Installation No. 550" (*Atomnyi proekt SSSR*, Vols. 1 and 2, Moscow: Nauka, 1998–2002).

2. G. A. Goncharov, "Termoyaderny proekt SSSR: predystoriya i desyat' let puti k termoyadernoy bombe," *Istoriya sovetskogo Atomnogo proekta. Dokumenty, vospominaniya, issledovaniya* (V. P. Vizgin, ed.), Saint Petersburg: Russkiy hristianskiy gumanitarnyy institut, 2002.

3. M. D. Frank-Kamenetsky interview, March 30, 1996.

4. L. A. Vernaya interview, May 25, 1996.

5. V. F. Sennikov interview, December 4, 1992.

6. D. Holloway, *Stalin and the Bomb: The Soviet Union and Atomic Energy, 1939–1956*, New Haven, CT: Yale University Press, 1994, ch. 10.

7. Goncharov, "Termoyaderny proekt SSSR."

8. V. L. Ginzburg, *O fizike i astrofizike*, Moscow: Kvantum, 1995.

9. E. S. Fradkin interview, Jauary 21, 1995.

10. L. P. Feoktistov interview, February 24, 1995

11. V. L. Ginzburg, *O nauke, o sebe i o drugikh*, Moscow: Fizmatlit, 2003.

12. In addition to Sakharov and Romanov, Tamm's group included Y. N. Babayev, G. A. Goncharov, V. G. Zagrafov, R. Y. Zaidel', B. N. Kozlov, I. A. Kurilov, V. I. Ritus, and M. P. Shumayev. Zeldovich's group consisted of V. B. Adamskii, G. M. Gandel'man, N. A. Dmitriev, E. I. Zababakhin, E. A. Negin, V. N. Rodigin, Y. A. Trutnev, and L. P. Feoktistov (Y. A. Romanov interview, November 11, 1992).

13. M. M. Agrest interview, March 23, 1996.

14. M. M. Agrest, "Izgnanie," *Khimiya i zhizn'*, no. 1, 1993.

15. "Nikolai Nikolaevich Bogolubov: Matematik, mekhanik, fizik," Dubna, 1994.

16. L. D. Landau, *Sobranie trudov*, Moscow: Nauka, 1969.

17. I. M. Khalatnikov interview, March 17, 1993. What follows is based on this interview.

18. I. M. Khalatnikov interview, March 17, 1993.

19. Y. B. Khariton and Y. N. Smirnov, "O nekotorykh mifakh i legendakh vokrug sovetskikh atomnogo i termoyadernogo proektov," *Materialy ubileinoi sessii Uchenogo soveta RNTs "Kurchatovskii institut" 12 yanvarya 1993*, Moscow: RNTs "Kurchatovskii institut," 1993.

20. O. A. Lavrentiev interview, June 30, 1994 (conducted by Y. N. Ranyuk, American Institute of Physics, Center for the History of Physics).

21. Ibid.

22. A. D. Sakharov, "Otzyv na rabotu O.A. Lavrent'eva," *Uspekhi Fizicheskikh Nauk*, no. 8, 2001.

23. V. L. Ginzburg interview, September 25, 1990.

24. V. I. Kogan interview, April 30, 1993.

Chapter 11

1. V. I. Ritus, "Esli ne ya, to kto?" *Priroda*, no. 8, 1990; *On mezhdu nami zhil . . . Vospominaniya o Sakharove* (B. L. Altshuler et al., eds.), Moscow: Praktika, 1996.

2. Arkhiv RAN (Archives of the Russian Academy of Sciences) 411–3–533.

3. Government report about the hydrogen bomb test in the Soviet Union, *Pravda*, August 20, 1953.

4. Arkhiv RAN 411-3-533.

5. V. L. Ginzburg interview, September 25, 1990.

6. G. A. Goncharov, "Termoyaderny proekt SSSR: predystoriya i desyat' let puti k termoyadernoy bombe," *Istoriya sovetskogo Atomnogo proekta. Dokumenty, vospominaniya, issledovaniya* (V. P. Vizgin, ed.), Saint Petersburg: Russkiy hristianskiy gumanitarnyy institut, 2002.

7. Ibid.

8. P. Galison and B. Bernstein, "In Any Light: Scientists and the Decision to Build the Superbomb," *Historical Studies in the Physical and Biological Sciences*, Vol. 19, no. 2, 1989, p. 292.

9. Goncharov, "Termoyaderny proekt SSSR."

10. V. I. Ritus interview, July 7, 1992.

11. L. V. Altshuler, presentation at the conference titled "History of the Soviet Atomic Project," Dubna, May 1996.

12. Y. A. Romanov interview, November 11, 1992.

13. Unofficially, it is considered that the superdevice is surrounded by a cas-

ing that evaporates instantly from radiation heat "brighter than a thousand suns" and thereby compresses the filling through reactive force.

14. Goncharov, "Termoyaderny proekt SSSR."

15. Ibid.

16. A. Fitzpatrick, "Igniting the Light Elements: The Los Alamos Thermonuclear Weapon Project, 1942–1952" [Ph.D. dissertation], Blacksburg: Virginia Polytechnic Institute and State University, 1998.

17. A. Poleshchuk, "Sluzhba vneshnei razvedki Rossii ne imeet konkurentov v mire," *Nezavisimaya gazeta*, December 18, 1996.

18. J. Albright and M. Kunstel, *Bombshell: The Secret Story of America's Unknown Atomic Spy Conspiracy*, New York: Random House, 1997.

19. Ibid., pp. 186–187.

20. L. L. Strauss, *Men and Decisions*, Garden City, NY: Doubleday, 1962, p. 270; *No Sacrifice Too Great: The Life of Lewis L. Strauss*, Charlottesville: University Press of Virginia, 1984, p. 137; J. M. Holl, "Notice to Newsmen," *Washington Post*, December 29, 1975; J. A. Wheeler interview, May 23, 1988 (conducted by Finn Aaserud, American Institute of Physics, Center for the History of Physics), pp. 81–82; J. Wheeler and K. Ford, *Geons, Black Holes, and Quantum Foam: A Life in Physics*, New York: Norton, 1998, pp. 284–286.

21. Thomas C. Reed is the author of *At the Abyss: An Insider's History of the Cold War* (New York: Ballantine Books, 2004).

22. G. Herken, *Brotherhood of the Bomb: The Tangled Lives and Loyalties of Robert Oppenheimer, Ernest Lawrence, and Edward Teller*, New York: Henry Holt, 2002, pp. 259–260; T. Reed, personal communication.

23. H. Bethe and F. Seitz, *How Close Is the Danger? In One World or None* (D. Masters and K. Way, eds.), New York: McGraw-Hill, 1946, p. 100.

24. A. A. Yatskov, "Atom i razvedka," *VIET*, no. 3, 1992.

25. Albright and Kunstel, *Bombshell*, p. 156.

26. V. L. Ginzburg, *Atomnoye yadro i ego energiya*, Moscow: OGIZ, 1946.

27. "Policy and Progress in the H-Bomb Program: A Chronology of Leading Events," Joint Committee on Atomic Energy, January 1, 1953, p. 79; C. Hansen, *The Swords of Armageddon: U.S. Nuclear Weapons Development since 1945*, Vol. 3, Sunnyvale, CA: Chukelea Publications, 1995, p. 137.

28. B. L. Ioffe, "Koe-chto iz istorii atomnogo proekta v SSSR," *Sibirskii fizicheskii zhurnal*, no. 2, 1995; "Osobo sekretnoe zadanie," *Novyi mir*, no. 5, 1999; "Vliyanie polyarizatsii na rasprostranenie kvantov v ionizovannon gaze," *Pis'ma v astronomicheskii zhurnal*, Vol. 20, 1994, pp. 803–819.

29. I am relying on Goncharov's analysis, cited numerous times above, because he studied Fuchs's report with an "armed" eye—armed with personal participation in the implementation of the Third Idea and four decades of work at the Installation.

30. Goncharov, "Termoyaderny proekt SSSR."

31. L. P. Feoktistov, "Vodorodnaya bomba: Kto zhe vydal ee sekret?" *NG-Nauka*, September 2, 1997; *Oruzhie, kotoroe sebya ischerpalo*, Moscow: Rossiyskiy komitet VMPYaV, 1999.

32. Goncharov, "Termoyaderny proekt SSSR."

33. H. A. Bethe, memorandum on the history of the thermonuclear program, May 28, 1952 (available at http://www.fas.org/nuke/guide/usa/nuclear/bethe-52.htm).

34. Edward Teller, "Comments on Bethe's History of the Thermonuclear Program," August 14, 1952; "Policy and Progress in the H-Bomb Program: A

Chronology of Leading Events," Joint Committee on Atomic Energy, January 1, 1953, pp.78, 79; Hansen, *The Swords of Armageddon*, Vol. 3, pp. 35, 191.

35. The Fuchs–von Neumann patent of May 28, 1946, is noted in the declassified chronology of the major events in the history of the hydrogen bomb ("Policy and Progress in the H-Bomb Program: A Chronology of Leading Events") prepared by the Joint Committee on Atomic Energy, January 1, 1953, but its description is deleted.

36. A. Fitzpatrick, "Igniting the Light Elements," p. 163.

37. Testimony in the matter of J. Robert Oppenheimer (available at http://www.nuclearfires.org/redocuments/1954/54–opp-testimony.html). In 1997 Bethe repeated his opinion: "[T]he crucial invention was made in 1951, by Teller" (H. J. Bethe, "Robert Oppenheimer," *Biographical Memoirs of the National Academy of Science*, Vol. 71, 1997).

38. Hans Bethe's statement to the FBI on Klaus Fuchs, February 14, 1950 (available at http://www.pbs.org/wgbh/amex/bomb/filmmore/reference/primary/hansbethestate.html).

39. Goncharov, "Termoyaderny proekt SSSR."

40. Y. A. Romanov interview, November 11, 1992.

41. *Lyudi Ob'ekta. Ocherki i vospominaniya*, Sarov-Moscow, 1996.

42. E. G. Bonner interview, December 18, 1997.

43. *Znakomyi neznakomyi Zeldovich*, Moscow: Nauka, 1993.

44. V. L. Ginzburg, *O fizike i astrofizike*, Moscow: Kvantum, 1995.

Chapter 12

1. This is what he replied to his new acquaintance Elena Bonner when she asked him, "Who are your friends?" (E. G. Bonner interview, January 31, 1996).

2. This remark was included only in the English edition of Sakharov's memoirs [Andrei Sakharov, *Memoirs* (R. Lourie, trans.), New York: Knopf, 1990, p. 125]. When preparing the Russian edition after Zeldovich's death, he evidently chose to soften his tone.

3. Landau began with the A-bomb: He was awarded the Order of Lenin and Stalin Prize "for developing a theory of calculating the efficiency factor." (*Atomnyi proekt SSSR*, Vols. 1 and 2, Moscow: Nauka, 1998–2002).

4. S. S. Ilizarov, "Partapparat protiv Tamma," *Kapitsa. Tamm. Semenov. V ocherkakh i pis'makh*, Moscow: Vagrius-Priroda, 1998.

5. L. N. Bell interview, August 30, 1994.

6. A. D. Sakharov, *Nauchnye trudy*, Moscow: Tsentrkom, 1995.

7. Y. I. Krivonosov, "Landau i Sakharov v 'razrabotkakh' KGB," *VIET*, no. 3, 1993.

8. "Box 5" was the Soviet synonym of Jewish ethnicity because question 5 in the standard personnel questionnaire required information about an individual's ethnicity.

9. I. M. Khalatnikov interview, March 17, 1993.

10. Ibid. In the words of a KGB agent: "At the end of March [1955] LANDAU was summoned along with GINZBURG to ZAVENYAGIN regarding secret work. In a conversation with the source, LANDAU expressed himself very harshly about ZELDOVICH, 'the creator of all manner of dirty tricks.' LANDAU told the source that he would not agree to engage in secret work again and that he found talking about it unpleasant. On the way to the Ministry LANDAU warned GINZBURG not

to state that he needed LANDAU for the upcoming work" (Krivonosov, "Landau i Sakharov"). It's possible that the person with whom Landau was speaking was identified incorrectly. As V. L. Ginzburg told the author (in a letter dated December 27, 2000), he has no recollection whatsoever of this incident and thinks that it concerns someone else, since he was not involved in secret work in 1955.

11. E. L. Feinberg, *Epokha i lichnost. Fiziki. Ocherki i vospominaniya*, Moscow: Fizmatlit, 2003.

12. *Akademik M. A. Leontovich: uchenyy, uchitel, grazhdanin*, Moscow: Nauka, 2003.

13. Arkhiv RAN (Archives of the Russian Academy of Sciences) 532-1-182.

14. *Akademik M. A. Leontovich: uchenyy, uchitel, grazhdanin.*

15. Ibid.

16. D. A. Kirzhnits interview, July 5, 1995.

17. Ibid.

18. M. A. Leontovich, G. S. Landsberg, A. A. Andronov, and I. E. Tamm were on the publishing commission.

19. Arkhiv RAN 532-1-213, 532-1-198.

20. Aleksandr Danilovich Aleksandrov (1912–1999) graduated from Leningrad University in 1933. He received a Stalin Prize in 1942, was elected to Corresponding Member of the Academy of Sciences of the USSR in 1946, and was elected to academician in 1964.

21. Arkhiv RAN 532-1-213.

22. Arkhiv RAN 532-1-198.

23. V. A. Fock, "Protiv nevezhestvennoi kritiki sovremennykh fizicheskikh teorii," *Voprosy filosofii*, no. 1, 1953.

24. *Akademik L. I. Mandelshtam. K 100-iyu so dnya rozhdeniya*, Moscow: Nauka, 1979.

25. Describing Aleksandrov, outstanding mathematician and founder of the Leningrad Geometry School, outside mathematics is not an easy task. An avid mountain climber, on flat land his temperament revealed itself in his craving for adventures. In 1947 the Corresponding Member and Stalin Laureate was ready to join the FIAN expedition to Brazil in any capacity—even as a simple laborer. His love of dialectical materialism was coupled with a loathing of "professional diamats [dialectical materialists]" like Maximov and Terletsky, whom he did not spare in his articles. Devoted to socialism, he personally took young dissidents whom he knew under his wing. He was accused of both anti-Semitism and complicity in Zionism.

26. Excuses of the warrior with "Einsteinism" that immediately follow Fock's article were accompanied by a demeaning footnote: "The article by the editorial committee member, A. A. Maksimov, is published by way of general discussion.—Ed."

27. S. S. Ilizarov and L. I. Pushkarev, "Beria i teoriya otnositel'nosti," *Istoricheskii Arkhiv*, no. 3, 1994.

28. Arkhiv RAN 532-1-232.

29. RGANI (Rossiiskii gosudarstvennyi arkhiv noveishei istorii) [Russian State Archive of Contemporary History] 5-17-434. Documents from this file are further cited without footnotes. For a detailed history of Znoyko's laboratory, see G. Gorelik, "Enrico Fermi, A. P. Znoyko i Klim Voroshilov," *Priroda*, no. 2, 1994 (also *Znanie-Sila*, no. 1, 1998).

30. RGANI 5-17-434.

31. *Nikolai Nikolaevich Bogolubov: Matematik, mekhanik, fizik*, Dubna, 1994.

32. L. A. Vernaya, interviews on May 25, 1996, and May 25, 1998.

33. *Istoriya sovetskogo Atomnogo proekta. Dokumenty, vospominaniya, issledovaniya* (V. P. Vizgin, ed.), Saint Petersburg: Russkiy hristianskiy gumanitarnyy institut, 2002.

34. Krivonosov, "Landau i Sakharov."

35. *Istoriya sovetskogo Atomnogo proekta.*

36. V. B. Adamskii interview, February 28, 1995.

37. N. A. Dmitriev interview, June 26, 1995.

38. S. P. Shubin, *Izbrannye trudy po teoreticheskoy fizike. Ocherk zhizni. Vospominaniya,* Sverdlovsk, 1991. See G. Gorelik, "Ne uspevshie stat' akademikami," *Priroda,* no. 1, 1990 (also in *Repressirovannaya nauka,* Leningrad: Nauka, 1990).

39. Arkhiv RAN 2-4a/125.

40. D. A. Kirzhnits interview, July 5, 1995.

41. Sakharov, *Nauchnye trudy.*

42. M. G. Meshcheryakov interview, March 19, 1993.

43. *Priroda,* no. 8, 1990.

44. *Kapitsa. Tamm. Semenyonov. V ocherkakh i pis'makh,* Moscow: Vagrius-Priroda, 1998.

45. Feinberg, *Epokha i lichnost.*

Chapter 13

1. V. I. Ritus interview, July 7, 1992.

2. *The Effects of Atomic Weapons,* Washington, DC: United States Government Printing Office, 1950.

3. Y. S. Zamyatnin interview, March 18, 1993.

4. A. Schweitzer, "A Declaration of Conscience," *Saturday Review,* May 18, 1957; L. Wittner, "Blacklisting Schweitzer," *Bulletin of the Atomic Scientists,* May/ June 1995; H. Wasserman et al., *Killing Our Own: The Disaster of America's Experience with Atomic Radiation,* New York: Delacorte Press, 1982.

5. *Science,* Vol. 265, 1994, p. 1507; *Science,* Vol. 266, p. 1141.

6. E. Hiebert, "The Impact of Atomic Energy. A History of Responses by Governments, Scientists, and Religious Groups," Newton, KS: Faith and Life Press, 1961; A. H. Compton, "Hiroshima Revisited," *Science,* Vol. 134, 1961, pp. 1231–1233; K. Fuchs, [Book review], "Zeitschrift fuer Geschichte der Naturwissenschaften, Technik und Medizin, 1961," Heft 4, 1 Jahrgang, pp. 138–143.

7. President Eisenhower warned the country about the dangers of the military-industrial complex in his farewell address in January 1961.

8. E. Teller and A. Latter, *Our Nuclear Future: Facts, Dangers, and Opportunities,* New York: Criterion Books, 1958.

9. E. Teller, "The History of the American Hydrogen Bomb," presented at the international symposium titled "History of the Soviet Atomic Project," Dubna, May 14–18, 1996.

10. L. Tisza interviews, February 28, 1994, and May 28, 1999.

11. E. Teller, "Science and Morality," *Science,* May 22, 1998, pp. 1200–1201.

12. V. B. Adamskii interview, February 28, 1995.

13. The first test of the "clean bomb" took place a year earlier.

14. Wasserman et al., *Killing Our Own.*

15. Sakharov wrote: "Perhaps in some way I asked for this assignment." V. B. Adamskii suggested the same (interview, February 28, 1995).

16. A. D. Sakharov, "Radioaktivnyi uglerod yadernykh vzryvov i neporogovye biologicheskie effekty" [Radioactive Carbon in Nuclear Explosions and Non-Threshold Biological Effects] (*Atomnaya energiya*, 1958), *Nauchnye trudy*, Moscow: Tsentrkom, 1995. The first person in the Soviet scientific press to raise the question of the danger of atmospheric testing was O. I. Leipunsky, whom Sakharov thanks "for a valuable discussion" and whose article he cites.

17. Teller and Latter, *Our Nuclear Future*, p. 126.

18. A. D. Sakharov, "O radioaktivnoi opasnosti yadernykh ispytanii" [On the Radioactive Danger of Nuclear Testing] (manuscript, 1958), in *Nauchnye trudy* [corrected from the manuscript (Andrei Sakharov Archives, Moscow)].

19. Ibid.

20. V. B. Adamskii interview, February 28, 1995; L. P. Feoktistov interview, February 24, 1995.

21. Sakharov, *Nauchnye trudy*.

22. The Central Committee Archive has a draft of his letter dated June 27, 1958: "Reading in the press about your speech regarding the question of testing and using atomic weapons, I think that you might be interested by my article, 'On the Radioactive Danger of Nuclear Testing,' which I append to this letter" (Andrei Sakharov Archives, Moscow).

23. F. Von Hippel, preface, *Science and Global Security*, Vol. 1, 1990, p. 175.

24. An American scientist, Linus Pauling, had come to an analogous conclusion about the effect of radioactive carbon. See L. Pauling, "Genetic and Somatic Effects of Carbon-14," *Science*, Vol. 128, no. 3333, November 14, 1958.

25. *Biograficheskiy slovar' deyateley estestvoznaniya i tekhniki*, Moscow: BSE, 1959.

26. V. I. Ritus, "Esli ne ya, to kto?" *Priroda*, no. 8, 1990.

27. A detailed account is in D. Holloway, *Stalin and the Bomb: The Soviet Union and Atomic Energy, 1939–1956*, New Haven, CT: Yale University Press, 1994, chs. 15 and 16; V. Zubok and C. Pleshakov, *Inside the Kremlin's Cold War: From Stalin to Khrushchev*, Cambridge, MA: Harvard University Press, 1996, ch. 5.

28. "Man of the Year 1957: Nikita Khrushchev," *Time*, January 6, 1958.

29. Veteran American bomb-makers still have their suspicions—it's too different from their experience, even though they know that Khrushchev banged his shoe at the UN, which isn't often encountered in American practice, either.

30. It was not violated until the French tests of 1960.

31. A. I. Solzhenitsyn, *Bodalsya telenok s dubom*, Moscow: Soglasie, 1996.

32. Lydia Chukovskaya, *Zapiski ob Anna Akhmatovoi* (1952–1962), Vol. 2, Moscow: Soglasia, 1997.

33. V. B. Adamskii and Y. N. Smirnov, "50-megatonnyi vzryv nad Novoi Zemlei," *VIET*, no. 3, 1995.

34. Ibid.

35. Ibid.

36. A. B. Koldobskii, "Strategicheskii podvodnyi flot SSSR i Rossii (proshloe, nastoyashchee, budushchee)," *Fizika*, no. 1, 2001.

37. E. A. Shitikov, "Kak sozdavalos' morskoe yadernoe oruzhie," *Voenno-istoricheskii zhurnal*, no. 9, 1994.

38. Adamskii and Smirnov, "50-megatonnyi vzryv."

39. N. S. Khrushchev, *Vospominaniya. Izbrannye otryvki*, New York: Chalidze Publications, 1982.

40. S. N. Khrushchev, *Pensioner souznogo znacheniya*, Moscow: Novosti, 1991. Sakharov became the fifth physicist to be thrice awarded Hero of Socialist Labor,

after Kurchatov, Khariton, and Shelkin (1949, 1951, 1954) and Zeldovich (1949, 1954, 1956).

41. L. P. Feoktistov interview, February 24, 1995.

42. D. Eisenhower, *New York Times*, January 18, 1961.

43. Teller and Latter, *Our Nuclear Future*, p. 126.

44. The letter develops three conclusions: (1) harm to the environment comes from atmospheric tests; (2) atmospheric blasts are much more effective for perfecting weapons and training troops; and (3) underground explosions have potential peaceful applications.

45. V. B. Adamskii, "K istorii zaklucheniya Moskovskogo dogovora o zapreshchenii yadernykh isputanii v trekh sredakh" (manuscript), March 18, 1995. See also V. B. Adamskii, "Dear Mr. Khrushchev," *Bulletin of Atomic Scientists*, November/December 1995.

46. *Repressirovannaya nauka*, Leningrad: Nauka, 1990.

47. "N. V. Timofeev-Resovskii na seminare P. L. Kapitsy," *VIET*, nos. 3 and 4, 1990.

48. Arkhiv RAN (Archives of the Russian Academy of Sciences) 411-3-308.

49. *Repressirovannaya nauka*.

Chapter 14

1. A. M. Yaglom interview, February 9, 1998.

2. *On mezhdu nami zhil . . . Vospominaniya o Sakharove* (B. L. Altshuler et al., eds.), Moscow: Praktika, 1996.

3. An actual example of someone combining the talents of theoretical physicist and writer was Matvei Bronstein (1906–1938), author of the first profound analysis of quantum gravitation and three small masterpieces of scientific literature for children.

4. V. L. Ginzburg interview, September 25, 1990.

5. This problem was solved in the late 1940s by R. Feynman, J. Schwinger, and S. Tomonaga. They received the Nobel Prize in 1965.

6. According to colleagues, the construction of the "unlimitedly" large Tsar Bomb, even though it brought Sakharov his third star of the Hero of Socialist Labor, was incomparable in terms of creative heroism to the work that earned the first two stars.

7. E. G. Bonner interview, February 13, 1997.

8. *On mezhdu nami zhil.*

9. D. A. Frank-Kamenetsky, *Fizicheskie protsessy vnutri zvezd*, Moscow: Fizmatfiz, 1959.

10. See G. Gorelik, "Vladimir Fock: Philosophy of Gravity and Gravity of Philosophy," *The Attraction of Gravitation* (J. Earman et al., eds.), Boston: Birkhäuser, 1993.

11. S. Weinberg, *The First Three Minutes: A Modern View of the Origin of the Universe*, New York: Basic Books, 1977.

12. E. L. Feinberg, *Epokha i lichnost. Fiziki. Ocherki i vospominaniya*, Moscow: Fizmatlit, 2003.

13. Y. B. Zeldovich and I. D. Novikov, *Relyativistkaya astrofizika*, Moscow: Nauka, 1967; Y. B. Zeldovich and I. D. Novikov, *Stroenie i evolutsiya Vselennoi*, Moscow: Nauka, 1975.

14. *On mezhdu nami zhil.*

15. A. D. Sakharov, "Passivnye mezony. Otchet FIAN 1948" [Passive Mesons. FIAN Report 1948], *Nauchnye trudy*, Moscow: Tsentrkom, 1995.

16. *Znakomyi neznakomyi Zeldovich*, Moscow: Nauka, 1993.

17. *Lyudi Ob'ekta. Ocherki i vospominaniya*, Sarov-Moscow, 1996.

18. Zeldovich and Novikov, *Stroenie i evolutsiya Vselennoi*.

19. Y. B. Zeldovich, *Chastitsy, yadra, vselennaya: izbrannye trudy*, Moscow: Nauka, 1985.

20. K. Thorne, *Black Holes and Time Warps: Einstein's Outrageous Legacy*, New York: W. W. Norton, 1994, ch. 6.

21. The direction of Wheeler's early gravitational work found no echo in Zeldovich's writing. In his article "Geons," Wheeler constructed gravitational models of elementary particles.

22. I was working then as an editor at Moscow University's publishing house. I remember at one meeting a censor criticizing an editor for leaving a reference to an anthology of physics problems edited by Sakharov (the father of the disgraced academician). In 1981, the Russian translation of Steven Weinberg's popular book *The First Three Minutes* was published under Zeldovich's editorship (*Pervye tri minuty. Sovremmennyi vzglyad na proiskhozhdenie Vselennoi*, Moscow: Enegoizdat, 1981). By then, Sakharov's idea of baryon asymmetry of the universe had become cutting edge. The editor's comments mention Sakharov's pioneering work, but apparently his name was deleted from the foreword.

23. *Vospominaniya o I. E. Tamme*. The first edition of this book came out in 1981, when Sakharov's name was banned.

24. A. D. Sakharov, "Nachal'naya stadiya rasshireniya Vselennoi i vozniknovenie neodnorodnosti paspredeleniya veshchestva" (ZhETF 1965), *Nauchnye trudy*.

25. A. D. Sakharov, "Simmetriya Vselennoi" (1967), *Nauchnye trudy*.

26. Weinberg, *Pervye tri minuty*, p. 97.

27. A. D. Sakharov, "Narushenie CP-invariantnosti, C-asimmetriya i barionnaya symmetriya Vselennoi" [Violation of CP-invariability, C-asymmetry and Baryon Symmetry of the Universe] (Letters to ZhEFT, 1967), *Nauchnye trudy*.

28. Eight years later in a monograph on cosmology, Zeldovich expounded on Sakharov's hypothesis without any sympathy for it (Zeldovich and Novikov, *Stroenie i evolutsiya Vselennoi*).

29. Feinberg, *Epokha i lichnost*.

30. L. D. Landau, "O zakonakh sokhraneniya pri slabykh vzaimodeistviyakh" [On Laws of Conservation in Weak Interactions], *ZhETF, Sov. Phys JETP, Nucl. Phys.*, 1957.

31. S. Okubo, letter to Gennady Gorelik, December 4, 1996.

32. S. Okubo, "Decay of the Sigma-Plus-Hyperon and Its Antiparticle," *Physical Review*, Vol. 109, 1958, pp. 984–985.

33. Andrei Sakharov Archives, Moscow.

34. Y. I. Krivonosov, "Landau i Sakharov v 'razrabotkakh' KGB," *VIET*, no. 3, 1993.

35. A. D. Sakharov, "Narushenie CP-invariantnosti, C-asimmetriya i barionnaya symmetriya Vselennoi" [Violation of CP-invariability, C-asymmetry and Baryon Symmetry of the Universe] (Letters to ZhEFT, 1967), *Nauchnye trudy*.

36. Sakharov, *Nauchnye trudy*. According to Okun, Sakharov came to ITEF with a fully developed idea and got only the technical support he needed from them (L. B. Okun interview, April 24, 1998).

37. R. Oppenheimer, "Perspectives in Modern Physics"; E. Teller, "On a Theory

of Quasars," *Perspectives in Modern Physics: Essays in Honor of Hans A. Bethe on the Occasion of His 60th Birthday*, July, 1966 (R. E. Marshak, ed.), New York, Interscience Publishers, 1966, pp. 13, 461.

38. S. Okubo, letter to Gennady Gorelik, November 14, 1996.

39. H. Quinn and M. Witherell, "The Asymmetry between Matter and Antimatter," *Scientific American*, October 1998.

40. A. D. Sakharov, "Avtobiografiya," *Trevoga i nadezhda*, Moscow: Inter-Verso, 1991.

41. Y. B. Zeldovich, "Kosmologicheskaya postoyannaya i elementarnye chastitsy," *Pis'ma v ZhETF*, Vol. 6, no. 9, 1967, pp. 883–884.

42. A. Akhmatova, "Mne ni k chemu odicheskie rati . . ." (1940), *Sochineniya*, Moscow: Tsitadel', 1996.

43. A. D. Sakharov, "Vakuumnye kvantovye fluktuatsii v iskrivlennom prostranstve i teoriya gravitatsii" [Vacuum Quantum Fluctuations in Curved Space and the Theory of Gravitation] (Reports of the Academy of Sciences of the USSR, 1967), *Nauchnye trudy*.

44. C. Misner, K. Thorne, and J. Wheeler, *Gravitation*, San Francisco: W. H. Freeman, 1973, pp. 426–428; J. Wheeler, "Beyond the End of Time," in M. Rees, R. Ruffini, and J. A. Wheeler, *Black Holes, Gravitational Waves, and Cosmology: An Introduction to Current Research*, New York: Gordon and Breach, 1974; *The World of Physics: A Small Library of the Literature of Physics from Antiquity to the Present* (J. H. Weaver, ed.), New York: Simon and Schuster, 1987, pp. 675–694.

45. *On mezhdu nami zhil*; D. A. Kirzhnits interview, July 5, 1995.

46. A. D. Sakharov, "Za i protiv," *1973 god: Dokumnety, fakty, sobytiya*, Moscow: PIK, 1991.

Chapter 15

1. Academy of Sciences President M. V. Keldysh asked Minister Slavsky of the Ministry of Medium Machine Building to reassign Academician Zeldovich to work permanently at the Institute of Applied Mathematics of the Academy of Sciences of the USSR "for further development of research in astrophysics and cosmogony" [Arkhiv RAN (Archives of the Russian Academy of Sciences) 411–3–498].

2. *Istochnik*, no. 1, 1995.

3. D. Holloway, *Stalin and the Bomb: The Soviet Union and Atomic Energy, 1939–1956*, New Haven, CT: Yale University Press, 1994, pp. 360–362.

4. S. N. Khrushchev, *Pensioner souznogo znacheniya*, Moscow: Novosti, 1991.

5. V. V. Krylov, "Vybor ili vybory? K istorii izbraniya presidenta Akademii nauk SSSR. Iyul' 1945," *Istoricheskii Arkhiv*, no. 2, 1996.

6. Outraged that Nuzhdin's candidacy was not pushed through in the 1964 elections to the Academy of Sciences, Khrushchev threated to shut down the Academy (*Istochnik*, no. 2, 1998; Khrushchev, *Pensioner souznogo znacheniya*, p. 160).

7. A. D. Sakharov, "Radioaktivnyi uglerod," *Nauchnye trudy*, Moscow: Tsentrkom, 1995.

8. A. D. Sakharov, "Nauka budushchego (Prognoz perspektiv razvitya nauki)" (1966), *Nauchnye trudy*.

9. A. D. Sakharov, letter to the Central Committee of the CPSU, July 21, 1967, Andrei Sakharov Archives, Moscow.

10. O. V. Golubev, Y. A. Kamenskiy, M. G. Minasyan, and B. D. Pupkov, *Ros-*

siyskaya sistema protivoraketnoy oborony (proshloe i nastoyaschee—vzglyad iznutri), Moscow: Tehnokonsalt, 1994.

11. W. T. Lee, "The ABM Treaty Charade: A Study in Elite Illusion and Delusion," Council for Social and Economic Studies, 1997 (available at http://www.cdiss.org/colfeb97.htm); A. Dobrynin, *In Confidence*, New York: Random House, 1995; A. Dobrynin, "Kak my upustili NATO," *Ogonek*, July 28, 1997.

12. H. Bethe and R. Garvin, "Antiballistic-Missile Systems," *Scientific American*, March 1968.

13. E. G. Bonner interview, February 3, 1997.

14. *On mezhdu nami zhil . . . Vospominaniya o Sakharove* (B. L. Altshuler et al., eds.), Moscow: Praktika, 1996.

15. V. F. Sennikov interview, December 4, 1992; *On mezhdu nami zhil.*

16. This connection is visible in Sakharov's interview with Western journalists on August 23, 1973, when he said that his views "changed because of the danger of nuclear testing, the arms race, and the construction of offensive and anti-missile systems" and that he "wrote an article with a journalist (never published) on anti-missile defense and the role of the intelligentsia in the contemporary world. From here is a direct transition to the memorandum of 1968" (*A. Sakharov v bor'be za mir*, Frankfurt/Main: Posev, 1973.)

17. *An End to Silence: Uncensored Opinion in the Soviet Union from Roy Medvedev's Underground Magazine Political Diary* (S. F. Cohen, ed.), New York: W. W. Norton, 1982, p. 228; L. P. Petrovsky, "Tainyi front Ernsta Genri," *Kentavr*, no. 6, 1995.

18. N. M. Dolotova-Geilikman interview, May 2, 1997.

19. E. L. Kapitsa, *Petr Leonidovich Kapitsa: Vospominaniya, pis'ma, dokumenty* (P. E. Rubinin, comp.), Moscow: Nauka, 1994.

20. "Dissidenty o dissidentstve," Znamya-plus 97/98, 1997.

21. A. S. Esenin-Volpin interview, May 3, 1992.

22. Esenin-Volpin came up with this fruitful form of public activity, which did not require sacrifice and had constructive aims, thanks to his profession—mathematical logic—and his experience: he was arrested in the early 1950s for his poetry.

23. Sakharov, *Nauchnye trudy.*

24. Petition on the law "On Dissemination, Seeking and Obtaining Information" (October 26, 1967), Andrei Sakharov Archives, Moscow.

Chapter 16

1. "Nur die Fuelle fuehrt zur Klarheit" (L. P. Petrovsky, "Neizvestnyi Sakharov," *Vestnik Rossiiskoi Akademii Nauk*, no. 5, 1996).

2. A. D. Sakharov, "Razmyshleniya o progresse, mirnom sosushchestvovanii i intellektual'noi svobode" (1968), *Trevoga i nadezhda*, Moscow: Inter-Verso, 1991.

3. Ibid.

4. Ibid.

5. J. Galbraith, *The New Industrial State*, Boston: Houghton Mifflin, 1967.

6. Sakharov, "Razmyshleniya."

7. Ibid.

8. "Medvedev wholly approves of Sakharov's article, since in his opinion it calls for the democratization of spiritual life, but nevertheless, he notes its utopian character. Medvedev expresses concern over Sakharov's fate and feels that he shouldn't be "pressuring the government with his authority" (Petrovsky, "Neizvestnyi Sakharov").

9. Sakharov, *Trevoga i nadezhda*.

10. Sakharov, "Razmyshleniya."

11. Sakharov, *Trevoga i nadezhda*.

12. V. P. Kartsev, *Traktat o prityazhenii*, Moscow: Sovetskaya Rossiya, 1968.

13. A magnetic field of twenty-five million gauss created a force of twenty-five thousand tons per square centimeter.

14. A. D. Sakharov, "Rekordy magnitnykh polei," *Izvestiya*, April 29, 1966.

15. V. P. Kartsev interview, January 13, 1997.

16. Personal archives of Vladimir Kartsev.

17. V. P. Kartsev interview, January 13, 1997.

18. Ibid.

19. B. L. Altshuler interview, December 8, 1998. Boris Altshuler's defense of his dissertation, "General Covariant Boundary Conditions for Einstein's Equations, Quantum Gravitation, and Cosmology," took place in January 1969.

20. "Samaya vysokaya temperatura" [The Highest Temperature], *Priroda*, no. 11, 1966; "Simmetriya Vselennoi" [Symmetry of the Universe], *Nauchanya mysl', Vestnik APN*, no. 1, 1967; "Sushchestvuet li elementarnya dlina?" [Does Elementary Length Exist?], *Fizika v shkole*, no. 2, 1968.

21. *30 let "Razmyshleniy . . ." Andreya Sakharova*, Moscow: Prava cheloveka, 1998.

22. A. D. Sakharov, "Opasnost' termoyadernoi voiny. Otkrytoe pis'mo doktoru Sidneu Drellu," *Trevoga i nadezhda*; "The Danger of Thermonuclear War. An Open Letter to Dr. Sidney Drell," *Foreign Affairs*, Vol. 61, no. 5, 1983, pp. 1001-1016.

23. Sakharov, *Trevoga i nadezhda*.

Chapter 17

1. The first Western journalist to believe that Sakharov's article was authentic was the correspondent of the Dutch newspaper *Het Parool*, who reported on the article on July 6, 1968. Raymond Anderson, the *New York Times* Moscow correspondent, got a copy of the manuscript from him (*30 let "Razmyshleniy . . ." Andreya Sakharova*, Moscow: Prava cheloveka, 1998).

2. R. Anderson, "Soviet Expert Asks Intellectual Liberty," *New York Times*, July 11, 1968, p. 1.

3. *30 let "Razmyshleniy . . ." Andreya Sakharova*.

4. A. D. Sakharov, *Progress, Coexistence, and Intellectual Freedom*, New York: W. W. Norton, 1968, p. 12.

5. Ibid., p. 94.

6. Y. B. Zeldovich and A. D. Sakharov, "Nuzhny estestvenno-matematicheskie shkoly," *Pravda*, November 19, 1958.

7. T. A. Shabad, "A Russian Physicist's Plan: US-Soviet Collaboration," *New York Times*, July 22, 1968, p. 16.

8. *World Scope Encyclopedia Yearbook, Events of 1969*, New York: Gache, 1970, p. 175.

9. Ibid., p. 410.

10. Ibid., p. 446.

11. E. L. Feinberg, *Epokha i lichnost. Fiziki. Ocherki i vospominaniya*, Moscow: Fizmatlit, 2003.

12. We can see here how entrenched in Sakharov's milieu were Lenin's expressions "American efficiency" and "Russian revolutionary sweep," abbreviated

in "Reflections" as AME and RRS. (A. D. Sakharov, "Razmyshleniya o progresse, mirnom sosushchestvovanii i intellektual'noi svobode" (1968), *Trevoga i nadezhda*, Moscow: Inter-Verso, 1991.)

13. A. I. Solzhenitsyn, *Bodalsya telenok s dubom*, Moscow: Soglasie, 1996.

14. According to Sakharov. When it was published (in the anthology *Under the Rubble*), Solzhenitsyn changed the title to "As Breathing and Consciousness Return."

15. V. I. Ritus interview, July 7, 1992.

16. *A. Sakharov v bor'be za mir*, Frankfurt/Main: Posev, 1973.

17. A. V. Korotkov et al., *Kremlevsky samosud*, Moscow: Rodina, 1994.

18. A. D. Sakharov, *Trevoga i nadezhda*, Moscow: Inter-Verso, 1991.

19. Ibid.

20. A. I. Solzhenitsyn, "Ugodilo zernyshko promezh dvukh zhernovov," *Novyi mir*, no. 9, 1998.

21. Sakharov, "Razmyshleniya."

22. A. I. Solzhenitsyn, *Publitsistika*, Yaroslavl: Verkhne-Volzhskoye izdatel'stvo, 1995.

23. *Slovo probivaet sebe dorogu* (V. Glotser and E. Chukovskaya, comp.), Moscow: Russkii put', 1998.

24. E. T. Chukovskaya interview, April 28, 1998.

25. L. Chukovskaya, Kakim on zapomnilsya, *Aprel'*, no. 10, 1998.

26. A. Akhmatova, "Tvorchestvo," *Sochineniya*, Moscow: Tsitadel', 1996.

27. In Govorukhin's documentary film, Solzhenitsyn explained his ability to work on the multivolume epic *Red Wheel* by a mathematical organization of labor. His wife, a mathematician by education, also speaks of this: "The fact that we're both mathematicians helps us deal with the mountain of material" (interview by Natalya Solzhenitsyn in *Izvestiya*, May 25, 1992). Solzhenitsyn explains the architectural concept of *Red Wheel* in mathematical terms.

28. D. Shturman and S. Titkin, *Sovremenniki*, Jerusalem: Lira, 1998.

29. Solzhenitsyn, *Publitsistika*.

30. Shturman and Titkin, *Sovremenniki*.

31. "Tamm v dnevnikakh i pis'makh," *Kapitsa. Tamm. Semenov. V ocherkakh i pis'makh*, Moscow: Vagrius-Priroda, 1998.

32. Solzhenitsyn, *Publitsistika*.

33. A. D. Sakharov, interview with Olle Stenholm (July 1973), in Sakharov, *Vospominaniya*, v. 2, Moscow: Prava cheloveka, 1996.

34. *Sovetskii entsiklopedicheskii slovar*, Moscow: Sovetskaya entsiklopediya, 1989.

35. Sakharov, "Razmyshleniya."

36. Sakharov, interview with Olle Stenholm (July 1973).

37. Solzhenitsyn, *Bodalsya telenok s dubom*.

38. V. I. Ritus interview, July 7, 1992.

39. Solzhenitsyn, *Publitsistika*.

Chapter 18

1. John A. Wheeler, "Sakharov: Man of Humility, Understanding and Leadership," *Andrei Sakharov: Facets of a Life*, Gif-sur-Yvette, France: Editions Frontières, 1991, pp. 647–649.

2. Personal archives of Boris Altshuler.

3. I. Y. Barit interview, October 31, 1992.

4. M. L. Levin, *Zhizn', vospominaniya, tvorchestvo*, Nizhny Novgorod: IPF RAN, 1998.

5. L. A. Vernaya interview, March 19, 2000.

6. Sakharov, *Nauchnye trudy*, Moscow: Tsentrkom, 1995.

7. My thanks to P. E. Rubinin for the text of this letter (P. L. Kapitsa Archive at the Institute of Physical Problems of the Russian Academy of Sciences).

8. P. L. Kapitsa, *Nauchnye trudy. Nauka i sovremennoe obshchestvo* (P. E. Rubinin, ed.), Moscow: Nauka, 1998.

9. Ibid.

10. A. D. Sakharov, *Vospominaniya*, 2 vols., Moscow: Prava cheloveka, 1996.

11. Andrei Sakharov Archives, Moscow.

12. L. P. Feoktistov interview, February 24, 1995.

13. Ibid.

14. V. F. Sennikov interview, December 4, 1992. In the 1960s, V. F. Sennikov was the KGB official assigned to FIAN.

15. A. D. Sakharov, "Za i protiv," *1973 god: Dokumnety, fakty, sobytiya*, Moscow: PIK, 1991.

16. B. M. Bolotovsky, *Andrei Dmitrievich Sakharov. Nauchnaya deyatel'nost'. Prepodavanie fiziki v vysshei shkole*, Moscow: Prometei, 1997.

17. This was said by Grishin, Moscow's party boss (A. V. Korotkov et al., *Kremlevsky samosud*, Moscow: Rodina, 1994).

18. Korotkov et al., *Kremlevsky samosud*.

19. Ibid.

20. "Glavlit i literatura v period 'literaturno-politicheskogo brozheniya v Sovetskom Souze," *Voprosy literatury*, no. 5, 1998.

21. "Dissidenty o dissidentstve," *Znamya-plus* 97/98, 1997

22. And in the *samizdat* publication the date was given incorrectly—March instead of July 1967 [*Politichesky dnevnik* (R. Medvedev, ed.), Amsterdam: Fond im. Gertsena, 1972].

23. Sakharov, *Vospominaniya*.

24. L. Litinsky, "Dokumenty i kommentarii," *Troitskii variant*, December 12, 1997.

25. E. L. Feinberg, *Epokha i lichnost. Fiziki. Ocherki i vospominaniya*, Moscow: Fizmatlit, 2003.

26. G. S. Podyapolsky, *Zolotomu veku ne byvat'*, Moscow: Zven'ya, 2003.

27. From the working minutes of the Politburo of the Central Committee CPSU on September 17, 1973. Korotkov et al., *Kremlevsky samosud*.

Chapter 19

1. E. L. Feinberg, *Epokha i lichnost. Fiziki. Ocherki i vospominaniya*, Moscow: Fizmatlit, 2003; S. E. Babenysheva interview, September 13, 1998.

2. E. G. Bonner interview, March 7, 1999.

3. E. G. Bonner interview, February 7, 1997.

4. E. G. Bonner, "Bessonnoi noch'iu v kanun ubileya," *Literaturnaya gazeta*, May 9, 1995.

5. *Vsevolod Bagritsky, Dnevniki, pis'ma, stikhi* (L. G. Bagritskaya and E. G. Bonner, comp.), Moscow: Sovetskii pisatel', 1964.

6. E. G. Bonner-Alikhanova, "V Irake (Glazami druga)," *Neva*, nos. 3 and 4, 1961.

7. The Nobel Peace Prize 1975 presentation speech by Mrs. Aase Lionaes, Chairman of the Nobel Committee of the Norwegian Storting (available at http://www.nobel.se/peace/laureates/1975/press.html).

8. Sergei Kovalev was accused of slandering the Soviet state, more precisely, with editing *The Chronicle of Current Events*, which monitored abuse of human rights.

9. In a letter to the magazine *Znamya* (no. 8, 1991, p. 228), in an attempt to discredit the story, Terletsky called Sakharov's *Memoirs* "a publication by Elena Bonner."

10. A. W. Bouis, letter to Gennady Gorelik, July 5, 2001.

11. Andrei Sakharov Archives, Moscow.

12. A. I. Solzhenitsyn, *Bodalsya telenok s dubom*, Moscow: Soglasie, 1996.

13. E. G. Bonner interview, February 13, 1997.

14. E. G. Bonner, "Nikakoi Sakharov im ne nuzhen," *Ogonek*, no. 50–51, 1994.

15. A. D. Sakharov, "The Danger of Thermonuclear War. An Open Letter to Dr. Sidney Drell," *Foreign Affairs*, Vol. 61, no. 5, 1983, pp. 1001–1016.

16. U.S. Declaration of Independence.

Chapter 20

1. A. I. Solzhenitsyn, *Bodalsya telenok s dubom*, Moscow: Soglasie, 1996.

2. *Akademik L. I. Mandelshtam. K 100-tiyu so dnya rozhdeniya*, Moscow: Nauka, 1979.

3. M. L. Levin, *Zhizn', vospominaniya, tvorchestvo*, Nizhny Novgorod: IPF RAN, 1998.

4. A. D. Sakharov, *Trevoga i nadezhda*, Moscow: Inter-Verso, 1991.

5. Kapitsa's attitude toward Sakharov's strategic ideas is clear in his article "Contemporary Goals in the Struggle for Peace," written in 1970, which he was not permitted to publish (P. L. Kapitsa, *Nauchnye trudy. Nauka i sovremennoe obshchestvo* [P. E. Rubinin, ed.], Moscow: Nauka, 1998).

6. L. Chukovskaya, *Protsess iskluchemiya*, Moscow: Gorizont, 1990.
Her article "Wrath of the People" served as the excuse for expelling Chukovskaya from the Writers' Union of the USSR.

7. V. L. Ginzburg interview, September 25, 1990.

8. A. D. Sakharov, *Vospominaniya*, 2 vols., Moscow: Prava cheloveka, 1996.

9. V. L. Ginzburg interview, September 25, 1990.

10. V. Bukovsky, Soviet Archives (available at http://psi.ece.jhu.edu/~kaplan/IRUSS/BUK/GBARC/pdfs/sakharov/sakh-r.html).

11. Only two of the forty academicians who signed the first letter later made a public apology—Sergei Vonsovsky and Ilya Frank.

12. R. Sagdeev, "Donesti slovo v zashchitu nauki," *Priroda*, no. 1, 1992.

13. R. Sagdeev, *The Making of a Soviet Scientist: My Adventures in Nuclear Fusion and Space—From Stalin to Star Wars*, New York: John Wiley & Sons, 1994, pp. 142–143.

14. *Nauka i obshchestvo: Sovetskie i zarubezhnyi uchenye otvechaut na anketu "Literaturnoi gazety"* (O. Moroz and A. Lepikhov, comp.), Moscow: Znanie, 1977.

15. A. D. Sakharov, "Razmyshleniya o progresse, mirnom sosushchestvovanii i intellektual'noi svobode" (1968), *Trevoga i nadezhda*.

16. V. I. Ritus, "Esli ne ya, to kto?" *Priroda*, no. 8, 1990.

17. Leo Tolstoy wrote an article under that title in 1908 about state violence, an epidemic of executions.

18. Chukovskaya, *Protsess isklucheniya*.

19. Andrei Sakharov Archives, Moscow.

20. Chukovskaya, *Protsess isklucheniya*.

21. Personal archives of Elena Chukovskaya.

22. Andrei Sakharov's diary, entry dated April 27, 1978. Personal archives of Elena Bonner.

23. Andrei Sakharov's diary, entry dated April 29, 1978. Personal archives of Elena Bonner. (Lyusha is Elena Chukovskaya, Lydia Chukovskaya's daughter.)

24. Personal archives of Elena Bonner.

25. P. L. Kapitsa, *Pis'ma o nauke* (P. E. Rubinin, comp.), Moscow: Moskovskii rabochii, 1989.

26. A. D. Sakharov, *Nauchnye trudy*, Moscow: Tsentrkom, 1995.

27. Sakharov, *Trevoga i nadezhda*.

28. Seventy-year-old Tamm concluded his article "On the Threshold of a New Theory" with these words: "My deepest wish is to live long enough to see the creation of a new physical theory, and still be in a state to understand it" (I. E. Tamm, *Sobranie nauchnykh trudov*, Moscow: Nauka, 1975).

29. Andrei Sakharov, letter to Lydia Chukovskaya, dated March 10, 1986. Personal archives of Elena Chukovskaya.

30. *On mezhdu nami zhil . . . Vospominaniya o Sakharove* (B. L. Altshuler et al., eds.), Moscow: Praktika, 1996.

31. A. D. Sakharov, "The World in Fifty Years," May 17, 1974. Published as "Tomorrow: The View from Red Square," *Saturday Review/World*, August 24, 1974, and as "Mir cherez polveka," *Trevoga i nadezhda*.

32. On December 4, 1981, on the twelfth day of Sakharov's hunger strike for permission for Liza Alexeyeva to leave the Soviet Union, Kapitsa, then eighty-seven years old, sent a very short note to Brezhnev: "I am a very old man, and life has taught me that generous deeds are never forgotten. Save Sakharov. Yes, he has great flaws and a difficult personality, but he is a great scientist of our country. Respectfully, P. L. Kapitsa" (P. E. Rubinin, "K istorii odnogo pis'ma P. L. Kapitsy," *Kommunist*, no. 7, 1991).

33. A. A. Kapitsa interview, February 16, 1990.

34. Rubinin, *K istorii odnogo pis'ma P. L. Kapitsy.*

35. B. L. Altshuler, "Po tu storonu okna," *Aprel'*, no. 3, 1990.

36. S. N. Khrushchev, *Pensioner souznogo znacheniya*, Moscow: Novosti, 1991.

37. R. Lourie, "The Smuggled Manuscript: Translating Sakharov's Memoirs," New York Times, June 3, 1990.

38. "Akademik Sakharov: 'Ya pytalsya byt' na urovne svoey sud'by," *Molodezh Estonii*, October 11, 1988 (*Zvezda*, no. 5, 1991).

39. M. G. Petrenko interview, March 28, 1999.

40. Levin, *Zhizn', vospominaniya, tvorchestvo*.

41. Andrei Sakharov, letter to Lydia Chukovskaya, dated December 30, 1981. Personal archives of Elena Chukovskaya.

42. *On mezhdu nami zhil.*

43. Levin, *Zhizn', vospominaniya, tvorchestvo.*

44. In the expression of Tatiana Velikanova (in the 1991 documentary *In the Shadow of Sakharov* by Sherry Jones).

45. A. Tolkunov, "Mister Teller and Co.," *Pravda*, November 12, 1980.

46. "In some sort of crude sense which no vulgarity, no humor, no overstatement can quite extinguish, the physicists have known sin; and this is a knowledge

which they cannot lose." Oppenheimer made this famous statement in November 1947 at MIT. Quoted in L. Badash, *Scientists and the Development of Nuclear Weapons from Fission to the Limited Test Ban Treaty, 1939–1963* (Atlantic Highlands, NJ: Humanities Press, 1995), p. 57. See G. Gorelik, "Paralleli mezhdu perpendikulyarami: Andrei Sakharov, Edward Teller i Robert Oppenheimer," *VIET*, no. 2, 2002.

47. Sakharov, *Trevoga i nadezhda*.

48. Sakharov, "Razmyshleniya."

49. Andrei Sakharov Archives, Moscow.

50. *Moskovskie Novosti*, November 8, 1987.

51. Sakharov, *Vospominaniya*.

52. Bukovsky, Soviet Archives.

53. A. and R. McGowan, "A Conversation with Andrei Sakharov and Elena Bonner," SIPIscope, Vol. 15, no. 2, June–July 1987.

54. A. W. Bouis, letter to Gennady Gorelik, dated July 5, 2001.

55. S. Weinberg, *The First Three Minutes: A Modern View of the Origin of the Universe*, New York: Basic Books, 1977, p. 154.

56. Sakharov, *Vospominaniya*.

57. Andrei Sakharov's diary, entry dated April 27, 1978. Personal archives of Elena Bonner.

58. From Sakharov's letter to Boris Altshuler, dated May 10, 1982. In *Andrei Dmitrievich. Vospominaniya o Sakharove*, Moscow: Terra, Knizhnoe obozrenie, 1990.

59. Quoted from Kolakowski, *Pokhvala neposledovatel'nosti* [In Praise of Inconsistency], Firenze: Edizioni Aurora, 1974, p. 24.

60. Sakharov, *Trevoga i nadezhda*.

61. Andrei Sakharov's acceptance speech for the peace prize of the Einstein Foundation, dated November 15, 1988. Andrei Sakharov Archives, Moscow.

SUGGESTIONS FOR
FURTHER READING

The following is a list of primary publications on the world of Andrei Sakharov.

30 let "Razmyshleniy . . . " Andreya Sakharova. Moscow: Prava cheloveka, 1998.

Akademik L. I. Mandelshtam. K 100-tiyu so dnya rozhdeniya. Moscow: Nauka, 1979.

Akademik M. A. Leontovich: uchenyy, uchitel, grazhjdanin. Moscow: Nauka, 2003.

Altshuler, B. L., et al., eds. *On mezhdu nami zhil . . . Vospominaniya o Sakharove*. Moscow: Praktika, 1996.

The American Institute of Physics. *Andrei Sakharov: Soviet Physics, Nuclear Weapons, and Human Rights*. Accessed January 3, 2005, at www.aip.org/history/sakharov/

Andrei Sakharov: Facets of a Life. Gif-sur-Yvette, France: Éditions Frontières, 1991.

Bonner, E. *Alone Together*. New York: Knopf, 1986.

Bonner, E. *Vol'nye zametki k rodoslovnoi Andreya Sakharova*. Moscow: Prava cheloveka, 1996.

Drell, S. D., & Kapitsa, S. P., eds. *Sakharov Remembered : A Tribute by Friends and Colleagues*. New York: American Institute of Physics, 1991.

Feinberg, E. L. *Epokha i lichnost. Fiziki. Ocherki i vospominaniya*. Moscow: Fizmatlit, 2003.

Ginzburg, V. L. *About Science, Myself and Others*. Bristol: IOP Publishing, 2004.

Ginzburg, V. L. *O nauke, o sebe i o drugikh*. Moscow: Fizmatlit, 2003.

Gorelik, G. "Vladimir Fock: Philosophy of Gravity and Gravity of Philosophy." In *The Attraction of Gravitation* (J. Earman et al., eds.). Boston: Birkhauser, 1993.

Gorelik, G. *"Meine antisowjetische Taetigkeit . . ." Russische Physiker unter Stalin*. Braunschweig/Wiesbaden: Vieweg, 1995.

Holloway, D. *Stalin and the Bomb: The Soviet Union and Atomic Energy, 1939–1956*. New Haven, CT: Yale University Press, 1994.

Kapitsa. Tamm. Semenov. V ocherkakh i pis'makh. Moscow: Vagrius-Priroda, 1998.

Sakharov, A. D. *Alarm and Hope*. New York: Alfred Knopf, 1978.

Sakharov, A. D. *Collected Scientific Works*. New York: Marcel Dekker, 1982.

Sakharov, A. D. *My Country and the World*. New York: Alfred Knopf, 1975.

Sakharov, A. D. *Sakharov Speaks.* New York: Alfred Knopf, 1974.

Sakharov, A. D. *Vospominaniya.* 2 v. Moscow: Prava cheloveka, 1996. English editions: *Memoirs* (R. Lourie, trans.), New York: Knopf, 1990; *Moscow and Beyond, 1986 to 1989* (A. Bouis, trans.), New York: Knopf, 1991.

The Sakharov Archives in Moscow. *A.D. Sakharov, stranitsy zhizni.* Accessed January 3, 2005, at www.sakharov-center.ru/sakharov-chronology/

Sergei Ivanovich Vavilov. Ocherki i vospominaniya. Moscow: Nauka, 1991.

Thorne, K. *Black Holes and Time Warps: Einstein's Outrageous Legacy.* New York: W. W. Norton, 1994.

Vospominaniya o I. Ye. Tamme. Moscow: Izdat, 1995.

INDEX